Textbook of

Integral Calculus AND Differential Equations

Khalil Ahmad

Currently a Professor in the Department of Mathematics, Jamia Millia Islamia, New Delhi, has been teaching undergraduate and postgraduate classes for the last 32 years. He had also taught in DAV (PG) College, Kanpur University, Kanpur and AMU, Aligarh. He was Visiting Professor in the Department of Mathematics, Yarmouk University, Irbid, Jordan for one year.

Dr. Khalil Ahmad has to his credit 33 research papers in Functional Analysis and Wavelet Analysis in various national and international journals, authored two books, coauthored four books, coedited three books and produced eight PhD's.

Textbook of

Integral Calculus AND Differential Equations

KHALIL AHMAD

Anamaya Publishers
New Delhi

Dr. Khalil Ahmad
Professor
Department of Mathematics
Jamia Millia Islamia
New Delhi - 110 025, India

ANAMAYA PUBLISHERS
F-230, Lado Sarai, New Delhi - 110 030, India
e-mail: anamayapub@vsnl.net
anamayapub@indiatimes.com

ISBN 81-88342-17-3

Published by Manish Sejwal for Anamaya Publishers,
F-230, Lado Sarai, New Delhi - 110 030. Printed in India.

Preface

The present book, which is in sequel to our book "Differential Calculus", written in accordance with the UGC model syllabus prepared by a team of experts on the initiative of UGC, covers the syllabus of calculus and differential equations for B.A./B.Sc. (Hons) and Pass Course students of Indian Universities. This book will also be most useful for engineering students and others preparing for competitive examinations like IAS, IES, PCS etc.

The contents of the book have been drawn from various streams of courses given to students by the author at different institutes during the last 33 years. The text begins with the concepts of integration and is followed by the study of integration of rational functions, irrational algebraic functions and trigonometric functions, definite integrals, reduction formulae, quadrature, rectification, volumes and surfaces of solids of revolution, differential equations of the first order and the first degree, differential equations of the first order and higher degree, trajectories, linear differential equations with constant coefficients, homogeneous linear differential equations, linear differential equations of second order, simultaneous differential equations and an appendix on Beta and Gamma functions.

Examples, exercises, remarks and notes at different places would help the reader in understanding the subject. Exercises, covering a wide range from simple numerical problems to the development of the theory, have been provided at the end of each section. Each chapter ends with a miscellaneous set of exercises. Answers to the problems in the set of exercises are given at the end of the book. Care has been taken to grade the exercises on the difficulty scale.

My debt to a large number of authors who have contributed to the discipline in the past is immense. I have not tried to attribute various theorems and proofs to their original discoverers. However, I take it as a duty to place on record my gratitude to them.

I am thankful to the generations of students who gave me the feedback during the course of discussions of many of these topics in the classroom. I have profited from their discussion of these topics.

I am also thankful to my associates in the Department of Mathematics, Jamia Millia Islamia and fellow mathematicians in India. They tempered the ideas and results in the book by valuable discussions from time to time. In particular, I owe a special debt to my esteemed friend Prof. V.N. Dixit of Jamia Millia Islamia, who encouraged me constantly to complete the project.

I am grateful to my family members for their cooperation and patience during the preparation of the book, in particular, to my daughter Miss Humaira, who helped me in solving the problems. Finally, I thank Mr Manish Sejwal, M/s Anamaya Publishers and Mr M.S. Sejwal, for doing an excellent job in bringing out the book in the present form.

January 1, 2005

KHALIL AHMAD

Contents

Textbook of

Integral Calculus AND Differential Equations

1

Elementary Integration

1.1 Introduction

Integral calculus deals mainly with two problems. First, inverse of the problem of the differential calculus. In differential calculus we are given a function and we are required to find its derivative or differential. In integral calculus we are to find a function whose derivative or differential is given.

The second problem is totally of different nature. In this we find the limit of the sum of a large number of small quantities when their number increases indefinitely while each quantity tends to zero. This problem was known long before Newton and Leibnitz discovered Differential Calculus. From this point of view, the Integral Calculus is very much older than the Differential Calculus. It will be seen later that under suitable conditions the solution of this problem can be made to depend on that of the first. This chapter confines to the study of indefinite integrals including some techniques of integration.

1.2 The Indefinite Integral

A function $F(x)$ which has $f(x)$ for its derivative or $f(x)\,dx$ for its differential is called an *integral* or a *primitive function* of $f(x)$ denoted by $\int f(x)\,dx$. Thus

$$F(x) = \int f(x)dx$$

The symbol $\int$ is just elongated S which is first letter of the word sum just as in the symbol dy/dx, the d in the numerator and denominator is the first letter of the word difference.

The given function $f(x)$ is called the *integrand* and the process of obtaining $F(x)$ from $f(x)$ is called *integration*. It is clear from the definition that in order to integrate a function $f(x)$ we have to look for a function which, on differentiation, gives $f(x)$. Thus if $F(x)$ is an integral of $f(x)$, then the function $F(x) + C$, where C is an arbitrary constant, is the most general function which has $f(x)$ for its derivative. The function $F(x) + C$ is called *indefinite* or the *general integral* of $f(x)$, and we write

$$\int f(x)dx = F(x) + C$$

Hence, the process of integration leads to a function which is indefinite to the extent of an additive arbitrary constant C, called the *constant of integration*. This constant is not written down usually and is understood with every indefinite integral.

1.3 Some Theorems

I. The integral of the product of a constant and a function is equal to the product of the constant and the integral of the function.

If a is constant and $f(x)$ is a function, then

$$\int af(x)dx = a\int f(x)dx$$

II. The integral of the sum or difference of two functions is equal to the sum or difference of their integrals.

If $f(x)$ and $g(x)$ be the functions, then

$$\int (f(x) \pm g(x))dx = \int f(x)dx \pm \int g(x)dx$$

1.4 Standard Forms

The integrals of numerous functions can be obtained from the standard results in differential calculus. Some of these are:

1. $\int x^n dx = \frac{x^{n+1}}{n+1}$, $(n \neq -1)$
2. $\int \frac{1}{x}dx = \log|x|$
3. $\int e^x dx = e^x$
4. $\int a^x dx = \frac{a^x}{\log a}$, $a > 0$
5. $\int \sin x\,dx = -\cos x$
6. $\int \cos x\,dx = \sin x$
7. $\int \sec^2 x\,dx = \tan x$
8. $\int \operatorname{cosec}^2 x\,dx = -\cot x$
9. $\int \sec x \tan x\,dx = \sec x$
10. $\int \operatorname{cosec} x \cot x\,dx = -\operatorname{cosec} x$
11. $\int \frac{1}{\sqrt{a^2 - x^2}}dx = \sin^{-1}\frac{x}{a}$ or $-\cos^{-1}\frac{x}{a}$

12. $\int \frac{1}{\sqrt{a^2 + x^2}} dx = \frac{1}{a} \tan^{-1} \frac{x}{a}$ or $-\frac{1}{a} \cot^{-1} \frac{x}{a}$

13. $\int \frac{1}{x\sqrt{x^2 - a^2}} dx = \frac{1}{a} \sec^{-1} \frac{x}{a}$ or $-\frac{1}{a} \operatorname{cosec}^{-1} \frac{x}{a}$

14. $\int \sinh x \, dx = \cosh x$

15. $\int \cosh x \, dx = \sinh x$

16. $\int \operatorname{sech}^2 x \, dx = \tanh x$

17. $\int \operatorname{cosech}^2 x \, dx = -\coth x$

18. $\int \frac{1}{\sqrt{x^2 + a^2}} dx = \sinh^{-1} \frac{x}{a}$ or $\log \left| \frac{x + \sqrt{x^2 + a^2}}{a} \right|$

19. $\int \frac{1}{\sqrt{x^2 - a^2}} dx = \cosh^{-1} \frac{x}{a}$ or $\log \left| \frac{x + \sqrt{x^2 - a^2}}{a} \right|$

20. $\int \frac{1}{a^2 - x^2} dx = \frac{1}{a} \tanh^{-1} \frac{x}{a}$ or $\frac{1}{2a} \log \left| \frac{a + x}{a - x} \right|, (x^2 < a^2)$

21. $\int \frac{1}{x^2 - a^2} dx = -\frac{1}{a} \coth^{-1} \frac{x}{a}$ or $-\frac{1}{2a} \log \left| \frac{x + a}{x - a} \right|, (x^2 > a^2)$

22. $\int \frac{1}{x\sqrt{a^2 - x^2}} dx = -\frac{1}{a} \operatorname{sech}^{-1} \frac{x}{a}$

23. $\int \frac{1}{x\sqrt{a^2 + x^2}} dx = -\frac{1}{a} \operatorname{cosech}^{-1} \frac{x}{a}$

1.5 Integration by Substitution

Functions to be integrated do not always have the standard form to make use of the rules of indefinite integration. Frequently, it may be necessary to choose a new variable, related somehow to the original one, in order that the integration may be performed. Such a procedure of substitution is based upon the following:

Theorem. If $x = \phi(t)$, where ϕ is any differentiable function, then

$$\int f(x) dx = \int f(\phi(t)) \, \phi'(t) dt$$

Following are some integrals where the method of substitution is used.

1. $\int f'(ax + b) dx = \frac{1}{a} f(ax + b)$

2. $\int (f(x))^n f'(x)\,dx = \frac{(f(x))^{n+1}}{n+1}, \quad (n \neq -1)$

3. $\int \frac{f'(x)}{f(x)}\,dx = \log|f(x)|$

4. $\int \tan x\,dx = \log|\sec x|$

5. $\int \cot x\,dx = \log|\sin x|$

6. $\int \sec x\,dx = \log|\sec x + \tan x| = \log\left|\tan\left(\frac{\pi}{4} + \frac{x}{2}\right)\right|$

7. $\int \operatorname{cosec} x\,dx = \log|\operatorname{cosec} x - \cot x| = \log\left|\tan\frac{x}{2}\right|$

8. $\int \frac{1}{a^2 + x^2}\,dx = \frac{1}{a}\tan^{-1}\frac{x}{a}$

9. $\int \frac{1}{\sqrt{a^2 - x^2}}\,dx = \sin^{-1}\frac{x}{a}$

10. $\int \frac{1}{\sqrt{a^2 + x^2}}\,dx = \sinh^{-1}\frac{x}{a} = \log\left|\frac{x + \sqrt{a^2 + x^2}}{a}\right|$

11. $\int \frac{1}{\sqrt{x^2 - a^2}}\,dx = \cosh^{-1}\frac{x}{a} = \log\left|\frac{x + \sqrt{x^2 - a^2}}{a}\right|$

12. $\int \sqrt{a^2 - x^2}\,dx = \frac{a^2}{2}\sin^{-1}\frac{x}{a} + \frac{x}{2}\sqrt{a^2 - x^2}$

13. $\int \sqrt{a^2 + x^2}\,dx = \frac{a^2}{2}\sinh^{-1}\frac{x}{a} + \frac{x}{2}\sqrt{a^2 + x^2}$

$$= \frac{a^2}{2}\log\left|\frac{x + \sqrt{a^2 + x^2}}{a}\right| + \frac{x}{2}\sqrt{a^2 + x^2}$$

14. $\int \sqrt{x^2 - a^2}\,dx = \frac{x}{2}\sqrt{x^2 - a^2} - \frac{a^2}{2}\cosh^{-1}\frac{x}{a}$

$$= \frac{x}{2}\sqrt{x^2 - a^2} - \frac{a^2}{2}\log\left|\frac{x + \sqrt{x^2 - a^2}}{a}\right|$$

Remark. Now we give some trigonometrical substitutions so that the students may choose the proper substitution. It should be seen that the substitution be such that the integrand may be free from fractional power.

Integrands involving	Substitutions
1. $[a^2 + (f(x))^2]$	$f(x) = a\tan\theta$, $a\cot\theta$ or $a\sinh\theta$

2. $[a^2 - (f(x))^2]$ $f(x) = a \sin\theta$ or $a \cos\theta$
3. $[(f(x))^2 - a^2]$ $f(x) = a \sec\theta$ or $a\ \mathrm{cosec}\ \theta$, $a \cosh\theta$
4. $[a + f(x)]$ $f(x) = a \tan^2\theta$, $a \cot^2\theta$ or $a \sinh^2\theta$
5. $[a - f(x)]$ $f(x) = a \sin^2\theta$ or $a \cos^2\theta$
6. $[f(x) - a]$ $f(x) = a \sec^2\theta$, $a\ \mathrm{cosec}^2\theta$ or $a \cosh^2\theta$
7. $[ax - x^2]$ $x = a \sin^2\theta$ or $a \cos^2\theta$
8. $(a - x)(x - b)$ $x = a \cos^2\theta + b \sin^2\theta$
9. $\dfrac{a^n - x^n}{a^n + x^n}$ $x^n = a^n \cos\theta$

1.6 Integration by Parts

Let f and g be any two differentiable functions of a single variable x. Then

$$\int f(x)\, g(x)\, dx = f(x) \int g(x)\, dx - \int [f'(x) \int g(x)\, dx]\, dx$$

Remark. There is no general rule to choose $f(x)$ and $g(x)$, but for simplicity the function $g(x)$ should be chosen in such a way that it may be integrated easily and the remaining factor of the integrand as $f(x)$.

Following are some integrals in which the method of integration by parts is used:

1. $\int \log x\, dx = x \log x - x$

2. $\int \sec^3 x\, dx = \frac{1}{2}\ [\sec x \tan x + \log|\sec x + \tan x|]$

3. $\int \tan^{-1} x\, dx = x \tan^{-1} x - \frac{1}{2} \log(1 + x^2)$

4. $\int \sin^{-1} x\, dx = x \sin^{-1} x + \sqrt{1 - x^2}$

5. $\int e^{ax} \sin bx\, dx = \dfrac{e^{ax}}{a^2 + b^2}(a \sin bx - b \cos bx)$

6. $\int e^{ax} \cos bx\, dx = \dfrac{e^{ax}}{a^2 + b^2}(a \cos bx + b \sin bx)$

7. $\int e^x (f(x) + f'(x))\, dx = e^x f(x)$

1.7 Definite Integrals

If $F(x)$ be any integral of $f(x)$, then the difference $F(b) - F(a)$ between the values of $F(x)$ at $x = a$ and $x = b$ is called the *definite integral* of $f(x)$ between the limits a and b denoted by

$$\int_a^b f(x)\,dx$$

The quantity $F(b) - F(a)$ is usually denoted by $[F(x)]_a^b$. Thus, we can write

$$\int_a^b f(x)\,dx = [F(x)]_a^b = F(b) - F(a)$$

The interval $[a, b]$ is called the *interval* or *range of integration*, a is called the *lower* or *inferior limit* and b the *upper* or *superior limit* of integration for the definite integral.

Remark. The value of the definite integral does not depend upon the choice of constant of integration. It is the same whatever indefinite integral is used to evaluate it. For, if $F(x) + C$ be any other indefinite integral of $f(x)$, then

$$\int_a^b f(x)\,dx = [F(x) + C]_a^b = F(b) + C - F(a) - C = F(b) - F(a)$$

Examples

1. Evaluate $\displaystyle\int \frac{\sin(2 + 3\log x)}{x}\,dx$.

Solution. Let $2 + 3\log x = t$. Then $\dfrac{3}{x}\,dx = dt$.

Therefore $\displaystyle\int \frac{\sin(2 + 3\log x)}{x}\,dx = \int \sin t \cdot \frac{1}{3}\,dt = \frac{1}{3}(-\cos t)$

$$= -\frac{1}{3}\cos(2 + 3\log x)$$

2. Evaluate $\displaystyle\int \frac{x^2 \tan^{-1} x^3}{1 + x^6}\,dx$.

Solution. Let $\tan^{-1} x^3 = t$. Then $\dfrac{3x^2}{1 + x^6}\,dx = dt$.

Therefore $\displaystyle\int \frac{x^2 \tan^{-1} x^3}{1 + x^6}\,dx = \int t \cdot \frac{1}{3}\,dt = \frac{1}{6}t^2 = \frac{1}{6}(\tan^{-1} x^3)^2$

3. Evaluate $\displaystyle\int \frac{x^{e-1} + e^{x-1}}{x^e + e^x}\,dx$.

Solution. Let $x^e + e^x = t$. Then $(ex^{e-1} + e^x)\,dx = dt$, i.e. $e(x^{e-1} + e^{x-1})\,dx = dt$. Therefore

$$\int \frac{x^{e-1} + e^{x-1}}{x^e + e^x}\,dx = \int \frac{1}{t}\cdot\frac{1}{e}\,dt = \frac{1}{e}\log|t| = \frac{1}{e}\log|x^e + e^x|$$

4. Evaluate $\int \frac{xe^x}{(x+1)^2}\,dx$.

Solution. Taking $f(x) = xe^x$ and $g(x) = \frac{1}{(x+1)^2}$ and integrating by parts, we have

$$\int \frac{xe^x}{(x+1)^2}\,dx = xe^x\left(-\frac{1}{x+1}\right) - \int (xe^x + e^x)\left(-\frac{1}{x+1}\right)dx$$

$$= -\frac{xe^x}{x+1} + \int e^x\,dx = -\frac{xe^x}{x+1} + e^x = \frac{e^x}{x+1}$$

5. Evaluate $\int e^x\left(\frac{1+\sin x}{1+\cos x}\right)dx$.

Solution. We have

$$\frac{1+\sin x}{1+\cos x} = \frac{1 + 2\sin\frac{x}{2}\cos\frac{x}{2}}{2\cos^2\frac{x}{2}} = \frac{1}{2}\sec^2\frac{x}{2} + \tan\frac{x}{2}$$

Therefore

$$\int e^x\left(\frac{1+\sin x}{1+\cos x}\right)dx = \int e^x\left(\frac{1}{2}\sec^2\frac{x}{2} + \tan\frac{x}{2}\right)dx$$

$$= \int e^x\cdot\frac{1}{2}\sec^2\frac{x}{2}\,dx + \int e^x\tan\frac{x}{2}\,dx$$

$$= e^x\tan\frac{x}{2} - \int e^x\tan\frac{x}{2}\,dx + \int e^x\tan\frac{x}{2}\,dx$$

$$= e^x\tan\frac{x}{2}$$

6. Evaluate $\int \sin^{-1}\left(\frac{2x}{1+x^2}\right)dx$.

Solution. Let $x = \tan\theta$. Then $dx = \sec^2\theta\, d\theta$. Also

$$\frac{2x}{1+x^2} = \frac{2\tan\theta}{1+\tan^2\theta} = \sin 2\theta$$

Therefore

$$\int \sin^{-1}\left(\frac{2x}{1+x^2}\right) dx = \int \sin^{-1}(\sin 2\theta)\sec^2\theta\, d\theta = \int 2\theta\sec^2\theta\, d\theta$$

$$= 2\theta\tan\theta - \int 2\tan\theta\, d\theta = 2\theta\tan\theta - 2\log\sec\theta$$

$$= 2x\tan^{-1}x - \log(1+x^2)$$

7. Evaluate $\displaystyle\int_0^1 \frac{x}{\sqrt{1+x^2}}\, dx$.

Solution. Let $x = \tan\theta$. Then $dx = \sec^2\theta\, d\theta$.
When $x = 0$, then $\tan\theta = 0 \Rightarrow \theta = 0$

When $x = 1$, then $\tan\theta = 1 \Rightarrow \theta = \dfrac{\pi}{4}$

Therefore $\displaystyle\int_0^1 \frac{x}{\sqrt{1+x^2}}\, dx = \int_0^{\frac{\pi}{4}} \frac{\tan\theta}{\sqrt{1+\tan^2\theta}} \cdot \sec^2\theta\, d\theta$

$$= \int_0^{\frac{\pi}{4}} \sec\theta\tan\theta\, d\theta = [\sec\theta]_0^{\pi/4}$$

$$= \sec\frac{\pi}{4} - \sec 0 = \sqrt{2} - 1$$

EXERCISES

Evaluate the following integrals:

1. $\displaystyle\int (1+\sin x)^3 \cos x\, dx$

2. $\displaystyle\int \frac{\cot x}{\log\sin x}\, dx$

3. $\displaystyle\int \frac{\tan(\sin^{-1}x)}{\sqrt{1-x^2}}\, dx$

4. $\displaystyle\int \frac{1}{\sqrt{1-\cos 2x}}\, dx$

5. $\displaystyle\int \frac{1}{a\cos x + b\sin x}\, dx$

6. $\displaystyle\int \frac{1}{\sqrt{9+4x^2}}\, dx$

7. $\displaystyle\int \frac{e^x}{\sqrt{e^{2x}-1}}\, dx$

8. $\displaystyle\int \operatorname{cosec}^3 x\, dx$

9. $\int \frac{\cot(\log x)}{x}\,dx$

10. $\int \frac{1}{(1+x^2)\tan^{-1} x}\,dx$

11. $\int \frac{e^{m \sin^{-1} x}}{\sqrt{1-x^2}}\,dx$

12. $\int x^2 \log x\,dx$

13. $\int x^2 (\log x)^2\,dx$

14. $\int x \sin^{-1} x\,dx$

15. $\int x \tan^{-1} x\,dx$

16. $\int_1^2 x \log x\,dx$

17. $\int_0^{\pi/2} x^2 \sin x\,dx$

18. $\int_0^{\infty} \frac{x \tan^{-1} x}{(1+x^2)^2}\,dx$

19. $\int_0^1 \sin^{-1} x\,dx$

20. $\int_0^{\pi/2} \sqrt{1+\sin x}\,dx$

21. $\int_0^{\infty} \frac{\sin(\tan^{-1} x)}{1+x^2}\,dx$

22. $\int_0^{\pi/4} \cos 3x \cos 5x\,dx$

23. $\int_0^{\pi/2} e^x (\sin x + \cos x)\,dx$

2

Integration of Rational Functions

2.1 Rational Functions

The quotient of one polynomial in x divided by another polynomial in x is called a *rational function.* Thus, a rational function is an expression of the form

$$\frac{a_0 x^m + a_1 x^{m-1} + \ldots + a_{m-1} x + a_m}{b_0 x^n + b_1 x^{n-1} + \ldots + b_{n-1} x + b_n} \tag{1}$$

where m, n are positive integers and $a_0, a_1, \ldots a_m$; $b_0, b_1, \ldots, b_n$ are constants. It may be supposed that the numerator and denominator of Eq. (1) have no common factor, for, if there is any, it may be cancelled.

A rational function is said to be *proper* if the degree m of the numerator is less than the degree n of the denominator. If a rational function is not proper, i.e. $m \geq n$, then by division it can be reduced to the sum of a polynomial and a proper rational function. In this case if we divide the numerator by the denominator, we shall obtain a quotient $Q(x)$ which will be a polynomial in x and a remainder which will be of lower degree than n. If in Eq. (1) $f(x)$ be the remainder and $g(x)$ the denominator, then it can be written as

$$Q(x) + \frac{f(x)}{g(x)}$$

where $f(x)/g(x)$ is a proper rational function.

The integration of such proper rational functions can be obtained by decomposing it into fractions with simpler denominators called the *partial fractions* of the given function. For this purpose, the denominator $g(x)$, which is a polynomial in x, is decomposed into its simplest factors which must be linear or quadratic and one or more of which may be repeated several times. Thus, $g(x)$ is expressed as the product of a series of factors of the form

$$(x - a),\ (x - a)^r,\ (x^2 + px + q),\ (x^2 + px + q)^s$$

where r and s are positive integers. We shall consider these four kinds of factors of the denominator separately.

2.2 When the Denominator Consists of Unrepeated Linear Factors

We explain the method by solving the problems.

Examples

1. Evaluate $\int \frac{x^2}{(x+1)(x-2)(x+3)} dx$.

Solution. We have

$$\frac{x^2}{(x+1)(x-2)(x+3)} = \frac{A}{x+1} + \frac{B}{x-2} + \frac{C}{x+3}$$

$$\Rightarrow \quad x^2 = A(x-2)(x+3) + B(x+1)(x+3) + C(x+1)(x-2)$$

Putting $x = -1, 2, -3$, respectively, we get

$$A = -\frac{1}{6}, \quad B = \frac{4}{15}, \quad C = \frac{9}{10}$$

Therefore

$$\int \frac{x^2}{(x+1)(x-2)(x+3)} dx = -\frac{1}{6}\int \frac{1}{x+1} dx + \frac{4}{15}\int \frac{1}{x-2} dx + \frac{9}{10}\int \frac{1}{x+3} dx$$

$$= -\frac{1}{6}\log|x+1| + \frac{4}{15}\log|x-2| + \frac{9}{10}\log|x+3|$$

2. Evaluate $\int \frac{x^5}{x^3 - 2x^2 - 5x + 6} dx$.

Solution. We have

$$\frac{x^5}{x^3 - 2x^2 - 5x + 6} = x^2 + 2x + 9 + \frac{22x^2 + 33x - 54}{x^3 - 2x^2 - 5x + 6}$$

Now

$$\frac{22x^2 + 33x - 54}{x^3 - 2x^2 - 5x + 6} = \frac{22x^2 + 33x - 54}{(x-1)(x+2)(x-3)} = \frac{A}{x-1} + \frac{B}{x+2} + \frac{C}{x-3}$$

$$\Rightarrow \quad 22x^2 + 33x - 54 = A(x+2)(x-3) + B(x-1)(x-3) + C(x-1)(x+2)$$

Putting $x = 1, -2, 3$, respectively, we get

$$A = -\frac{1}{6},\ B = -\frac{32}{15},\ C = \frac{243}{10}$$

Therefore

$$\int \frac{x^5}{x^3 - 2x^2 - 5x + 6}\,dx = \int (x^2 + 2x + 9)\,dx - \frac{1}{6}\int \frac{1}{x-1}\,dx$$

$$- \frac{32}{15}\int \frac{1}{x+2}\,dx + \frac{243}{10}\int \frac{1}{x-3}\,dx$$

$$= \frac{1}{3}x^3 + x^2 + 9x - \frac{1}{6}\log|x-1| - \frac{32}{15}\log|x+2| + \frac{243}{10}\log|x-3|$$

EXERCISES

Evaluate

1. $\int \frac{1}{(x+1)(x+2)(x+3)}\,dx$
2. $\int \frac{x^2}{(x+1)(x+2)(x+3)}\,dx$
3. $\int \frac{1}{(x^2 - 3x + 2)}\,dx$
4. $\int \frac{(x^2 + x + 2)}{(x-1)(x-2)}\,dx$
5. $\int \frac{x^3}{(x-1)(x-2)(x-3)}\,dx$
6. $\int \frac{(x^2 + 1)}{(x-1)(x+1)(2x+1)}\,dx$
7. $\int \frac{(x^2 + x - 1)}{x^3 + x^2 - 6x}\,dx$
8. $\int \frac{x}{x^3 - 6x^2 + 11x - 6}\,dx$
9. $\int \frac{x^2}{(x-1)(3x-1)(3x-2)}\,dx$
10. $\int \frac{(x^2 + 5x + 41)}{(x-1)(x+3)(2x-1)}\,dx$

2.3 When the Denominator Consists of Repeated and Unrepeated Linear Factors

We now consider the case where the denominator $g(x)$ has a factor of the form $(x - a)^r$ so that $g(x) = (x - a)^r h(x)$, $g(x)$ and $h(x)$ being of the nth and $(n - r)$th degrees, respectively. We may assume

$$\frac{f(x)}{g(x)} = \frac{f(x)}{(x-a)^r h(x)} = \frac{A_1}{x-a} + \frac{A_2}{(x-a)^2} + \ldots + \frac{A_r}{(x-a)^r} + \frac{f_1(x)}{h(x)}$$

where $f_1(x)/h(x)$ represents the sum of all the partial fractions of $f(x)/g(x)$ except $\frac{A_1}{x-a}, \frac{A_2}{(x-a)^2}, \ldots, \frac{A_r}{(x-a)^r}$. Also $f_1(x)$ is an expression of at most $(n - r - 1)$th degree.

The methods for finding the values of $A_1, A_2, \ldots, A_r$ are illustrated in the solved examples below.

Examples

1. Evaluate $\int \frac{x^2}{(x-1)^3(x-2)}\,dx.$

Solution. Let

$$\frac{x^2}{(x-1)^3(x-2)} = \frac{A}{x-1} + \frac{B}{(x-1)^2} + \frac{C}{(x-1)^3} + \frac{D}{x-2}$$

$$\Rightarrow \quad x^2 \equiv A(x-1)^2(x-2) + B(x-1)(x-2) + C(x-2) + D(x-1)^3$$

$$\Rightarrow \quad x^2 = A(x^3 - 4x^2 + 5x - 2) + B(x^2 - 3x + 2) + C(x-2) + D(x^3 - 3x^2 + 3x - 1) \qquad (1)$$

Putting $x = 1, 2$ in Eq. (1), we get $C = -1$, $D = 4$.

To find A, B we equate the coefficients of x^3, x^2 on both sides of Eq. (1) and get

$$0 = A + D$$

$$1 = -4A + B - 3D$$

Solving these equations, we get

$$A = -4, B = -3$$

Therefore

$$\int \frac{x^3}{(x-1)^3(x-2)}\,dx = -4\int \frac{1}{x-1}dx - 3\int \frac{1}{(x-1)^2}dx - \int \frac{1}{(x-1)^3}dx + 4\int \frac{1}{x-2}dx$$

$$= -4\log|x-1| + \frac{3}{x-1} + \frac{1}{2(x-1)^2} + 4\log|x-2|$$

2. Evaluate $\int \frac{(x^3+2)}{(x-1)(x-2)^3}\,dx.$

Solution. Let

$$\frac{x^3+2}{(x-1)(x-2)^3} = \frac{A}{x-1} + \frac{B}{x-2} + \frac{C}{(x-2)^2} + \frac{D}{(x-2)^3}$$

$$\Rightarrow x^3 + 2 = A(x-2)^3 + B(x-1)(x-2)^2 + C(x-1)(x-2) + D(x-1)$$

$$\Rightarrow \quad x^3 + 2 = A(x^3 - 6x^2 + 12x + 8) + B(x^3 - 5x^2 + 8x - 4) + C(x^2 - 3x + 2) + D(x-1) \qquad (1)$$

Putting $x = 1, 2$ in Eq. (1), we get $A = -3$, $D = 10$.

To find B, C we equate the coefficients of x^3, x^2 on both sides of Eq. (1) and get

$$1 = A + B$$

$$0 = -6A - 5B + C$$

Solving these equations, we get

$$B = 4, \quad C = 2$$

Therefore

$$\int \frac{x^3 + 2}{(x-1)(x-2)^3} dx = -3 \int \frac{1}{x-1} dx + 4 \int \frac{1}{x-2} dx$$

$$+ 2 \int \frac{1}{(x-2)^2} dx + 10 \int \frac{1}{(x-2)^3} dx$$

$$= -3 \log |x-1| + 4 \log |x-2| - \frac{2}{x-2} - \frac{5}{(x-2)^2}$$

EXERCISES

Evaluate

1. $\int \frac{(x^2+1)}{(x+1)^3 (x-2)} dx$
2. $\int \frac{(x+1)}{x^4 (x-1)} dx$
3. $\int \frac{(3x+1)}{(x+1)(x-1)^3} dx$
4. $\int \frac{x}{(x-1)^3 (x-2)} dx$
5. $\int \frac{(x+1)}{(x-1)^2 (x+2)^2} dx$
6. $\int \frac{1}{x^2 (x^2-1)^2} dx$
7. $\int \frac{(x^2+1)}{(x-1)(x+2)^3} dx$
8. $\int \frac{(x^3 - 4x^2 + 5x - 2)}{x^3 + 4x^2 + 5x + 2} dx$
9. $\int \frac{1}{x^3 (x-1)^2 (x+1)} dx$
10. $\int_0^{1/2} \frac{1}{(1-x^2)^2} dx$

2.4 When the Denominator Contains Unrepeated Quadratic Factors

Let $f(x)/g(x)$ be the given fraction and let $g(x)$ have an unrepeated factor of the form $x^2 + px + q$ which cannot be further resolved into real linear factors. Then, we can write

$$\frac{f(x)}{g(x)} = \frac{Cx + D}{x^2 + px + q} + \frac{h(x)}{\phi(x)} \tag{1}$$

where $h(x)/\phi(x)$ stands for the sum of all the partial fractions exclusive of $(Cx + D)/(x^2 + px + q)$. Multiplying both sides of Eq. (1) by $g(x)$, we get

$$f(x) = (Cx + D)\,\phi(x) + h(x)(x^2 + px + q) \tag{2}$$

The values of C and D can be obtained by equating the coefficients of two suitable powers of x on the two sides of Eq. (2).

Examples

1. Evaluate $\displaystyle\int \frac{1}{x^3 - 1}\,dx$.

Solution. Let

$$\frac{1}{x^3 - 1} = \frac{1}{(x - 1)(x^2 + x + 1)} = \frac{A}{x - 1} + \frac{Bx + C}{x^2 + x + 1}$$

$$\Rightarrow \qquad 1 = A(x^2 + x + 1) + (Bx + C)(x - 1)$$

$$\Rightarrow \qquad 1 = A(x^2 + x + 1) + Bx^2 - Bx + Cx - C \tag{1}$$

Putting $x = 1$ in Eq. (1), we get $A = \frac{1}{3}$.

To find B and C we equate the coefficients of x^2, x on both sides of Eq. (1) and get

$$0 = A + B$$

$$0 = A - B + C$$

Solving these equations, we get $B = -\frac{1}{3}$, $C = -\frac{2}{3}$.

Therefore

$$\int \frac{1}{x^3 - 1}\,dx = \frac{1}{3}\int \frac{1}{x - 1}\,dx - \frac{1}{3}\int \frac{(x + 2)}{x^2 + x + 1}\,dx$$

$$= \frac{1}{3}\log|x - 1| - \frac{1}{6}\int \frac{(2x + 1)}{x^2 + x + 1}\,dx - \frac{1}{2}\int \frac{1}{x^2 + x + 1}\,dx$$

$$= \frac{1}{3}\log|x - 1| - \frac{1}{6}\log|x^2 + x + 1| - \frac{1}{2}\int \frac{1}{\left(x + \frac{1}{2}\right)^2 + \left(\frac{\sqrt{3}}{2}\right)^2}\,dx$$

$$= \frac{1}{3}\log|x - 1| - \frac{1}{6}\log|x^2 + x + 1| - \frac{1}{\sqrt{3}}\tan^{-1}\left(\frac{2x + 1}{\sqrt{3}}\right)$$

2. Evaluate $\int \frac{1}{(x+1)^2(x^2+1)}\,dx.$

Solution. Let

$$\frac{1}{(x+1)^2(x^2+1)} = \frac{A}{x+1} + \frac{B}{(x+1)^2} + \frac{Cx+D}{x^2+1}$$

$$\Rightarrow \quad 1 = A(x+1)(x^2+1) + B(x^2+1) + (Cx+D)(x+1)^2$$

$$\Rightarrow \quad 1 = A(x^3+x^2+x+1) + B(x^2+1) + Cx^3 + (D+2C)x^2$$

$$+ (2D+C)\,x + D \qquad (1)$$

Putting $x = -1$ in Eq. (1), we get $B = \frac{1}{2}$.

To find A, C and D we equate the coefficients of x^3, x^2, x, 1 on both sides of Eq. (1) and get

$$0 = A + C$$

$$0 = A + B + D + 2C$$

$$0 = A + 2D + C$$

$$1 = A + B + D$$

Solving these equations, we get

$$A = \frac{1}{2},\ C = -\frac{1}{2},\ D = 0$$

Therefore

$$\int \frac{1}{(x+1)^2(x^2+1)}\,dx = \frac{1}{2}\int \left[\frac{1}{x+1} + \frac{1}{(x+1)^2} - \frac{x}{x^2+1}\right]dx$$

$$= \frac{1}{2}\int \frac{1}{x+1}\,dx + \frac{1}{2}\int \frac{1}{(x+1)^2}\,dx - \frac{1}{4}\int \frac{2x}{x^2+1}\,dx$$

$$= \frac{1}{2}\log|x+1| - \frac{1}{2(x+1)} - \frac{1}{4}\log|x^2+1|$$

EXERCISES

Evaluate

1. $\int \frac{x}{x^3+1}\,dx$

2. $\int \frac{x}{x^3+x^2+x+1}\,dx$

3. $\int \frac{x^3}{x^4+3x^2+2}\,dx$

4. $\int \frac{1}{(x-1)^2(x^2+4)}\,dx$

5. $\int \frac{1}{x^4 - 1} dx$

6. $\int \frac{x^4}{(x-1)^2(x^2+4)} dx$

7. $\int \frac{1}{(x^2+1)(x^2+x+1)} dx$

8. $\int \frac{x^2}{(x-1)^2(x^2+1)} dx$

9. $\int \frac{1}{x^4+x^2+1} dx$

10. $\int_0^1 \frac{x^3}{(x^2+1)(x^2+7x+12)} dx$

2.5 Miscellaneous Methods

The general method of integrating a rational function has been described in detail in the preceding sections. We now proceed to illustrate some methods by means of which rational function of certain forms can be integrated more expeditiously. Very often a given rational function can be reduced to a simpler one by a suitable substitution, which can then be integrated by the methods given above. No general rules can be laid down for the substitution to be tried in any particular case, but one or two cases are of special importance and we proceed to solve some problems of these types.

Case I. If either the numerator or the denominator of a rational function contains only odd powers of x while the other contains even powers of x, the substitution $x^2 = t$ will always simplify the given function considerably.

EXAMPLES

1. Evaluate $\int \frac{1}{x^3(1+x^2)} dx$.

Solution. We put $x^2 = t$. Then $2x\,dx = dt$.

Therefore

$$\int \frac{1}{x^3(1+x^2)} dx = \frac{1}{2}\int \frac{2x}{x^4(1+x^2)} dx = \frac{1}{2}\int \frac{1}{t^2(1+t)} dt$$

$$= \frac{1}{2}\int \left(\frac{1}{t^2} - \frac{1}{t} + \frac{1}{1+t}\right) dt = \frac{1}{2}\left(-\frac{1}{t} - \log t + \log|1+t|\right)$$

$$= -\frac{1}{2x^2} - \frac{1}{2}\log x^2 + \frac{1}{2}\log|1+x^2| = -\frac{1}{2x^2} - \log|x| + \frac{1}{2}\log|1+x^2|$$

2. Evaluate $\int \frac{(x^2+1)}{x^4+1} dx$.

Solution. On dividing numerator and denominator by x^2, we get

$$\int \frac{(x^2+1)}{x^4+1}dx = \int \frac{\left(1+\frac{1}{x^2}\right)}{x^2+\frac{1}{x^2}}dx = \int \frac{\left(1+\frac{1}{x^2}\right)}{\left(x-\frac{1}{x}\right)^2+2}dx$$

Putting $x - \frac{1}{x} = t$, we get $\left(1+\frac{1}{x^2}\right)dx = dt$. Therefore, the given integral

$$\int \frac{1}{t^2+2}dt = \int \frac{1}{t^2+(\sqrt{2})^2}dt = \frac{1}{\sqrt{2}}\tan^{-1}\frac{t}{\sqrt{2}} = \frac{1}{\sqrt{2}}\tan^{-1}\frac{\left(x-\frac{1}{x}\right)}{\sqrt{2}}$$

$$= \frac{1}{\sqrt{2}}\tan^{-1}\frac{(x^2-1)}{x\sqrt{2}}$$

3. Evaluate $\int \frac{x}{x^4+x^2+1}dx.$

Solution. Substituting $x^2 = t$, we have $2x\,dx = dt$. Therefore

$$\int \frac{x}{x^4+x^2+1}dx = \frac{1}{2}\int \frac{2x}{x^4+x^2+1}dx = \frac{1}{2}\int \frac{1}{t^2+t+1}dt$$

$$= \frac{1}{2}\int \frac{1}{\left(t+\frac{1}{2}\right)^2+\left(\frac{\sqrt{3}}{2}\right)^2}dt = \frac{1}{2}\cdot\frac{1}{\frac{\sqrt{3}}{2}}\tan^{-1}\frac{\left(t+\frac{1}{2}\right)}{\frac{\sqrt{3}}{2}}$$

$$= \frac{1}{\sqrt{3}}\tan^{-1}\frac{(2x^2+1)}{\sqrt{3}}$$

4. Evaluate $\int \frac{(x^2-1)}{x^4+1}dx.$

Solution. On dividing numerator and denominator by x^2, we get

$$\int \frac{(x^2-1)}{x^4+1}dx = \int \frac{\left(1-\frac{1}{x^2}\right)}{x^2+\frac{1}{x^2}}dx = \int \frac{\left(1-\frac{1}{x^2}\right)}{\left(x+\frac{1}{x}\right)^2-2}dx$$

Let $x + \frac{1}{x} = t$. Then $\left(1 - \frac{1}{x^2}\right) dx = dt$. Therefore

$$\int \frac{(x^2 - 1)}{x^4 + 1} dx = \int \frac{1}{t^2 - 2} dt = \int \frac{1}{t^2 - (\sqrt{2})^2} dt = \frac{1}{2\sqrt{2}} \log \left| \frac{t - \sqrt{2}}{t + \sqrt{2}} \right|$$

$$= \frac{1}{2\sqrt{2}} \log \left| \frac{x + \frac{1}{x} - \sqrt{2}}{x + \frac{1}{x} + \sqrt{2}} \right| = \frac{1}{2\sqrt{2}} \log \left| \frac{x^2 - x\sqrt{2} + 1}{x^2 + x\sqrt{2} + 1} \right|$$

Case II. If the numerator and the denominator of the integrand are functions of x^2, then we may regard x^2 as a new variable, i.e. we may first put t for the purpose of resolution into partial fractions (but not for the purpose of integration) and then integrate after replacing t by x^2.

Examples

1. Evaluate $\int \frac{x^2}{(x^2 + 1)(2x^2 + 1)} dx$.

Solution. Since the integrand is a function of x^2 only, putting $x^2 = t$, we get

$$\frac{x^2}{(x^2 + 1)(2x^2 + 1)} = \frac{t}{(t + 1)(2t + 1)} = \frac{1}{t + 1} - \frac{1}{2t + 1} = \frac{1}{x^2 + 1} - \frac{1}{2x^2 + 1}$$

Therefore $\int \frac{x^2}{(x^2 + 1)(2x^2 + 1)} dx = \int \frac{1}{x^2 + 1} dx - \int \frac{1}{2x^2 + 1} dx$

$$= \int \frac{1}{x^2 + 1} dx - \frac{1}{2} \int \frac{1}{x^2 + \frac{1}{2}} dx = \tan^{-1} x - \frac{1}{2} \cdot \frac{1}{\frac{1}{\sqrt{2}}} \tan^{-1} \frac{x}{\frac{1}{\sqrt{2}}}$$

$$= \tan^{-1} x - \frac{1}{\sqrt{2}} \tan^{-1} (x\sqrt{2})$$

2. Evaluate $\int \frac{(x^2 + 1)(x^2 + 2)}{(x^2 + 3)(x^2 + 4)} dx$.

Solution. Since the integrand is a function of x^2 only, putting $x^2 = t$, we get

$$\frac{(x^2 + 1)(x^2 + 2)}{(x^2 + 3)(x^2 + 4)} = \frac{(t + 1)(t + 2)}{(t + 3)(t + 4)} = \frac{t^2 + 3t + 2}{t^2 + 7t + 12} = 1 + \frac{-4t - 10}{(t + 3)(t + 4)}$$

$$= 1 + \frac{2}{t + 3} - \frac{6}{t + 4} = 1 + \frac{2}{x^2 + 3} - \frac{6}{x^2 + 4}$$

Therefore

$$\int \frac{(x^2+1)(x^2+2)}{(x^2+3)(x^2+4)}\,dx = \int dx + 2\int \frac{1}{x^2+3}\,dx - 6\int \frac{1}{x^2+4}\,dx$$

$$= x + 2\cdot\frac{1}{\sqrt{3}}\tan^{-1}\frac{x}{\sqrt{3}} - 6\cdot\frac{1}{2}\tan^{-1}\frac{x}{2}$$

$$= x + \frac{2}{\sqrt{3}}\tan^{-1}\frac{x}{\sqrt{3}} - 3\tan^{-1}\frac{x}{2}$$

EXERCISES

Evaluate the following:

1. $\displaystyle\int \frac{2x}{(x^2+1)(x^2+3)}\,dx$

2. $\displaystyle\int \frac{1}{x(x^2+1)(x^2+2)^2}\,dx$

3. $\displaystyle\int \frac{1}{x(x^5+1)}\,dx$

4. $\displaystyle\int \frac{1}{x(x^2+1)^3}\,dx$

5. $\displaystyle\int \frac{1}{x(1+x^6)}\,dx$

6. $\displaystyle\int \frac{(x^2-2)}{(x^2+2)^3}\,dx$

7. $\displaystyle\int_2^3 \frac{x^2}{(x^2-1)(x^2+2)}\,dx$

8. $\displaystyle\int \frac{(x^2+4)}{(x^2+1)\,(x^2+3)}\,dx$

9. $\displaystyle\int \frac{(x^2-1)}{x^4+x^2+1}\,dx$

   $\left[\text{Put } x+\frac{1}{x}=t\right]$

10. $\displaystyle\int \frac{(x^2+1)}{x^4-x^2+1}\,dx$

11. $\displaystyle\int \frac{(x^2+1)}{(x^2-1)^2}\,dx$

2.6 Integration of Rational Function of e^x

A rational function of e^x is transformed into a rational function of t by the substitution $e^x = t$. Therefore, the integral of such a function is obtained by using previous methods.

Examples

1. Evaluate $\displaystyle\int \frac{1}{(1+e^x)(1+e^{-x})}\,dx.$

Solution. Let $e^x = t$. Then $dx = \frac{1}{t}\,dt$. Therefore

$$\int \frac{1}{(1+e^x)(1+e^{-x})}\,dx = \int \frac{1}{(1+t)\left(1+\frac{1}{t}\right)}\cdot\frac{1}{t}\,dt = \int \frac{1}{(1+t)^2}\,dt$$

$$= -\frac{1}{1+t} = -\frac{1}{1+e^x}$$

2. Evaluate $\int \frac{1}{(e^x - 1)(e^x + 3)} dx.$

Solution. Let $e^x = t$. Then $dx = \frac{1}{t} dt$. Therefore

$$\int \frac{1}{(e^x - 1)(e^x + 3)} dx = \int \frac{1}{(t - 1)(t + 3)} \cdot \frac{1}{t} dt = \int \frac{1}{t(t - 1)(t + 3)} dt$$

$$= \int \left[-\frac{1}{3t} + \frac{1}{4(t - 1)} + \frac{1}{12(t + 3)} \right] dt$$

$$= -\frac{1}{3} \log |t| + \frac{1}{4} \log |t - 1| + \frac{1}{12} \log |t + 3|$$

$$= -\frac{1}{3} \log e^x + \frac{1}{4} \log |e^x - 1| + \frac{1}{12} \log |e^x + 3|$$

$$= -\frac{1}{3} x + \frac{1}{4} \left[\log |e^x - 1| + \frac{1}{3} \log |e^x + 3| \right]$$

3. Evaluate $\int \frac{1}{1 + e^x - 2e^{2x}} dx.$

Solution. Let $e^x = t$. Then $dx = \frac{1}{t} dt$. Therefore

$$\int \frac{1}{1 + e^x - 2e^{2x}} dx = \int \frac{1}{(1 + t - 2t^2)} \cdot \frac{1}{t} dt = \int \frac{1}{t(1 - t)(1 + 2t)} dt$$

$$= \int \left[\frac{1}{t} + \frac{1}{3(1 - t)} - \frac{4}{3(1 + 2t)} \right] dt$$

$$= \log |t| - \frac{1}{3} \log |1 - t| - \frac{4}{3} \cdot \frac{1}{2} \log |1 + 2t|$$

$$= \log e^x - \frac{1}{3} \log |1 - e^x| - \frac{2}{3} \log |1 + 2e^x|$$

$$= x - \frac{1}{3} \log |1 - e^x| (1 + 2e^x)^2$$

4. Evaluate $\int \frac{1}{1 + \cosh x} dx.$

Solution. We know that $\cosh x = \frac{e^x + e^{-x}}{2}$. Therefore

$$\int \frac{1}{1 + \cosh x} dx = \int \frac{1}{1 + \frac{e^x + e^{-x}}{2}} dx = \int \frac{2}{2 + e^x + e^{-x}} dx$$

Let $e^x = t$. Then $dx = \frac{1}{t}\,dt$. Now

$$\int \frac{1}{1+\cosh x}\,dx = \int \frac{2}{\left(2+t+\frac{1}{t}\right)}\cdot\frac{1}{t}\,dt = \int \frac{2}{t^2+2t+1}\,dt = \int \frac{2}{(t+1)^2}\,dt$$

$$= -\frac{2}{t+1} = -\frac{2}{e^x+1}$$

EXERCISES

Evaluate the following:

1. $\int \frac{1}{1+3e^x+2e^{2x}}\,dx$
2. $\int \frac{1}{e^x+e^{2x}}\,dx$
3. $\int \frac{1}{(e^x-1)^2}\,dx$
4. $\int_0^1 \frac{1}{2e^x-1}\,dx$
5. $\int \frac{(4e^x+6e^{-x})}{9e^x-4e^{-x}}\,dx$

MISCELLANEOUS EXERCISES

Evaluate

1. $\int \frac{1}{x-x^3}\,dx$
2. $\int \frac{x}{(x-a)(x-b)(x-c)}\,dx$
3. $\int \frac{(x^2+1)}{(x^2-1)(x^2-4)}\,dx$
4. $\int \frac{(x^2-3x+3)}{x^3-4x^2-7x+10}\,dx$
5. $\int \frac{(x^3+1)}{x^4-3x^3+3x^2-x}\,dx$
6. $\int \frac{(x^2+1)}{(x^2-1)^2}\,dx$
7. $\int \frac{x^3}{(x-1)(x+2)(x+1)^4}\,dx$
8. $\int \frac{1}{x^4+1}\,dx$
9. $\int \frac{1}{(x-1)^3(x^3+1)}\,dx$
10. $\int \frac{1}{(x-1)^2(x^2+1)}\,dx$
11. $\int \frac{1}{1-x^6}\,dx$
12. $\int \frac{1}{(x^2+a^2)^2}\,dx$
13. $\int \frac{x^2}{x^4+1}\,dx$
14. $\int \frac{x^2}{x^4+x^2+1}\,dx$
15. $\int \frac{(x+1)^2}{x^4+x^2+1}\,dx$
16. $\int \frac{(x^2+1)}{(x^2+2)(2x^2+1)}\,dx$
17. $\int \frac{1}{(x^2+a^2)(x^2+b^2)}\,dx$
18. $\int \frac{1}{(x-1)^2(x-2)(x^2+4)}\,dx$

19. $\displaystyle\int \frac{(x^2+1)}{x^4+x^2+1}\,dx$

20. $\displaystyle\int \frac{(3x^2-5x+4)}{x^3+2x^2+3x+6}\,dx$

21. $\displaystyle\int \frac{(3x^2+x-2)}{(x+1)^3(x^2+1)}\,dx$

22. $\displaystyle\int \frac{(57x^3-25x^2+9x-1)}{(x-1)^2(2x-1)(5x-1)}\,dx$

23. $\displaystyle\int_0^{\pi/2} \frac{\cos x}{(1+\sin x)(2+\sin x)}\,dx$

24. $\displaystyle\int_0^{\pi/4} \sqrt{\tan\theta}\,d\theta$

[Put $\tan\theta = t^2$]

25. $\displaystyle\int_0^{\pi/4} \sqrt{\cot\theta}\,d\theta$

Show that

26. $\displaystyle\int_2^3 \frac{(x^2+1)}{(2x+1)(x-1)(x+1)}\,dx = \frac{7}{3}\log 2 - \frac{5}{6}\log\frac{7}{5} - \log 3$

27. $\displaystyle\int_0^\infty \frac{x^2}{(x^2+a^2)(x^2+b^2)(x^2+c^2)}\,dx = \frac{\pi}{2(a+b)(b+c)(c+a)}$

3

Integration of Irrational Algebraic Functions

3.1 Introduction

This chapter discusses the methods of integration of some simple irrational algebraic functions. These are usually integrated by reducing them to rational or other more easily integrable functions by suitable substitutions.

3.2 Integration of $f(x)\sqrt{ax+b}$ or $f(x)/\sqrt{ax+b}$, where $f(x)$ is a Rational Function of x

Integral of such functions can be evaluated in all cases by substituting $\sqrt{ax+b} = t$.

Examples

1. Evaluate $\displaystyle\int \frac{1}{(x-1)\sqrt{x+2}}\,dx.$

Solution. Let $x + 2 = t^2$. Then $dx = 2t\,dt$ and $x - 1 = t^2 - 3.$

Now

$$\int \frac{1}{(x-1)\sqrt{x+2}}\,dx = \int \frac{2t}{(t^2-3)t}\,dt = 2\int \frac{1}{t^2-3}\,dt$$

$$= \frac{1}{\sqrt{3}} \log \left|\frac{t-\sqrt{3}}{t+\sqrt{3}}\right| = \frac{1}{\sqrt{3}} \log \left|\frac{\sqrt{x+2}-\sqrt{3}}{\sqrt{x+2}+\sqrt{3}}\right|$$

2. Evaluate $\displaystyle\int \frac{(x^2-x+1)}{(x+1)\sqrt{x-2}}\,dx.$

Solution. Let $x - 2 = t^2$. Then

$$dx = 2t\,dt,\ x + 1 = t^2 + 3 \quad\text{and}\quad x^2 - x + 1 = t^4 + 3t^2 + 3$$

Now

$$\int \frac{(x^2 - x + 1)}{(x + 1)\sqrt{x - 2}}\, dx = \int \frac{(t^4 + 3t^2 + 3)\cdot 2t}{(t^2 + 3)\,t}\, dt = 2\int \left(t^2 + \frac{3}{t^2 + 3}\right) dt$$

$$= 2\left[\frac{t^3}{3} + 3\cdot\frac{1}{\sqrt{3}}\tan^{-1}\frac{t}{\sqrt{3}}\right] = 2\left[\frac{(x-2)^{3/2}}{3} + \sqrt{3}\tan^{-1}\sqrt{\frac{x-2}{3}}\right]$$

EXERCISES

Evaluate

1. $\int \frac{(x^2 + 1)}{(3x + 2)\sqrt{x - 1}}\, dx$
2. $\int \frac{(x^2 + 2x + 1)}{(x - 1)\sqrt{x + 2}}\, dx$
3. $\int \frac{1}{x^2\sqrt{x + 1}}\, dx$
4. $\int \frac{1}{(4x + 3)\sqrt{x + 1}}\, dx$
5. $\int \frac{(2 - 3x)}{x\sqrt{1 + x}}\, dx$
6. $\int \frac{x^3}{(x - 1)\sqrt{x + 2}}\, dx$
7. $\int \frac{(x + 3)}{(x - 3)^{\frac{1}{3}}}\, dx$
8. Show that $\int_8^{15} \frac{1}{(x - 3)\sqrt{x + 1}}\, dx = \frac{1}{2}\log\frac{5}{3}$.
9. Show that $\int \frac{1}{(1 + x)\sqrt{1 - x}}\, dx = \sqrt{2}\tan^{-1}\sqrt{\frac{1 - x}{2}}$.

3.3 Integration of $\frac{1}{(ax + b)\sqrt{px^2 + qx + r}}$ or $\frac{1}{(ax + b)^n\sqrt{px^2 + qx + r}}$

Integrals of such functions can be evaluated in all cases by substituting $ax + b = \frac{1}{t}$.

Examples

1. Evaluate $\int \frac{1}{(1 + x)\sqrt{1 + 2x - x^2}}\, dx$.

Solution. Let $1 + x = \frac{1}{t}$. Then

$$dx = -\frac{1}{t^2}\, dt \quad \text{and} \quad x = \frac{1 - t}{t}$$

Now

$$\int \frac{1}{(1+x)\sqrt{1+2x-x^2}}\,dx = \int \frac{1}{\frac{1}{t}\sqrt{1+\frac{2(1-t)}{t}-\left(\frac{1-t}{t}\right)^2}}\cdot\left(-\frac{1}{t^2}\right)dt$$

$$= -\int \frac{1}{\sqrt{t^2+2t(1-t)-(1-t)^2}}\,dt = -\int \frac{1}{\sqrt{-2t^2+4t-1}}\,dt$$

$$= -\frac{1}{\sqrt{2}}\int \frac{1}{\sqrt{\frac{1}{2}-(t-1)^2}}\,dt = -\frac{1}{\sqrt{2}}\sin^{-1}((t-1)\sqrt{2})$$

$$= -\frac{1}{\sqrt{2}}\sin^{-1}\left(\frac{-x\sqrt{2}}{1+x}\right)$$

2. Evaluate $\displaystyle\int \frac{1}{(x-3)^2\sqrt{x^2-6x+8}}\,dx.$

Solution. Let $x-3=\frac{1}{t}$. Then

$$dx = -\frac{1}{t^2}\,dt \quad \text{and} \quad x = \frac{3t+1}{t}$$

Now $\displaystyle\int \frac{1}{(x-3)^2\sqrt{x^2-6x+8}}\,dx = \int \frac{1}{\frac{1}{t^2}\sqrt{\frac{1-t^2}{t^2}}}\left(-\frac{1}{t^2}\right)dt$

$$= \int \frac{(-t)}{\sqrt{1-t^2}}\,dt = \sqrt{1-t^2} = \sqrt{1-\frac{1}{(x-3)^2}} = \sqrt{\frac{x^2-6x+8}{x-3}}$$

EXERCISES

Evaluate the following:

1. $\displaystyle\int \frac{1}{x\sqrt{x^2+2x-1}}\,dx$

2. $\displaystyle\int \frac{1}{(x+2)\sqrt{x^2+6x+7}}\,dx$

3. $\displaystyle\int \frac{1}{(x-1)\sqrt{x^2+x+1}}\,dx$

4. $\displaystyle\int \frac{1}{(2-x)\sqrt{1-2x+3x^2}}\,dx$

5. $\displaystyle\int \frac{1}{(x+1)\sqrt{2x^2+3x+4}}\,dx$

6. $\displaystyle\int \frac{1}{(x+2)^2\sqrt{3x^2+4x-5}}\,dx$

7. Show that $\displaystyle\int_1^2 \frac{1}{(x+1)\sqrt{x^2-1}}\,dx = \frac{1}{\sqrt{3}}$

8. Show that $\displaystyle\int_1^2 \frac{1}{(x+1)\sqrt{x^2+x+1}}\,dx = \log\frac{\sqrt{3}\,(1+2\sqrt{7})}{9}$

3.4 Integration of $\dfrac{1}{(ax^2+bx+c)\sqrt{px^2+qx+r}}$

Integrals of such functions can be evaluated in all cases by substituting $\dfrac{px^2+qx+r}{ax^2+bx+c} = t^2$.

Examples

1. Evaluate $\displaystyle\int \frac{1}{(1+x^2)\sqrt{2+x^2}}\,dx.$

Solution. Let $\dfrac{2+x^2}{1+x^2} = t^2$. Then

$$-\frac{x}{(1+x^2)^2}\,dx = t\,dt \quad \text{and} \quad x^2 = \frac{t^2-2}{1-t^2}$$

Now

$$\int \frac{1}{(1+x^2)\sqrt{2+x^2}}\,dx = \int \frac{1}{(1+x^2)\sqrt{2+x^2}}\,\frac{(1+x^2)^2}{(-x)}\,t\,dt$$

$$= -\int \frac{(1+x^2)\,t\,dt}{x\sqrt{2+x^2}} = -\int \sqrt{\frac{1+x^2}{2+x^2}}\cdot\frac{\sqrt{1+x^2}}{x}\,t\,dt$$

$$= -\int \frac{1}{t}\cdot\sqrt{1+\frac{1}{x^2}}\cdot t\,dt = -\int \sqrt{1+\frac{1-t^2}{t^2-2}}\,dt$$

$$= -\int \frac{1}{\sqrt{2-t^2}}\,dt = \cos^{-1}\frac{t}{\sqrt{2}} = \cos^{-1}\left(\sqrt{\frac{2+x^2}{2(1+x^2)}}\right)$$

2. Evaluate $\displaystyle\int \frac{1}{(2x^2+3)\sqrt{3x^2-4}}\,dx.$

Solution. We put $\dfrac{3x^2-4}{2x^2+3} = t^2$. Then

$$dx = \frac{(2x^2+3)^2}{17x}\cdot t\,dt \quad \text{and} \quad x^2 = \frac{3t^2+4}{3-2t^2}$$

Now

$$\int \frac{1}{(2x^2+3)\sqrt{3x^2-4}}\,dx = \int \frac{1}{(2x^2+3)\sqrt{3x^2-4}}\cdot\frac{(2x^2+3)^2}{17x}\cdot t\,dt$$

$$= \frac{1}{17}\int \sqrt{\frac{2x^2+3}{3x^2-4}}\cdot\sqrt{\frac{2x^2+3}{x^2}}\cdot t\,dt = \frac{1}{17}\int \frac{1}{t}\cdot\sqrt{2+\frac{3}{x^2}}\cdot t\,dt$$

$$= \frac{1}{17}\int \sqrt{2+3\left(\frac{3-2t^2}{3t^2+4}\right)}\,dt = \frac{\sqrt{17}}{17}\int \frac{1}{\sqrt{3t^2+4}}\,dt$$

$$= \frac{1}{\sqrt{51}}\int \frac{1}{\sqrt{t^2+\left(\frac{2}{\sqrt{3}}\right)^2}}\,dt = \frac{1}{\sqrt{51}}\sinh^{-1}\frac{t}{\frac{2}{\sqrt{3}}} = \frac{1}{\sqrt{51}}\sinh^{-1}\sqrt{\frac{9x^2-12}{8x^2+12}}$$

EXERCISES

Evaluate the following:

1. $\int \frac{1}{(1+x^2)\sqrt{1-x^2}}\,dx$
2. $\int \frac{1}{(x^2-1)\sqrt{x^2+1}}\,dx$
3. $\int \frac{1}{(x^2+1)\sqrt{x^2-1}}\,dx$
4. $\int \frac{1}{(2x^2+3)\sqrt{x^2-4}}\,dx$
5. $\int \frac{(x+1)}{(x^2+4)\sqrt{x^2+9}}\,dx$
6. $\int \frac{1}{(x^2+x+1)\sqrt{x^2+x+2}}\,dx$
7. Show that $\int_0^1 \frac{1}{(1+x^2)\sqrt{1-x^2}}\,dx = \frac{\pi}{2\sqrt{2}}$
8. Show that $\int_0^{1/2} \frac{1}{(1-2x^2)\sqrt{1-x^2}}\,dx = \frac{1}{2}\log(2+\sqrt{3})$

MISCELLANEOUS EXERCISES

Evaluate the following:

1. $\int \frac{x^4}{(x-1)\sqrt{x+2}}\,dx$
2. $\int \frac{1}{(2x+3)\sqrt{4x+5}}\,dx$

3. $\int \frac{1}{(x^2 - 4)\sqrt{x + 1}}\,dx$

4. $\int \frac{(x + 2)}{(x^2 + 3x + 3)\sqrt{x + 1}}\,dx$

5. $\int \frac{1}{(x - 1)\sqrt{x^2 + 1}}\,dx$

6. $\int_0^4 \frac{1}{(1 + 2x)\sqrt{2 + x^2}}\,dx$

7. $\int \frac{(x + 1)}{(x + 2)\sqrt{2x^2 - 3x + 1}}\,dx$

8. $\int \frac{(x^2 + 2x - 1)}{(x + 2)\sqrt{2x^2 + 3x - 4}}\,dx$

9. $\int \frac{1}{(2 + 3x)\sqrt{4 - x^2}}\,dx$

10. $\int \frac{1}{(x - 1)\sqrt{x^2 + 1}}\,dx$

11. $\int \frac{1}{(1 + x)\sqrt{1 + x - x^2}}\,dx$

12. $\int \frac{1}{x(x + 1)\sqrt{x^2 + x - 1}}\,dx$

13. $\int \frac{1}{(x^2 - 1)\sqrt{1 + x^2}}\,dx$

14. $\int \frac{(x^2 + 2x + 3)}{(x^2 + 1)\sqrt{x^2 - 2}}\,dx$

15. $\int \frac{(x^3 + x^2 + x + 1)}{(x^2 + 2)\sqrt{x^2 - 3}}\,dx$

16. $\int \frac{(1 - x^2)}{(1 + x^2)\sqrt{1 + x^4}}\,dx$

17. $\int \frac{(x^4 - 1)}{x^2\sqrt{x^4 + x^2 + 1}}\,dx$

4

Integration of Trigonometric Functions

4.1 Introduction

In this chapter, we consider the integrals of some simple trigonometric functions. In many cases the integration of trigonometrical expressions can be reduced to those of algebraic expressions by means of suitable substitutions. We now proceed to discuss the integration of trigonometric functions.

4.2 Integration of a Rational Function of sin x and cos x

When the integrand is a rational function of sin x and cos x, the suitable substitution $\tan \frac{x}{2} = t$ will always transform the integrand to a rational function of t. Then, this function may be evaluated by the methods given in Chapter 2. We have

$$dx = \frac{2}{1 + t^2}\, dt$$

Also

$$\sin x = \frac{2 \tan \frac{x}{2}}{1 + \tan^2 \frac{x}{2}} = \frac{2t}{1 + t^2}$$

and

$$\cos x = \frac{1 - \tan^2 \frac{x}{2}}{1 + \tan^2 \frac{x}{2}} = \frac{1 - t^2}{1 + t^2}$$

Examples

1. Evaluate $\int \frac{1}{a + b \cos x}\, dx$.

Solution. Put $\tan \frac{x}{2} = t$. Then

$$dx = \frac{2}{1+t^2}\,dt,\ \cos x = \frac{1-t^2}{1+t^2}$$

Now

$$\int \frac{1}{a + b\cos x}\,dx = \int \frac{1}{a + b\left(\frac{1-t^2}{1+t^2}\right)} \cdot \frac{2\,dt}{(1+t^2)}$$

$$= \int \frac{2}{a(1+t^2) + b(1-t^2)}\,dt = \int \frac{2}{(a+b) + (a-b)t^2}\,dt$$

$$= \frac{2}{(a-b)} \int \frac{1}{t^2 + \frac{a+b}{a-b}}\,dt$$

We see that $\frac{a+b}{a-b}$ is positive or negative according as

$$a^2 > b^2 \quad \text{or} \quad a^2 < b^2$$

Case I. Let $a^2 > b^2$. Then $\frac{a+b}{a-b}$ is positive and we have

$$\int \frac{1}{a + b\cos x}\,dx = \frac{2}{(a-b)} \int \frac{1}{t^2 + \left(\sqrt{\frac{a+b}{a-b}}\right)^2}\,dt$$

$$= \frac{2}{(a-b)} \cdot \frac{1}{\sqrt{\frac{a+b}{a-b}}} \tan^{-1} \frac{t}{\sqrt{\frac{a+b}{a-b}}} = \frac{2}{\sqrt{a^2-b^2}} \tan^{-1}\left(\sqrt{\frac{a-b}{a+b}}\tan\frac{x}{2}\right)$$

Case II. Let $a^2 < b^2$. Then $\frac{a+b}{a-b}$ is negative and we have

$$\int \frac{1}{a + b\cos x}\,dx = \frac{2}{(a-b)} \int \frac{1}{t^2 - \left(\frac{b+a}{b-a}\right)}\,dt$$

$$= \frac{2}{(a-b)} \int \frac{1}{t^2 - \left(\sqrt{\frac{b+a}{b-a}}\right)^2}\,dt = \frac{2}{(b-a)} \int \frac{1}{\left(\sqrt{\frac{b+a}{b-a}}\right)^2 - t^2}\,dt$$

$$= \frac{2}{(b-a)} \cdot \frac{1}{2\sqrt{\frac{b+a}{b-a}}} \log \left| \frac{\sqrt{\frac{b+a}{b-a}} + t}{\sqrt{\frac{b+a}{b-a}} - t} \right|$$

$$= \frac{1}{\sqrt{b^2 - a^2}} \log \left| \frac{\sqrt{b+a} + \sqrt{b-a} \tan \frac{x}{2}}{\sqrt{b+a} - \sqrt{b-a} \tan \frac{x}{2}} \right|$$

Case III. Let $a^2 = b^2$. Then, either $a = b$ or $a = -b$. If $a = b$, then

$$\int \frac{1}{a + b\cos x}\, dx = \int \frac{2}{2a}\, dt = \frac{1}{a} t = \frac{1}{a} \tan \frac{x}{2}$$

Further, if $a = -b$, then

$$\int \frac{1}{a + b\cos x}\, dx = \frac{1}{a} \int \frac{1}{t^2}\, dt = -\frac{1}{at} = -\frac{1}{a} \cot \frac{x}{2}$$

2. Evaluate $\int \frac{1}{a + b\sin x}\, dx$.

Solution. We put $\tan \frac{x}{2} = t$. Then

$$dx = \frac{2}{1+t^2}\, dt, \ \sin x = \frac{2t}{1+t^2}$$

Now, assuming $a \neq 0$, we have

$$\int \frac{1}{a + b\sin x}\, dx = \int \frac{1}{a + b \cdot \frac{2t}{1+t^2}} \cdot \frac{2\,dt}{(1+t^2)} = \frac{2}{a} \int \frac{1}{t^2 + \frac{2bt}{a} + 1}\, dt$$

$$= \frac{2}{a} \int \frac{1}{\left(t + \frac{b}{a}\right)^2 + \frac{a^2 - b^2}{a^2}}\, dt$$

There arise three cases according as $a^2 - b^2$ is positive, zero or negative.

Case I. Let $a^2 > b^2$. Then, $a^2 - b^2$ is positive and we have

$$\int \frac{1}{a + b\sin x}\, dx = \frac{2}{a} \int \frac{1}{\left(t + \frac{b}{a}\right)^2 + \left(\frac{\sqrt{a^2 - b^2}}{a}\right)^2}\, dt$$

$$= \frac{2}{a} \cdot \frac{1}{\frac{\sqrt{a^2 - b^2}}{a}} \tan^{-1}\left(\frac{t + \frac{b}{a}}{\frac{\sqrt{a^2 - b^2}}{a}}\right) = \frac{2}{\sqrt{a^2 - b^2}} \tan^{-1}\left(\frac{a \tan \frac{x}{2} + b}{\sqrt{a^2 - b^2}}\right)$$

Case II. Let $a^2 = b^2$. Then, $a^2 - b^2$ is zero and we have

$$\int \frac{1}{a + b \sin x}\, dx = \frac{2}{a} \int \frac{1}{\left(t + \frac{b}{a}\right)^2}\, dt = -\frac{2}{a\left(t + \frac{b}{a}\right)} = -\frac{2}{a \tan \frac{x}{2} + b}$$

Case III. Let $a^2 < b^2$. Then, $a^2 - b^2$ is negative and we have

$$\int \frac{1}{a + b \sin x}\, dx = \frac{2}{a} \int \frac{1}{\left(t + \frac{b}{a}\right)^2 - \frac{b^2 - a^2}{a^2}}\, dt$$

$$= \frac{2}{a} \int \frac{1}{\left(t + \frac{b}{a}\right)^2 - \left(\frac{\sqrt{b^2 - a^2}}{a}\right)^2}\, dt$$

$$= \frac{2}{a} \cdot \frac{1}{\frac{2\sqrt{b^2 - a^2}}{a}} \log \left| \frac{t + \frac{b}{a} - \frac{\sqrt{b^2 - a^2}}{a}}{t + \frac{b}{a} + \frac{\sqrt{b^2 - a^2}}{a}} \right|$$

$$= \frac{1}{\sqrt{b^2 - a^2}} \log \left| \frac{a \tan \frac{x}{2} + b - \sqrt{b^2 - a^2}}{a \tan \frac{x}{2} + b + \sqrt{b^2 - a^2}} \right|$$

3. Show that $\int_0^{\pi} \frac{1}{3 + 2 \sin x + \cos x}\, dx = \frac{\pi}{4}$.

Solution. We put $\tan \frac{x}{2} = t$. Then

$$dx = \frac{2}{1 + t^2}\, dt, \quad \sin x = \frac{2t}{1 + t^2} \text{ and } \cos x = \frac{1 - t^2}{1 + t^2}$$

Now

$$\int \frac{1}{3 + 2 \sin x + \cos x}\, dx = \int \frac{1}{3 + 2 \cdot \frac{2t}{1 + t^2} + \frac{1 - t^2}{1 + t^2}} \cdot \frac{2}{(1 + t^2)}\, dt$$

$$= \int \frac{2}{2t^2 + 4t + 4}\, dt = \int \frac{1}{t^2 + 2t + 2}\, dt$$

$$= \int \frac{1}{(t+1)^2 + 1}\, dt = \tan^{-1}(t+1) = \tan^{-1}\left(\tan\frac{x}{2} + 1\right)$$

Therefore $\displaystyle\int_0^{\pi} \frac{1}{3 + 2\sin x + \cos x}\, dx = \left[\tan^{-1}\left(\tan\frac{x}{2} + 1\right)\right]_0^{\pi}$

$$= \tan^{-1}\infty - \tan^{-1}1 = \frac{\pi}{2} - \frac{\pi}{4} = \frac{\pi}{4}$$

EXERCISES

Evaluate the following:

1. $\displaystyle\int \frac{1}{3 + 4\cos x}\, dx$
2. $\displaystyle\int \frac{1}{5 + 4\cos x}\, dx$
3. $\displaystyle\int \frac{1}{5 + 4\sin x}\, dx$
4. $\displaystyle\int_0^{\pi/2} \frac{1}{1 + 2\cos x}\, dx$
5. $\displaystyle\int_0^{\pi/2} \frac{1}{4 + 5\sin x}\, dx$
6. $\displaystyle\int_0^{\pi} \frac{1}{5 + 3\cos x}\, dx$
7. $\displaystyle\int_0^{\pi/2} \frac{1}{1 + 2\sin x + \cos x}\, dx$
8. $\displaystyle\int_0^{\pi} \frac{1}{1 - 2a\cos x + a^2}\, dx$

Show that

9. $\displaystyle\int_0^{\pi} \frac{1}{a + b\cos x}\, dx = \frac{\pi}{\sqrt{a^2 - b^2}}$ if $a^2 > b^2$
10. $\displaystyle\int_0^{\pi/2} \frac{1}{5 + 4\cos x}\, dx = \frac{2}{3}\tan^{-1}\frac{1}{3}$
11. $\displaystyle\int_{-\pi/2}^{\pi/2} \frac{1}{5 + 7\cos x + \sin x}\, dx = \frac{1}{5}\log 6$
12. $\displaystyle\int_0^{\alpha} \frac{1}{1 - \cos\alpha\cos x}\, dx = \frac{\pi}{2}\operatorname{cosec}\alpha$
13. $\displaystyle\int_0^{\alpha} \frac{1}{\cos\alpha + \cos x}\, dx = \operatorname{cosec}\alpha \log\sec\alpha$
14. $\displaystyle\int_0^{\pi/2} \frac{1}{4 + 5\cos x}\, dx = \frac{1}{3}\log 2$

4.3 Integration of $\dfrac{a\cos x + b\sin x}{c\sin x + d\cos x}$

Integral of such functions may also be obtained by introducing two constants λ and μ and determine their values such that

$a \cos x + b \sin x \equiv \lambda \times$ Denominator $+ \mu \times$ Derivative of denominator

$$\equiv \lambda(c \sin x + d \cos x) + \mu(c \cos x - d \sin x)$$

Equating the coefficients of $\cos x$ and $\sin x$ on both sides, we get

$$\lambda d + \mu c = a \quad \text{and} \quad \lambda c - \mu d = b$$

$$\Rightarrow \qquad \lambda = \frac{ad + bc}{c^2 + d^2}, \quad \mu = \frac{ac - bd}{c^2 + d^2}$$

Therefore

$$\int \frac{(a \cos x + b \sin x)}{c \sin x + d \cos x}\, dx = \lambda \int dx + \mu \int \frac{(c \cos x - d \sin x)}{c \sin x + d \cos x}\, dx$$

$$= \lambda x + \mu \log |c \sin x + d \cos x|$$

where λ and μ have the values found above.

4.4 Integration of $\dfrac{a \cos x + b \sin x + c}{p \cos x + q \sin x + r}$

Integral of such functions may be obtained by introducing three constants λ, μ and ν and determine their values such that

$$a \cos x + b \sin x + c \equiv \lambda \times \text{Denominator}$$

$$+ \mu \times \text{Derivative of denominator} + \nu$$

$$\equiv \lambda(p \cos x + q \sin x + r) + \mu(-p \sin x + q \cos x) + \nu$$

$$\Rightarrow \qquad p\lambda + q\mu = a,\ q\lambda - p\mu = b,\ r\lambda + \nu = c$$

With these values of λ, μ and ν, we have

$$\int \frac{(a \cos x + b \sin x + c)}{(p \cos x + q \sin x + r)}\, dx = \int \lambda\, dx + \mu \int \frac{(-p \sin x + q \cos x)}{p \cos x + q \sin x + r}\, dx$$

$$+ \nu \int \frac{1}{p \cos x + q \sin x + r}\, dx$$

where λ, μ and ν have the values found above.

Examples

1. Evaluate $\displaystyle\int \frac{(2 \sin x + 3 \cos x)}{3 \sin x + 4 \cos x}\, dx.$

Solution. We introduce two constants λ and μ and determine their values such that

$$2 \sin x + 3 \cos x \equiv \lambda(3 \sin x + 4 \cos x) + \mu(3 \cos x - 4 \sin x)$$

$$\Rightarrow \qquad 3\lambda - 4\mu = 2,\ 4\lambda + 3\mu = 3$$

On solving these equations, we get

$$\lambda = \frac{18}{25},\ \mu = \frac{1}{25}$$

Therefore

$$\int \frac{(2 \sin x + 3 \cos x)}{3 \sin x + 4 \cos x}\, dx = \frac{18}{25} \int dx + \frac{1}{25} \int \frac{(3 \cos x - 4 \sin x)}{3 \sin x + 4 \cos x}\, dx$$

$$= \frac{18}{25} x + \frac{1}{25} \log |3 \sin x + 4 \cos x|.$$

2. Evaluate $\displaystyle\int \frac{(\cos x + 2 \sin x + 3)}{4 \cos x + 5 \sin x + 6}\, dx.$

Solution. We introduce three constants λ, μ and v and determine their values such that

$$\cos x + 2 \sin x + 3 \equiv \lambda(4 \cos x + 5 \sin x + 6) + \mu(-4 \sin x + 5 \cos x) + v$$

$$\Rightarrow \qquad 4\lambda + 5\mu = 1, \quad 5\lambda - 4\mu = 2, \quad 6\lambda + v = 3$$

On solving these equations, we get

$$\lambda = \frac{14}{41}, \quad \mu = -\frac{3}{41}, \quad v = \frac{39}{41}$$

Therefore

$$\int \frac{(\cos x + 2 \sin x + 3)}{4 \cos x + 5 \sin x + 6}\, dx = \frac{14}{41} \int dx - \frac{3}{41} \int \frac{(-4 \sin x + 5 \cos x)}{4 \cos x + 5 \sin x + 6}\, dx$$

$$+ \frac{39}{41} \int \frac{1}{4 \cos x + 5 \sin x + 6}\, dx$$

$$= \frac{14x}{41} - \frac{3}{41} \log |\, 4 \cos x + 5 \sin x + 6 \,|$$

$$+ \frac{39}{41} \int \frac{1}{4 \cos x + 5 \sin x + 6}\, dx$$

Now, we evaluate $\displaystyle\int \frac{1}{4 \cos x + 5 \sin x + 6}\, dx$ separately. Putting $\tan \frac{x}{2} = t$, we get

$$dx = \frac{2}{1+t^2}\,dt, \quad \sin x = \frac{2t}{1+t^2} \quad \text{and} \quad \cos x = \frac{1-t^2}{1+t^2}$$

Then

$$\int \frac{1}{4\cos x + 5\sin x + 6}\,dx = \int \frac{1}{4\left(\frac{1-t^2}{1+t^2}\right) + \frac{5.2t}{1+t^2} + 6} \cdot \frac{2}{(1+t^2)}\,dt$$

$$= \int \frac{2}{2t^2 + 10t + 10}\,dt = \int \frac{1}{t^2 + 5t + 5}\,dt = \int \frac{1}{\left(t+\frac{5}{2}\right)^2 - \left(\frac{\sqrt{5}}{2}\right)^2}\,dt$$

$$= \frac{1}{2\frac{\sqrt{5}}{2}} \log \left| \frac{t + \frac{5}{2} - \frac{\sqrt{5}}{2}}{t + \frac{5}{2} + \frac{\sqrt{5}}{2}} \right| = \frac{1}{\sqrt{5}} \log \left| \frac{2\tan\frac{x}{2} + 5 - \sqrt{5}}{2\tan\frac{x}{2} + 5 + \sqrt{5}} \right|$$

Thus

$$\int \frac{(\cos x + 2\sin x + 3)}{4\cos x + 5\sin x + 6}\,dx = \frac{14x}{41} - \frac{3}{41} \log |\, 4\cos x + 5\sin x + 6 \,|$$

$$+ \frac{39}{41} \cdot \frac{1}{\sqrt{5}} \log \left| \frac{2\tan\frac{x}{2} + 5 - \sqrt{5}}{2\tan\frac{x}{2} + 5 + \sqrt{5}} \right|$$

3. Evaluate $\displaystyle\int \frac{1}{a + b\tan x}\,dx.$

Solution. The given integral can be written as

$$\int \frac{\cos x}{a\cos x + b\sin x}\,dx$$

We now introduce two constants λ and μ and determine their values such that

$$\cos x \equiv \lambda(a\cos x + b\sin x) + \mu(-a\sin x + b\cos x)$$

$$\Rightarrow \qquad a\lambda + b\mu = 1, \; b\lambda - a\mu = 0$$

On solving these equations, we get

$$\lambda = \frac{a}{a^2 + b^2}, \quad \mu = \frac{b}{a^2 + b^2}$$

Therefore

$$\int \frac{1}{a + b \tan x} dx = \frac{a}{a^2 + b^2} \int dx + \frac{b}{a^2 + b^2} \int \frac{(-a \sin x + b \cos x)}{a \cos x + b \sin x} dx$$

$$= \frac{ax}{a^2 + b^2} + \frac{b}{a^2 + b^2} \log |a \cos x + b \sin x|$$

EXERCISES

Evaluate the following:

1. $\int \frac{1}{1 + \tan x} dx$
2. $\int_0^{\frac{\pi}{2}} \frac{1}{1 + \cot x} dx$
3. $\int \frac{(3 \cos x + 4 \sin x)}{4 \cos x + 5 \sin x} dx$
4. $\int \frac{(3 \cos x + 2)}{\sin x + 2 \cos x + 3} dx$
5. $\int \frac{\cos x}{3 \cos x + 4 \sin x} dx$
6. $\int \frac{(\sin x + 2 \cos x)}{2 \sin x + \cos x} dx$
7. $\int \frac{\cos x}{2 \sin x + 3 \cos x} dx$
8. $\int \frac{(3 \sin x + 4 \cos x)}{\sin x + \cos x} dx$

MISCELLANEOUS EXERCISES

Evaluate the following:

1. $\int_0^{\pi} \frac{1}{2 + \cos x} dx$
2. $\int \frac{\operatorname{cosec} x}{2 + \operatorname{cosec} x} dx$
3. $\int \frac{1}{2 + 4 \sin x} dx$
4. $\int \frac{(\sin x + \cos x)}{3 \sin x + 4 \cos x + 1} dx$
5. $\int \frac{\sqrt{2}}{2\sqrt{2} + \sin x + \cos x} dx$
6. $\int \frac{1}{\cos x (5 + 3 \cos x)} dx$
7. $\int_0^{\pi/2} \frac{1}{4 + 3 \cos x} dx$
8. $\int \frac{1}{3 \cos x - 4 \sin x + 5} dx$
9. $\int \frac{(5 \cos x + 6)}{2 \cos x + \sin x + 3} dx$
10. $\int \frac{1}{3 \sin x + 4 \cos x} dx$
11. $\int \frac{1}{5 \sin x + 12 \cos x} dx$
12. $\int \frac{1}{13 + 3 \cos x + 4 \sin x} dx$
13. $\int \frac{(2 \cos x + 3 \sin x)}{3 \cos x + 2 \sin x} dx$
14. $\int_0^{\pi} \frac{(3 + 4 \sin x + 2 \cos x)}{3 + 2 \sin x + \cos x} dx$

5

Definite Integrals

5.1 Introduction

We have so far considered the process of integration as the inverse of differentiation and, in Sec. 1.7, defined the definite integral $\int_a^b f(x)\,dx$ as the difference $F(b) - F(a)$, where $F(x)$ is an indefinite integral of $f(x)$. We shall now define the definite integral as the limit of a sum which not only does not depend upon our previous knowledge of $F(x)$ the primitive of $f(x)$, but actually may be used to define $F(x)$ if $f(x)$ is continuous. We shall also prove the *Fundamental Theorem* of Integral Calculus which establishes the equivalence of the two definitions.

5.2 Definite Integral as the Limit of a Sum

Let $[a, b]$ be a closed interval. Divide $[a, b]$ into n sub-intervals and let the points of division be $x_0, x_1, x_2, \ldots, x_n$, where

$$a = x_0 \le x_1 \le x_2 \le \ldots \le x_n = b$$

Let $\Delta x_1 = x_1 - x_0$, $\Delta x_2 = x_2 - x_1, \ldots, \Delta x_n = x_n - x_{n-1}$. Now, let t_i be any value of x in the ith subinterval so that

$$x_{i-1} \le t_i \le x_i \ (i = 1, 2, \ldots, n)$$

Let f be a bounded real valued function on $[a, b]$. Then, from the sum

$$S_n = \sum_{i=1}^{n} f(t_i)\Delta x_i = f(t_1)\Delta x_1 + f(t_2)\Delta x_2 + \ldots + f(t_n)\Delta x_n$$

The limit of S_n if it exists, when the number of sub-intervals tends to infinity in such a manner that the length of each sub-interval tends to zero, is defined to be the *definite integral* of $f(x)$ over the interval $[a, b]$ and is denoted by

$$\int_a^b f(x)\,dx$$

As in Sec. 1.7, a is called the *lower* or *inferior limit* and b the *upper* or *superior limit.*

Note. Let the sub-intervals Δx_i be of equal length h. Then $h = (b - a)/n$, so that $h \to 0$ as $n \to \infty$. Choose the points t_i such that

$$t_1 = a + h,\ t_2 = a + 2h,\ \ldots,\ t_n = a + nh$$

Then $$\int_a^b f(x)\,dx = \lim_{h\to 0} h[f(a+h) + f(a+2h) + \ldots + f(a+nh)]$$
$$= \lim_{h\to 0} h \sum_{r=1}^{n} f(a+rh)$$

Examples

1. Evaluate $\int_a^b x^3 dx$ as the limit of a sum.

Solution. Here $f(x) = x^3$, so dividing the interval $[a, b]$ into n equal sub-intervals each of length h. Then

$$S_n = h[f(a+h) + f(a+2h) + \ldots + f(a+nh)]$$
$$= h[(a+h)^3 + (a+2h)^3 + \ldots + (a+nh)^3]$$
$$= h[na^3 + 3a^2h(1 + 2 + 3 + \ldots + n) + 3ah^2(1^2 + 2^2 + \ldots + n^2)$$
$$+ h^3(1^3 + 2^3 + \ldots + n^3)]$$
$$= h\left[na^3 + 3a^2h \cdot \frac{n(n+1)}{2} + 3ah^2 \cdot \frac{n(n+1)(2n+1)}{6} + \frac{h^3 \cdot n^2(n+1)^2}{4}\right]$$
$$= nha^3 + 3a^2h^2 \cdot \frac{n(n+1)}{2} + 3ah^3 \cdot \frac{n(n+1)(2n+1)}{6} + h^4 \cdot \frac{n^2(n+1)^2}{4}$$
$$= a^3(b-a) + \frac{3}{2}a^2(b-a)^2 \cdot \frac{(n+1)}{n} + \frac{1}{2}a(b-a)^3 \cdot \frac{(n+1)(2n+1)}{n^2}$$
$$+ \frac{1}{4}(b-a)^4 \cdot \frac{(n+1)^2}{n^2}$$

replacing h by $(b - a)/n$. Now

$$\int_a^b x^3 dx = \lim_{n\to\infty} S_n \qquad (\text{when } n \to \infty,\ h \to 0)$$
$$= \lim_{n\to\infty}\left[a^3(b-a) + \frac{3}{2}a^2(b-a)^2 \cdot \frac{(n+1)}{n}\right.$$
$$\left. + \frac{1}{2}a(b-a)^3 \cdot \frac{(n+1)(2n+1)}{n^2} + \frac{1}{4}(b-a)^4 \cdot \frac{(n+1)^2}{n^2}\right]$$

$$= a^3(b-a) + \frac{3}{2}a^2(b-a)^2 + a(b-a)^3 + \frac{1}{4}(b-a)^4$$

$$= \frac{1}{4}(b^4 - a^4)$$

2. Evaluate $\int_a^b \sin x\, dx$ as the limit of a sum.

Solution. We have

$$\int_a^b \sin x dx = \lim_{h\to 0} h[f(a+h) + f(a+2h) + \ldots + f(a+nh)]$$

$$= \lim_{h\to 0} h[\sin(a+h) + \sin(a+2h) + \cdots + \sin(a+nh)]$$

Let $\quad S_n = h[\sin(a+h) + \sin(a+2h) + \cdots + \sin(a+nh)]$.

Multiplying both sides by $2\sin\frac{h}{2}$, we get

$$2\sin\frac{h}{2}\cdot S_n = h\left[2\sin(a+h)\sin\frac{h}{2} + 2\sin(a+2h)\sin\frac{h}{2}\right.$$

$$\left. + \ldots + \sin(a+nh)\sin\frac{h}{2}\right]$$

$$= h\left[\cos\left(a+\frac{h}{2}\right) - \cos\left(a+\frac{3h}{2}\right) + \cos\left(a+\frac{3h}{2}\right) - \cos\left(a+\frac{5h}{2}\right)\right.$$

$$\left. + \cdots + \cos\left(a+\frac{(2n-1)h}{2}\right) - \cos\left(a+\frac{(2n+1)h}{2}\right)\right]$$

$$= h\left[\cos\left(a+\frac{h}{2}\right) - \cos\left(a+\frac{(2n+1)h}{2}\right)\right]$$

Therefore

$$\int_a^b \sin x\, dx = \lim_{h\to 0}\frac{h}{2\sin\frac{h}{2}}\left[\cos\left(a+\frac{h}{2}\right) - \cos\left(a+\frac{(2n+1)h}{2}\right)\right]$$

$$= \lim_{h\to 0}\frac{\frac{h}{2}}{\sin\frac{h}{2}}\left[\cos\left(a+\frac{h}{2}\right) - \cos\left(b+\frac{h}{2}\right)\right] \quad (\because nh = (b-a))$$

$$= \cos a - \cos b$$

EXERCISES

Evaluate the following integrals as the limit of a sum:

1. $\int_a^b k\,dx$
2. $\int_a^b x\,dx$
3. $\int_a^b x^2\,dx$
4. $\int_a^b \frac{1}{x^2}\,dx$
5. $\int_a^b \frac{1}{\sqrt{x}}\,dx$
6. $\int_a^b \sqrt{x}\,dx$
7. $\int_a^b e^x\,dx$
8. $\int_a^b \cos x\,dx$
9. $\int_a^b \sin^2 x\,dx$

5.3 Geometrical Interpretation of a Definite Integral: Area Under a Curve

Let CD be the graph of the function $y = f(x)$ which is finite, single-valued and continuous in the closed interval $[a, b]$. Further, suppose that $f(x)$ retains its sign in $[a, b]$. We shall prove that the definite integral $\int_a^b f(x)\,dx$ represents the area under the curve CD, i.e. the area bounded by the arc CD, the x-axis and the ordinates AC and BD (Fig. 5.1).

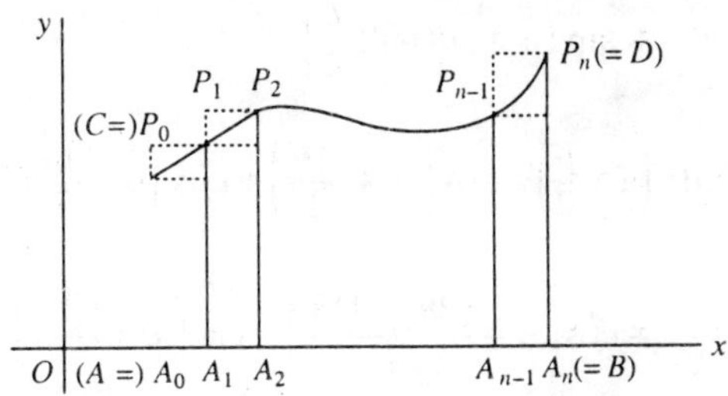

Fig. 5.1

In order to find this area, divide the interval AB on the x-axis into n subintervals. Let the points of division be $A_0(= A), A_1, A_2, \ldots, A_n(= B)$. Draw ordinates of the curve through each of these points of subdivision meeting the curve in $P_0(= C), P_1, P_2, \ldots, P_n(= D)$, respectively. Let $x_0(= a), x_1, x_2, \ldots, x_n (= b)$ be the abscissae of the points $A_0, A_1, \ldots, A_n$.

Further, let

$$A_0A_1 = x_1 - x_0 = \Delta x_1$$
$$A_1A_2 = x_2 - x_1 = \Delta x_2$$
$$\ldots \qquad \ldots \qquad \ldots$$
$$\ldots \qquad \ldots \qquad \ldots$$
$$A_{n-1}A_n = x_n - x_{n-1} = \Delta x_n$$

Let m_1 be the infimum and M_1 the supremum of the function as x changes from x_0 to x_1, m_2 the infimum and M_2 the supremum as x changes from x_1 to x_2 etc. Then, from Fig. 5.1, we have

$$m_1\Delta x_1 \le P_0A_0A_1P_1 \le M_1\,\Delta x_1$$
$$m_2\Delta x_2 \le P_1A_1A_2P_2 \le M_2\Delta x_2$$
$$\ldots \quad \ldots \quad \ldots \quad \ldots$$
$$\ldots \quad \ldots \quad \ldots \quad \ldots$$
$$m_n\Delta x_n \le P_{n-1}A_{n-1}A_nP_n \le M_n\,\Delta x_n$$

On adding, we get

$$\sum_{i=1}^{n} m_i\,\Delta x_i \le P_0A_0A_1P_1 + \ldots + P_{n-1}A_{n-1}A_nP_n \le \sum_{i=1}^{n} M_i\,\Delta x_i$$

$$\Rightarrow \qquad \sum_{i=1}^{n} m_i\,\Delta x_i \le A \le \sum_{i=1}^{n} M_i\Delta x_i \qquad (1)$$

where A denotes the required area.

Now, let the number of points of subdivision tend to infinity in such a manner that each subinterval Δx_i tends to zero. Then, from the definition of a finite integral, each of the extreme members of the inequalities (1) has the definite integral $\int_a^b f(x)dx$ as its limit, and since A lies between them, whatever the mode of subdivision, A must be equal to this limit.

Hence
$$A = \int_a^b f(x)dx$$

Note. The expression $\sum_{i=1}^{n} m_i\Delta x_i$ represents the area of the inner rectangles and tends to A from below and the expression $\sum_{i=1}^{n} M_i\Delta x_i$ represents the area of the outer rectangles and tends to A from above.

5.4 The Fundamental Theorem

In Sec. 5.3, it has been shown that the definite integral $\int_a^b f(x)dx$ represents the area under the graph CD of the curve $y = f(x)$ as x varies from a to b. We now proceed to find the same area by a different method (Fig. 5.2).

Let $A(x)$ denote the area bounded by the curve, the x-axis, the ordinate at $x = a$ and an ordinate NP through a variable point $P(x, y)$ on the curve. Then, clearly $A(x)$, the area $CANP$, is a function of x. Let x be increased by a small amount Δx and let ΔA, the area $PNLQ$, be the corresponding increment in $A(x)$. Let m and M be the infinum and supremum of the

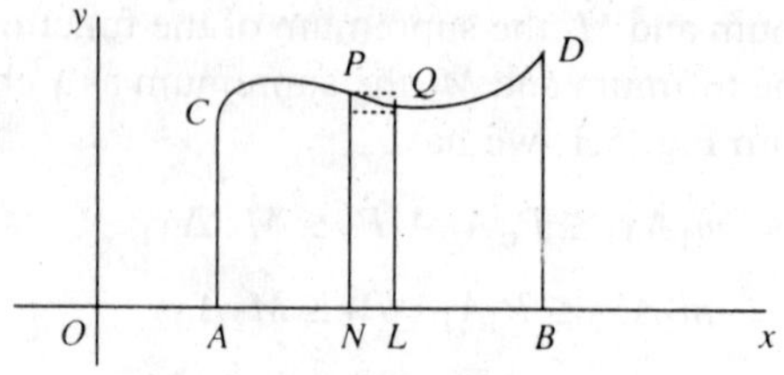

Fig. 5.2

function in the closed interval $[x, x + \Delta x]$. Then, ΔA lies between two rectangles whose base is Δx and whose heights are m and M, respectively. Thus

$$m\Delta x \leq \Delta A \leq M\Delta x \quad \left(\because m \leq \frac{\Delta A}{\Delta x} \leq M\right)$$

But the function f is continuous. Therefore, *m and M* tend to a common limit as $\Delta x \to 0$, the common limit being $f(x)$. Hence proceeding to the limit as $\Delta x \to 0$, and observing that the limit of the extreme members is $f(x)$, we have

$$\lim \frac{\Delta A}{\Delta x} = f(x) \Rightarrow \frac{dA}{dx} = f(x)$$

Thus, $A(x)$ turns out to be a function whose derivative is $f(x)$ and can therefore be determined except for an additive constant. Hence

$$A(x) = \int f(x)dx + C = F(x) + C \tag{1}$$

where $F(x)$ is an indefinite integral of $f(x)$.

To determine C, the constant of integration, we observe that for $x = a$, the area $CANP = 0$, i.e. $A(a) = 0$. Putting $x = a$ in Eq. (1), we get

$$A(a) = F(a) + C \Rightarrow 0 = F(a) + C$$

$$\Rightarrow C = -F(a)$$

Hence $\qquad A(x) = F(x) - F(a)$.

To calculate the area $CABD$, we put $x = b$. Therefore, the required area is $F(b) - F(a)$. Thus

$$\int_a^b f(x)dx = F(b) - F(a)$$

where $F(x)$ is an indefinite integral of $f(x)$.

Note. The above result which expresses the equivalence of the two definitions of a definite integral is called the *Fundamental Theorem* of Integral Calculus.

Remark. Earlier the evaluation of a definite integral as the inverse of a derivative was possible only after the integral function was known. But in view of Fundamental Theorem of Integral Calculus, it is now possible without any such knowledge. In fact, it is not always possible to find the integral function, yet it may be possible to evaluate the definite integral by considering it as the limit of a sum. At the same time, it may be observed that the process of summation is considerably more labourious, if not difficult, and is, therefore, replaced by the other method whenever convenient.

5.5 Summation of Series

There is a certain type of series the limits of whose sums (as the number of their terms increases indefinitely) can be expressed as definite integrals and can thus be evaluated easily with the help of the Fundamental Theorem. The characteristics of a series which can be so expressed will be evident if we consider the definite integral $\int_a^b f(x)\,dx$ expressed as the limit of a sum. If the interval $[a, b]$ is divided into n equal parts, each of length h, we have $b - a = nh$ and

$$\int_a^b f(x)\,dx = \lim_{h\to 0} h[f(a+h) + f(a+2h) + \ldots + f(a+nh)]$$

$$= \lim_{h\to 0} h \sum_{r=1}^{n} f(a+rh)$$

$$= (b-a) \lim_{n\to\infty} \frac{1}{n} \sum_{r=1}^{n} f\left(a + \frac{r(b-a)}{n}\right)$$

$$(\because nh = b - a \text{ and when } h \to 0,\ n \to \infty)$$

$$= (b-a) \int_0^1 f(a + (b-a)x)\,dx$$

Working Rule. The sum of the series can be expressed as definite integral by replacing r/n by x, the common difference $1/n$ by dx and $\lim\limits_{n\to\infty} \Sigma$ by $\int$.

Note. In particular

I. $$\lim_{n\to\infty} \frac{1}{n} \sum_{r=1}^{n} f\left(\frac{r}{n}\right) = \int_0^1 f(x)\,dx$$

Here $h = 1/n$; number of terms $= n$. Therefore, upper limit is $nh = 1$.

II. $$\lim_{n\to\infty} \frac{1}{n} \sum_{r=1}^{2n} f\left(\frac{r}{n}\right) = \int_0^2 f(x)\,dx$$

Here $h = 1/n$; number of terms $= 2n$. Therefore, upper limit is $2nh = 2$.

Remark. The lower limit can always be taken as zero since a definite integral can always be so transformed as to have its lower limit zero.

Examples

1. Evaluate $\lim_{n\to\infty}\left(\dfrac{1}{n+1}+\dfrac{1}{n+2}+\cdots+\dfrac{1}{2n}\right)$.

Solution. We have

$$\frac{1}{n+1}+\frac{1}{n+2}+\cdots+\frac{1}{n+n}=\frac{1}{n}\left(\frac{1}{1+\frac{1}{n}}+\frac{1}{1+\frac{2}{n}}+\cdots+\frac{1}{1+n\cdot\frac{1}{n}}\right)$$

$$=h\left(\frac{1}{1+h}+\frac{1}{1+2h}+\cdots+\frac{1}{1+nh}\right)$$

where $h = 1/n$ so that $h \to 0$ as $n \to \infty$. Since the number of terms is n, the upper limit is

$$nh = n\cdot\frac{1}{n} = 1$$

Therefore

$$\lim_{n\to\infty}\left(\frac{1}{n+1}+\frac{1}{n+2}+\cdots+\frac{1}{2n}\right)=\lim_{n\to\infty}\frac{1}{n}\sum_{r=1}^{n}\frac{1}{1+\frac{r}{n}}=\int_0^1\frac{1}{1+x}\,dx$$

$$=[\log|1+x|]_0^1=\log 2$$

2. Evaluate $\lim_{n\to\infty}\left(\dfrac{1}{n+1}+\dfrac{1}{n+2}+\cdots+\dfrac{1}{3n}\right)$.

Solution. We have

$$\frac{1}{n+1}+\frac{1}{n+2}+\cdots+\frac{1}{3n}=\frac{1}{n+1}+\frac{1}{n+2}+\cdots+\frac{1}{n+2n}$$

$$=\frac{1}{n}\left(\frac{1}{1+\frac{1}{n}}+\frac{1}{1+\frac{2}{n}}+\cdots+\frac{1}{1+2n\cdot\frac{1}{n}}\right)$$

$$=h\left(\frac{1}{1+h}+\frac{1}{1+2h}+\cdots+\frac{1}{1+2nh}\right)$$

where $h = 1/n$ so that $h \to 0$ as $n \to \infty$. Since the number of terms is $2n$, the upper limit is

$$2nh = 2n\cdot\frac{1}{n} = 2$$

Therefore

$$\lim_{n\to\infty}\left(\frac{1}{n+1}+\frac{1}{n+2}+\cdots+\frac{1}{3n}\right)=\lim_{n\to\infty}\frac{1}{n}\sum_{r=1}^{2n}\frac{1}{1+\frac{r}{n}}=\int_0^2\frac{1}{1+x}\,dx$$

$$=[\log|1+x|]_0^2=\log 3$$

3. Evaluate $\lim_{n\to\infty}\left(\frac{1}{n+1}+\frac{1}{n+2}+\cdots+\frac{1}{4n}\right)$.

Solution. We have

$$\frac{1}{n+1}+\frac{1}{n+2}+\cdots+\frac{1}{4n}=\frac{1}{n+1}+\frac{1}{n+2}+\cdots+\frac{1}{n+3n}$$

$$=\frac{1}{n}\left(\frac{1}{1+\frac{1}{n}}+\frac{1}{1+\frac{2}{n}}+\cdots+\frac{1}{1+3n\cdot\frac{1}{n}}\right)$$

$$=h\left(\frac{1}{1+h}+\frac{1}{1+2h}+\cdots+\frac{1}{1+3nh}\right)$$

where $h=1/n$, so that $h\to 0$ as $n\to\infty$. Since the number of terms is $3n$, the upper limit is

$$3nh=3n\cdot\frac{1}{n}=3$$

Therefore

$$\lim_{n\to\infty}\left(\frac{1}{n+1}+\frac{1}{n+2}+\cdots+\frac{1}{4n}\right)=\lim_{n\to\infty}\frac{1}{n}\sum_{r=1}^{3n}\frac{1}{1+\frac{r}{n}}=\int_0^3\frac{1}{1+x}\,dx$$

$$=[\log|1+x|]_0^3=\log 4$$

4. Evaluate $\lim_{n\to\infty}\left(\frac{n}{n^2+1^2}+\frac{n}{n^2+2^2}+\cdots+\frac{1}{2n}\right)$.

Solution. We have

$$\frac{n}{n^2+1^2}+\frac{n}{n^2+2^2}+\ldots+\frac{1}{2n}=\frac{n}{n^2+1^2}+\frac{n}{n^2+2^2}+\ldots+\frac{n}{n^2+n^2}$$

$$=\frac{1}{n}\left(\frac{1}{1+\left(\frac{1}{n}\right)^2}+\frac{1}{1+\left(\frac{2}{n}\right)^2}+\cdots+\frac{1}{1+\left(\frac{n}{n}\right)^2}\right)$$

$$=h\left(\frac{1}{1+h^2}+\frac{1}{1+(2h)^2}+\cdots+\frac{1}{1+(nh)^2}\right)$$

where $h = 1/n$, so that $h \to 0$ as $n \to \infty$. Since the number of terms is n, the upper limit is

$$nh = n \cdot \frac{1}{n} = 1$$

Therefore $$\lim_{n\to\infty}\left(\frac{n}{n^2+1^2}+\frac{n}{n^2+2^2}+\cdots+\frac{1}{2n}\right)$$

$$= \lim_{n\to\infty}\frac{1}{n}\sum_{r=1}^{n}\frac{1}{1+\left(\frac{r}{n}\right)^2} = \int_0^1 \frac{1}{1+x^2}\,dx$$

$$= [\tan^{-1}x]_0^1 = \tan^{-1}1 - \tan^{-1}0 = \frac{\pi}{4}$$

5. Evaluate $\lim\limits_{n\to\infty}\sum\limits_{r=1}^{n}\dfrac{n^3}{(n^2+r^2)(n^2+2r^2)}$.

Solution. Since the number of terms is n, the upper limit is 1. Therefore

$$\lim_{n\to\infty}\sum_{r=1}^{n}\frac{n^3}{(n^2+r^2)(n^2+2r^2)} = \lim_{n\to\infty}\frac{1}{n}\sum_{r=1}^{n}\frac{1}{\left(1+\frac{r^2}{n^2}\right)\left(1+\frac{2r^2}{n^2}\right)}$$

$$= \int_0^1 \frac{1}{(1+x^2)(1+2x^2)}\,dx = \int_0^1\left(\frac{2}{1+2x^2}-\frac{1}{1+x^2}\right)dx$$

$$= \int_0^1\left(\frac{1}{x^2+\frac{1}{2}}-\frac{1}{1+x^2}\right)dx = \left[\frac{1}{\frac{1}{\sqrt{2}}}\tan^{-1}\frac{x}{\frac{1}{\sqrt{2}}}-\tan^{-1}x\right]_0^1$$

$$= [\sqrt{2}\tan^{-1}x\sqrt{2}-\tan^{-1}x]_0^1 = \sqrt{2}\tan^{-1}\sqrt{2}-\frac{\pi}{4}$$

6. Evaluate $\lim\limits_{n\to\infty}\left(\dfrac{\lfloor n}{n^n}\right)^{\frac{1}{n}}$

Solution. Let $y = \left(\dfrac{\lfloor n}{n^n}\right)^{\frac{1}{n}}$. Then

$$\log y = \frac{1}{n}(\log \lfloor n - n\log n)$$

$$= \frac{1}{n}(\log 1 + \log 2 + \log 3 + \ldots + \log n - n\log n)$$

$$= \frac{1}{n}\,[(\log 1 - \log n) + (\log 2 - \log n) + \ldots + (\log n - \log n)]$$

$$= \frac{1}{n}\left[\log\frac{1}{n} + \log\frac{2}{n} + \cdots + \log\frac{n}{n}\right]$$

$$= h\,[\log h + \log 2h + \ldots + \log nh]$$

where $h = \frac{1}{n}$, so that $h \to 0$ as $n \to \infty$. Since the number of terms is n, the upper limit is

$$nh = n \cdot \frac{1}{n} = 1$$

Therefore

$$\lim_{n\to\infty} \log y = \lim_{n\to\infty} \frac{1}{n}\sum_{r=1}^{n} \log\frac{r}{n} = \int_0^1 \log x\, dx = [x \log x - x]_0^1 = -1$$

Hence
$$\lim_{n\to\infty} y = e^{-1} = \frac{1}{e}$$

7. Evaluate $\lim_{n\to\infty}\left(1+\frac{1}{n}\right)\left(1+\frac{2}{n}\right)^{\frac{1}{2}}\left(1+\frac{3}{n}\right)^{\frac{1}{3}}\cdots\left(1+\frac{n}{n}\right)^{\frac{1}{n}}$.

Solution. Let $y = \left(1+\frac{1}{n}\right)\left(1+\frac{2}{n}\right)^{\frac{1}{2}}\left(1+\frac{3}{n}\right)^{\frac{1}{3}}\cdots\left(1+\frac{n}{n}\right)^{\frac{1}{n}}$. Then

$$\log y = \log\left(1+\frac{1}{n}\right) + \frac{1}{2}\log\left(1+\frac{2}{n}\right) + \cdots + \frac{1}{n}\log\left(1+\frac{n}{n}\right)$$

$$= \sum_{r=1}^{n}\frac{1}{r}\log\left(1+\frac{r}{n}\right) = \sum_{r=1}^{n}\frac{1}{n}\cdot\frac{1}{\frac{r}{n}}\log\left(1+\frac{r}{n}\right)$$

Since the number of terms is n, the upper limit is 1. Therefore

$$\lim_{n\to\infty}\log y = \lim_{n\to\infty}\frac{1}{n}\sum_{r=1}^{n}\frac{1}{\frac{r}{n}}\log\left(1+\frac{r}{n}\right) = \int_0^1 \frac{1}{x}\log(1+x)dx$$

$$= \int_0^1 \frac{1}{x}\left(x - \frac{x^2}{2} + \frac{x^3}{3} - \frac{x^4}{4} + \cdots\right) dx$$

$$= \int_0^1 \left(1 - \frac{x}{2} + \frac{x^2}{3} - \frac{x^3}{4} + \cdots\right) dx$$

$$= \left[x - \frac{x^2}{2^2} + \frac{x^3}{3^2} - \frac{x^4}{4^2} + \cdots\right]_0^1 = 1 - \frac{1}{2^2} + \frac{1}{3^2} - \frac{1}{4^2} + \cdots$$

$$= \frac{\pi^2}{12} \qquad \left(\because\ 1 - \frac{1}{2^2} + \frac{1}{3^2} - \frac{1}{4^2} + \cdots = \frac{\pi^2}{12}\right)$$

Hence $\lim\limits_{n\to\infty} y = e^{\pi^2/12}$

EXERCISES

Find the limits of the following sums, when $n \to \infty$:

1. $\dfrac{n}{(n+1)^2} + \dfrac{n}{(n+2)^2} + \dfrac{n}{(n+3)^2} + \cdots + \dfrac{n}{(n+n)^2}$

2. $\dfrac{1}{n} + \dfrac{n^2}{(n+1)^3} + \dfrac{n^2}{(n+2)^3} + \cdots + \dfrac{1}{8n}$

3. $\dfrac{1}{1+n^3} + \dfrac{4}{8+n^3} + \dfrac{9}{27+n^3} + \dfrac{16}{64+n^3} + \cdots + \dfrac{1}{2n}$

4. $\dfrac{1}{n^2}\sec^2\dfrac{1}{n^2} + \dfrac{2}{n^2}\sec^2\dfrac{2^2}{n^2} + \dfrac{3}{n^2}\sec^2\dfrac{3^2}{n^2} + \cdots + \dfrac{n}{n^2}\sec^2\dfrac{n^2}{n^2}$

5. $\dfrac{1}{\sqrt{n^2}} + \dfrac{1}{\sqrt{n^2-1^2}} + \dfrac{1}{\sqrt{n^2-2^2}} + \cdots + \dfrac{1}{\sqrt{n^2-(n-1)^2}}$

6. $\dfrac{1}{\sqrt{2n-1^2}} + \dfrac{1}{\sqrt{4n-2^2}} + \dfrac{1}{\sqrt{6n-3^2}} + \cdots + \dfrac{1}{n}$

7. $\dfrac{n}{(n+1)\sqrt{2n+1}} + \dfrac{n}{(n+2)\sqrt{2(2n+2)}} + \dfrac{n}{(n+3)\sqrt{3(2n+3)}} + \cdots + \dfrac{n}{2n\sqrt{n\cdot 3n}}$

8. Evaluate $\lim\limits_{n\to\infty} \sum\limits_{r=1}^{n} \dfrac{r^3}{n^4+r^4}$

9. Evaluate $\lim\limits_{n\to\infty} \sum\limits_{r=0}^{n} \sqrt{\dfrac{n+r}{n-r}}$

10. Evaluate $\lim\limits_{n\to\infty} \sum\limits_{r=0}^{n-1} \sqrt{\dfrac{n^2-r^2}{n^2}}$

11. Find the limit, as $n \to \infty$, of the product

$$\left[\left(1+\frac{1}{n}\right)\left(1+\frac{2}{n}\right)\left(1+\frac{3}{n}\right)\cdots\left(1+\frac{n}{n}\right)\right]^{\frac{1}{n}}$$

12. Prove that

$$\lim_{n\to\infty}\left[\left(1+\frac{1}{n^2}\right)\left(1+\frac{2^2}{n^2}\right)\left(1+\frac{3^2}{n^2}\right)\cdots\left(1+\frac{n^2}{n^2}\right)\right]^{\frac{1}{n}} = 2e^{(\pi-4)/2}$$

5.6 Evaluation of Definite Integrals

In Sec. 5.2, we have studied the method of evaluating a definite integral as the limit of a sum. But this method can be applied only when the integrand is a simple function. If the integrand is not a simple function, then it may not be possible to find the limit of the sum. In such cases, we consider below some properties of the definite integrals which will be useful in evaluating the definite integrals more easily.

We recall that if

$$\int f(x)\,dx = F(x) \text{ then } \int_a^b f(x)dx = F(b) - F(a)$$

I. $$\int_a^b f(x)dx = -\int_b^a f(x)dx$$

Proof. We have

$$\int_a^b f(x)dx = [F(x)]_a^b = F(b) - F(a)$$

and $$-\int_b^a f(x)dx = -[F(x)]_b^a = -F(a) + F(b)$$

Hence $$\int_a^b f(x)dx = -\int_b^a f(x)dx$$

II. $$\int_a^b f(x)\,dx = \int_a^b f(t)\,dt$$

Proof. We have

$$\int_a^b f(x)\,dx = F(b) - F(a)$$

and $$\int_a^b f(t)\,dt = [F(t)]_a^b = F(b) - F(a)$$

Hence $$\int_a^b f(x)dx = \int_a^b f(t)\,dt$$

III. $$\int_a^b f(x)dx = \int_a^c f(x)dx + \int_c^b f(x)dx, \text{ where } a < c < b$$

Proof. We have

$$\begin{aligned} \text{R.H.S.} &= \int_a^c f(x)dx + \int_c^b f(x)dx \\ &= F(c) - F(a) + F(b) - F(c) \\ &= F(b) - F(a) = \int_a^b f(x)dx = \text{L.H.S.} \end{aligned}$$

IV. $$\int_a^b f(x)dx = \int_a^b f(a+b-x)dx$$

Proof. Let $a + b - x = t$. Then $dx = -dt$.
When $x = a$, $t = b$ and when $x = b$, $t = a$.
Therefore

$$\int_a^b f(a+b-x)dx = -\int_b^a f(t)dt = \int_a^b f(t)dt \quad \text{(by Property I)}$$

$$= \int_a^b f(x)dx \quad \text{(by Property II)}$$

V. $$\int_0^a f(x)dx = \int_0^a f(a-x)dx$$

Proof. Let $a - x = t$. Then $dx = -dt$.
When $x = 0$, $t = a$ and when $x = a$, $t = 0$.
Therefore

$$\int_0^a f(a-x)dx = -\int_a^0 f(t)dt = \int_0^a f(t)dt \quad \text{(by Property I)}$$

$$= \int_0^a f(x)dx \quad \text{(by Property II)}$$

VI. $$\int_0^{2a} f(x)dx = \begin{cases} 2\int_0^a f(x)dx, & \text{if } f(2a-x) = f(x) \\ 0 \quad , & \text{if } f(2a-x) = -f(x) \end{cases}$$

Proof. From Property III, we have

$$\int_0^{2a} f(x)dx = \int_0^a f(x)dx + \int_a^{2a} f(x)dx$$

In the second integral on the right, putting $x = 2a - t$, so that $dx = -dt$.
Also, when $x = a$, $t = a$ and when $x = 2a$, $t = 0$.
Therefore

$$\int_a^{2a} f(x)dx = -\int_a^0 f(2a-t)dt = \int_0^a f(2a-x)dx \text{ (by Properties I and II)}$$

Hence $$\int_0^{2a} f(x)dx = \int_0^a f(x)dx + \int_0^a f(2a-x)dx \tag{1}$$

Case 1. If $f(2a - x) = f(x)$, then from Eq. (1), we get

$$\int_0^{2a} f(x)dx = \int_0^a f(x)dx + \int_0^a f(x)dx = 2\int_0^a f(x)dx$$

Case 2. If $f(2a-x) = -f(x)$, then from Eq. (1), we get

$$\int_0^{2a} f(x)\,dx = \int_0^a f(x)\,dx - \int_0^a f(x)\,dx = 0$$

VII. $$\int_{-a}^{a} f(x)\,dx = \begin{cases} 2\int_0^a f(x)dx, & \text{if } f(x) \text{ is an even function} \\ 0 & \text{if } f(x) \text{ is an odd function} \end{cases}$$

Proof. From Property III, we have

$$\int_{-a}^{a} f(x)dx = \int_{-a}^{0} f(x)dx + \int_0^a f(x)dx$$

In the first integral on the right, putting $x = -t$, so that $dx = -dt$. Also, when $x = -a$, $t = a$ and when $x = 0$, $t = 0$.

Therefore

$$\int_{-a}^{0} f(x)dx = -\int_a^0 f(-t)dt = \int_0^a f(-x)dx \quad \text{(by Properties I and II)}$$

Hence $$\int_{-a}^{a} f(x)dx = \int_0^a f(-x)dx + \int_0^a f(x)dx \qquad (1)$$

Case 1. If $f(x)$ is an even function, then $f(-x) = f(x)$. Therefore, from Eq. (1), we get

$$\int_{-a}^{a} f(x)dx = \int_0^a f(x)dx + \int_0^a f(x)dx = 2\int_0^a f(x)dx$$

Case 2. If $f(x)$ is an odd function, then $f(-x) = -f(x)$. Therefore, from Eq. (1), we get

$$\int_{-a}^{a} f(x)dx = -\int_0^a f(x)dx + \int_0^a f(x)dx = 0$$

Examples

1. Evaluate $\displaystyle\int_0^{\pi/2} \frac{\sin x}{\sin x + \cos x}\,dx.$

Solution. Let

$$I = \int_0^{\pi/2} \frac{\sin x}{\sin x + \cos x}\,dx \qquad (1)$$

Also $$I = \int_0^{\pi/2} \frac{\sin\left(\frac{\pi}{2} - x\right)}{\sin\left(\frac{\pi}{2} - x\right) + \cos\left(\frac{\pi}{2} - x\right)}\,dx = \int_0^{\pi/2} \frac{\cos x}{\cos x + \sin x}\,dx \qquad (2)$$

On adding Eqs. (1) and (2), we get

$$2I = \int_0^{\pi/2} \frac{(\sin x + \cos x)}{\sin x + \cos x}\, dx = \int_0^{\pi/2} dx = [x]_0^{\pi/2} = \frac{\pi}{2}$$

Hence $I = \frac{\pi}{4}$.

2. Evaluate $\int_0^{\pi} \frac{x \sin x}{1 + \cos^2 x}\, dx$.

Solution. Let $$I = \int_0^{\pi} \frac{x \sin x}{1 + \cos^2 x}\, dx \tag{1}$$

Also $$I = \int_0^{\pi} \frac{(\pi - x) \sin (\pi - x)}{1 + \cos^2 (\pi - x)}\, dx = \int_0^{\pi} \frac{(\pi - x) \sin x}{1 + \cos^2 x}\, dx \tag{2}$$

On adding Eqs. (1) and (2), we get

$$2I = \int_0^{\pi} \frac{(x + \pi - x) \sin x}{1 + \cos^2 x}\, dx = \pi \int_0^{\pi} \frac{\sin x}{1 + \cos^2 x}\, dx$$

Putting $\cos x = t$, so that $\sin x\, dx = -\, dt$. Also, when $x = 0$, $t = 1$ and when $x = \pi$, $t = -1$.

Therefore $$2I = -\pi \int_1^{-1} \frac{1}{1 + t^2}\, dt = \pi \int_{-1}^{1} \frac{1}{1 + t^2}\, dt$$

$$= 2\pi \int_0^{1} \frac{1}{1 + t^2}\, dt \quad \left(\because f(t) = \frac{1}{1 + t^2} \text{ is an even function}\right)$$

$$= 2\pi \left[\tan^{-1} t\right]_0^1 = 2\pi \cdot \frac{\pi}{4}$$

Hence $I = \frac{\pi^2}{4}$.

3. Evaluate $\int_0^{\pi/4} \log (1 + \tan x)\, dx$.

Solution. Let

$$I = \int_0^{\pi/4} \log (1 + \tan x)\, dx \tag{1}$$

Also

$$I = \int_0^{\pi/4} \log\left(1 + \tan\left(\frac{\pi}{4} - x\right)\right) dx = \int_0^{\pi/4} \log\left(1 + \frac{\tan \frac{\pi}{4} - \tan x}{1 + \tan \frac{\pi}{4} \tan x}\right) dx$$

$$= \int_0^{\pi/4} \log\left(1 + \frac{1 - \tan x}{1 + \tan x}\right) dx = \int_0^{\pi/4} \log\left(\frac{2}{1 + \tan x}\right) dx$$

$$= \int_0^{\pi/4} \log 2\, dx - \int_0^{\pi/4} \log(1 + \tan x)\, dx = (\log 2)[x]_0^{\pi/4} - I \quad \text{(by (1))}$$

Therefore $$2I = \frac{\pi}{4} \log 2$$

$\Rightarrow$ $$I = \frac{\pi}{8} \log 2$$

4. Evaluate $\int_0^{\pi/2} \log \sin x\, dx$.

Solution. Let

$$I = \int_0^{\pi/2} \log \sin x\, dx \tag{1}$$

Also $$I = \int_0^{\pi/2} \log \sin\left(\frac{\pi}{2} - x\right) dx = \int_0^{\pi/2} \log \cos x\, dx \tag{2}$$

On adding Eqs. (1) and (2), we get

$$2I = \int_0^{\pi/2} (\log \sin x + \log \cos x)\, dx = \int_0^{\pi/2} \log \sin x \cos x\, dx$$

$$= \int_0^{\pi/2} \log\left(\frac{2 \sin x \cos x}{2}\right) dx = \int_0^{\pi/2} \log \sin 2x\, dx - \int_0^{\pi/2} \log 2\, dx$$

$$= \int_0^{\pi/2} \log \sin 2x\, dx - \frac{\pi}{2} \log 2$$

Putting $2x = t$, we get

$$2I = \frac{1}{2} \int_0^{\pi} \log \sin t\, dt - \frac{\pi}{2} \log 2 = \frac{1}{2} \int_0^{\pi} \log \sin x\, dx - \frac{\pi}{2} \log 2$$

$$= \frac{1}{2} \cdot 2 \int_0^{\pi/2} \log \sin x\, dx - \frac{\pi}{2} \log 2 \ (\because \log \sin(\pi - x) = \log \sin x)$$

$$= I - \frac{\pi}{2} \log 2$$

Hence $$I = -\frac{\pi}{2} \log 2$$

5. Show that $\int_0^{\pi} \frac{x \tan x}{\sec x + \cos x}\, dx = \frac{\pi^2}{4}$.

Solution. Let

$$I = \int_0^{\pi} \frac{x \tan x}{\sec x + \cos x}\, dx \tag{1}$$

Also $$I = \int_0^{\pi} \frac{(\pi - x) \tan (\pi - x)}{\sec (\pi - x) + \cos (\pi - x)}\, dx = \int_0^{\pi} \frac{(\pi - x) \tan x}{\sec x + \cos x}\, dx \tag{2}$$

On adding Eqs. (1) and (2), we get

$$2I = \int_0^{\pi} \frac{(x + \pi - x) \tan x}{\sec x + \cos x}\, dx = \pi \int_0^{\pi} \frac{\tan x}{\sec x + \cos x}\, dx$$

$$= \pi \int_0^{\pi} \frac{\sin x}{1 + \cos^2 x}\, dx = -\pi \left[\tan^{-1} (\cos x)\right]_0^{\pi}$$

$$= -\pi \left[\tan^{-1} (-1) - \tan^{-1} 1\right] = -\pi\left(-\frac{\pi}{4} - \frac{\pi}{4}\right) = \frac{\pi^2}{2}$$

Hence $I = \frac{\pi^2}{4}$.

EXERCISES

Evaluate the following definite integrals:

1. $\int_0^{\pi/2} \frac{\cos x}{\sin x + \cos x}\, dx$

2. $\int_0^{\pi/2} \frac{\sqrt{\sin x}}{\sqrt{\sin x} + \sqrt{\cos x}}\, dx$

3. $\int_0^{\pi/2} \frac{\sqrt{\cos x}}{\sqrt{\cos x} + \sqrt{\sin x}}\, dx$

4. $\int_0^{\pi/2} \frac{\sin^2 x}{\sin x + \cos x}\, dx$

5. $\int_0^{1} \frac{\log (1 + x)}{1 + x^2}\, dx$

6. $\int_0^{\pi/2} \frac{(\sin x - \cos x)}{1 + \sin x \cos x}\, dx$

7. $\int_0^{\pi/2} \log \tan x\, dx$

8. $\int_0^{\pi} x \log \sin x\, dx$

9. $\int_0^{\pi/2} \log (\tan x + \cot x)\, dx$

MISCELLANEOUS EXERCISES

Evaluate the following definite integrals as limits of sums:

1. $\int_a^b \sinh x\, dx$ 2. $\int_a^b \cosh 2x\, dx$

Find the limit, when $n \to \infty$, of the series:

3. $\frac{n+1}{n^2+1^2} + \frac{n+2}{n^2+2^2} + \frac{n+3}{n^2+3^2} + \ldots + \frac{1}{n}$

4. $\frac{n^2}{(n^2+1^2)^{3/2}} + \frac{n^2}{(n^2+2^2)^{3/2}} + \ldots + \frac{n^2}{(n^2+(n-1)^2)^{3/2}}$

5. $\frac{\sqrt{n}}{\sqrt{n^3}} + \frac{\sqrt{n}}{\sqrt{(n+4)^3}} + \frac{\sqrt{n}}{\sqrt{(n+8)^3}} + \frac{\sqrt{n}}{\sqrt{(n+12)^3}} + \ldots + \frac{\sqrt{n}}{\sqrt{[n+4(n-1)]^3}}$

6. $\frac{\sqrt{n}}{(3+4\sqrt{n})^2} + \frac{\sqrt{n}}{\sqrt{2}\,(3\sqrt{2}+4\sqrt{n})^2} + \frac{\sqrt{n}}{\sqrt{3}\,(3\sqrt{3}+4\sqrt{n})^2} + \ldots + \frac{1}{49n}$

7. $\frac{1}{n}\sum_{r=0}^{n} \frac{1}{\left(1+\frac{3r}{n}\right)^{3/2}}$ 8. $\frac{\sqrt{n+1}}{n\sqrt{n}} + \frac{\sqrt{n+2}}{n\sqrt{n}} + \frac{\sqrt{n+3}}{n\sqrt{n}} + \ldots + \frac{\sqrt{2n}}{n\sqrt{n}}$

9. $\frac{1}{n^3} + \frac{4}{n^3} + \frac{9}{n^3} + \ldots + \frac{n^2}{n^3}$ 10. $\left[\sin\frac{\pi}{2n}\sin\frac{2\pi}{2n}\sin\frac{3\pi}{2n}\ldots\sin\frac{n\pi}{2n}\right]^{\frac{1}{n}}$

11. $\left[\tan\frac{\pi}{2n}\tan\frac{2\pi}{2n}\tan\frac{3\pi}{2n}\ldots\tan\frac{n\pi}{2n}\right]^{\frac{1}{n}}$

12. Prove that $\int_0^{\pi/2} \frac{\sin^3 x}{\sin^3 x + \cos^3 x}\, dx = \frac{\pi}{4}$

13. Show that $\int_0^{\pi/2} \frac{\cos^n x}{\cos^n x + \sin^n x}\, dx = \frac{\pi}{4}$

14. Evaluate $\int_0^{\pi/2} \frac{\log\left(x+\frac{1}{x}\right)}{1+x^2}\, dx$ 15. Evaluate $\int_0^{\pi} \log(1+\cos x)\, dx$

16. Show that $\int_0^{\pi} \frac{x\tan x}{\sec x + \tan x}\, dx = \pi\left(\frac{\pi}{2} - 1\right)$ 17. Evaluate $\int_0^{\pi} \frac{x}{1+\sin x}\, dx$

6

Reduction Formulae

6.1 Introduction

A *reduction formula* in Integral Calculus is a formula by means of which any integral not directly obtainable is made to depend on a simpler integral of the same type. Repeated application of this formula will then, usually lead to an integral which is already known. Various methods are used to obtain reduction formulae, the most common being the method of integration by parts and the method of connection. The students are familiar with the first of these and we shall describe the second in the following sections. Throughout the chapter we consider m and n as positive integers.

6.2 Reduction Formulae for $\int \sin^n x dx$ and $\int \cos^n x dx$

Writing $\sin^n x = \sin^{n-1} x \cdot \sin x$ and integrating by parts, we have

$$\int \sin^n x\, dx = \int \sin^{n-1} x. \sin x\, dx$$

$$= \sin^{n-1} x\,(-\cos x) - \int (n-1) \sin^{n-2} x \cdot \cos x\,(-\cos x)\, dx$$

$$= -\sin^{n-1} x \cos x + (n-1) \int \sin^{n-2} x \cos^2 x\, dx$$

$$= -\sin^{n-1} x \cos x + (n-1) \int \sin^{n-2} x\,(1 - \sin^2 x)\, dx$$

$$= -\sin^{n-1} x \cos x + (n-1) \int \sin^{n-2} x\, dx - (n-1) \int \sin^n x\, dx$$

$$\Rightarrow \int \sin^n x dx + (n-1) \int \sin^n x dx = -\sin^{n-1} x \cos x + (n-1) \int \sin^{n-2} x\, dx$$

$$\Rightarrow n \int \sin^n x\, dx = -\sin^{n-1} x \cos x + (n-1) \int \sin^{n-2} x dx$$

$$\Rightarrow \int \sin^n x\,dx = -\frac{1}{n}\sin^{n-1} x\cos x + \frac{n-1}{n}\int \sin^{n-2} x\,dx$$

Similarly $\displaystyle\int \cos^n x\,dx = \frac{1}{n}\cos^{n-1} x\sin x + \frac{n-1}{n}\int \cos^{n-2} x\,dx$

These are the required reduction formulae.

Examples

1. Evaluate $\int \sin^5 x\,dx$.

Solution. We know that

$$\int \sin^n x\,dx = -\frac{1}{n}\sin^{n-1} x\cos x + \frac{n-1}{n}\int \sin^{n-2} x\,dx \tag{1}$$

Putting $n = 5, 3$ successively in Eq. (1), we get

$$\int \sin^5 x\,dx = -\frac{1}{5}\sin^4 x\cos x + \frac{4}{5}\int \sin^3 x\,dx \tag{2}$$

$$\int \sin^3 x\,dx = -\frac{1}{3}\sin^2 x\cos x + \frac{2}{3}\int \sin x\,dx = -\frac{1}{3}\sin^2 x\cos x - \frac{2}{3}\cos x \tag{3}$$

Hence, from Eqs. (2) and (3), we get

$$\int \sin^5 x\,dx = -\frac{1}{5}\sin^4 x\cos x + \frac{4}{5}\left[-\frac{1}{3}\sin^2 x\cos x - \frac{2}{3}\cos x\right]$$

$$= -\frac{1}{5}\sin^4 x\cos x - \frac{4}{15}\sin^2 x\cos x - \frac{8}{15}\cos x$$

2. Evaluate $\int \cos^6 x\,dx$.

Solution. We know that

$$\int \cos^n x\,dx = \frac{1}{n}\cos^{n-1} x\sin x + \frac{n-1}{n}\int \cos^{n-2} x\,dx \tag{1}$$

Putting $n = 6, 4, 2$ successively in Eq. (1), we get

$$\int \cos^6 x\,dx = \frac{1}{6}\cos^5 x\sin x + \frac{5}{6}\int \cos^4 x\,dx \tag{2}$$

$$\int \cos^4 x\,dx = \frac{1}{4}\cos^3 x\sin x + \frac{3}{4}\int \cos^2 x\,dx \tag{3}$$

$$\int \cos^2 x\,dx = \frac{1}{2}\cos x\sin x + \frac{1}{2}\int dx = \frac{1}{2}\cos x\sin x + \frac{1}{2}x \tag{4}$$

Hence, from Eqs. (2), (3) and (4), we get

$$\int \cos^6 x\,dx = \frac{1}{6}\cos^5 x \sin x$$

$$+\frac{5}{6}\left[\frac{1}{4}\cos^3 x \sin x + \frac{3}{4}\left(\frac{1}{2}\cos x \sin x + \frac{1}{2}x\right)\right]$$

$$= \frac{1}{6}\cos^5 x \sin x + \frac{5}{24}\cos^3 x \sin x + \frac{5}{16}\cos x \sin x + \frac{5}{16}x$$

EXERCISES

Evaluate

1. $\int \sin^3 x\,dx$
2. $\int \sin^4 x\,dx$
3. $\int \sin^6 x\,dx$
4. $\int \sin^7 x\,dx$
5. $\int \cos^3 x\,dx$
6. $\int \cos^4 x\,dx$
7. $\int \cos^7 x\,dx$

6.3 Evaluation of $\int_0^{\pi/2} \sin^n x\,dx$ and $\int_0^{\pi/2} \cos^n x\,dx$

First of all we obtain the reduction formulae

$$\int \sin^n x\,dx = -\frac{1}{n}\sin^{n-1} x \cos x + \frac{n-1}{n}\int \sin^{n-2} x\,dx$$

Now $$\int_0^{\frac{\pi}{2}} \sin^n x\,dx = \left[-\frac{1}{n}\sin^{n-1} x \cos x\right]_0^{\frac{\pi}{2}} + \frac{n-1}{n}\int_0^{\frac{\pi}{2}} \sin^{n-2} x\,dx$$

$$= \frac{(n-1)}{n}\int_0^{\frac{\pi}{2}} \sin^{n-2} x\,dx$$

Writing $I_n = \int_0^{\frac{\pi}{2}} \sin^n x\,dx$, we get

$$I_n = \frac{n-1}{n} I_{n-2}$$

In the above formulae putting $n = 2, 3, \ldots$, we get

$$I_2 = \frac{1}{2} I_0 = \frac{1}{2}\int_0^{\frac{\pi}{2}} dx = \frac{1}{2}\cdot\frac{\pi}{2}$$

$$I_3 = \frac{2}{3} I_1 = \frac{2}{3} \int_0^{\frac{\pi}{2}} \sin x \, dx = \frac{2}{3}[-\cos x]_0^{\frac{\pi}{2}} = \frac{2}{3}$$

$$I_4 = \frac{3}{4} I_2 = \frac{3}{4} \cdot \frac{1}{2} \cdot \frac{\pi}{2}$$

$$I_5 = \frac{4}{5} I_3 = \frac{4}{5} \cdot \frac{2}{3}$$

$$I_6 = \frac{5}{6} I_4 = \frac{5}{6} \cdot \frac{3}{4} \cdot \frac{1}{2} \cdot \frac{\pi}{2}$$

$$\ldots \qquad \ldots \qquad \ldots$$

$$\ldots \qquad \ldots \qquad \ldots$$

In general $I_n = \begin{cases} \dfrac{n-1}{n} \cdot \dfrac{n-3}{n-2} \cdot \dfrac{n-5}{n-4} \cdots \dfrac{4}{5} \cdot \dfrac{2}{3}, & \text{when } n \text{ is odd} \\ \dfrac{n-1}{n} \cdot \dfrac{n-3}{n-2} \cdot \dfrac{n-5}{n-4} \cdots \dfrac{3}{4} \cdot \dfrac{1}{2} \cdot \dfrac{\pi}{2}, & \text{when } n \text{ is even} \end{cases}$

Similarly

$$\int_0^{\pi/2} \cos^n x \, dx = \begin{cases} \dfrac{n-1}{n} \cdot \dfrac{n-3}{n-2} \cdot \dfrac{n-5}{n-4} \cdots \dfrac{4}{5} \cdot \dfrac{2}{3}, & \text{when } n \text{ is odd} \\ \dfrac{n-1}{n} \cdot \dfrac{n-3}{n-2} \cdot \dfrac{n-5}{n-4} \cdots \dfrac{3}{4} \cdot \dfrac{1}{2} \cdot \dfrac{\pi}{2}, & \text{when } n \text{ is even} \end{cases}$$

Examples

1. Evaluate $\int_0^{\pi/2} \sin^6 x \, dx$.

Solution. We have

$$\int_0^{\pi/2} \sin^6 x \, dx = \frac{5}{6} \cdot \frac{3}{4} \cdot \frac{1}{2} \cdot \frac{\pi}{2} = \frac{5\pi}{32}$$

2. Evaluate $\int_0^{\pi/4} \cos^5 2x \, dx$.

Solution. Putting $2x = t$. Then $dx = \frac{1}{2} dt$.

When $x = \frac{\pi}{4}, t = \frac{\pi}{2}$

When $x = 0, t = 0$

Therefore $\int_0^{\pi/4} \cos^5 2x \, dx = \frac{1}{2} \int_0^{\pi/2} \cos^5 t \, dt = \frac{1}{2} \cdot \frac{4}{5} \cdot \frac{2}{3} = \frac{4}{15}$

3. Evaluate $\int_0^1 \frac{x^5}{\sqrt{1-x^2}}\,dx.$

Solution. Let $x = \sin\theta$. Then $dx = \cos\theta\, d\theta$.

When $x = 1,\ \theta = \frac{\pi}{2}$

When $x = 0,\ \theta = 0$

Therefore

$$\int_0^1 \frac{x^5}{\sqrt{1-x^2}}\,dx = \int_0^{\pi/2} \frac{\sin^5\theta}{\sqrt{1-\sin^2\theta}} \cos\theta\, d\theta = \int_0^{\pi/2} \sin^5\theta\, d\theta$$

Now $$\int_0^{\pi/2} \sin^5\theta\, d\theta = \frac{4}{5}\cdot\frac{2}{3} = \frac{8}{15}$$

Hence $$\int_0^1 \frac{x^5}{\sqrt{1-x^2}}\,dx = \frac{8}{15}$$

EXERCISES

Evaluate

1. $\int_0^{\pi/2} \cos^6 x\,dx$
2. $\int_0^{\pi/2} \sin^7 x\,dx$
3. $\int_0^{\pi/2} \cos^8 x\,dx$
4. $\int_0^{\pi/2} \cos^9 x\,dx$
5. $\int_0^{\pi/2} \sin^{10} x\,dx$
6. $\int_0^{\pi/4} \cos^6 2x\,dx$
7. $\int_0^{\pi} \cos^8 \frac{x}{2}\,dx$
8. $\int_0^{\pi/6} \sin^6 3x\,dx$
9. $\int_0^{2\pi} \sin^7 \frac{x}{4}\,dx$
10. $\int_0^1 \frac{x^6}{\sqrt{1-x^2}}\,dx$
11. $\int_0^a \frac{x^4}{\sqrt{a^2-x^2}}\,dx$
12. $\int_0^1 x^5 \sin^{-1} x\,dx$
13. $\int_0^{2a} \frac{x^{9/2}}{\sqrt{2a-x}}\,dx$

6.4 Reduction Formulae for $\int \tan^n x\,dx$ and $\int \cot^n x\,dx$

We have

$$\int \tan^n x\,dx = \int \tan^{n-2} x \tan^2 x\,dx = \int \tan^{n-2} x\,(\sec^2 x - 1)\,dx$$

$$= \int \tan^{n-2} x \sec^2 x\,dx - \int \tan^{n-2} x\,dx$$

Therefore

$$\int \tan^n x dx = \frac{\tan^{n-1} x}{n-1} - \int \tan^{n-2} x dx$$

This is the required reduction formula.

Further, we have

$$\int \cot^n x dx = \int \cot^{n-2} x \cot^2 x \, dx = \int \cot^{n-2} x(\operatorname{cosec}^2 x - 1) dx$$

$$= \int \cot^{n-2} x \operatorname{cosec}^2 x \, dx - \int \cot^{n-2} x \, dx$$

Therefore $$\int \cot^n x \, dx = -\frac{\cot^{n-1} x}{n-1} - \int \cot^{n-2} x \, dx$$

This is the required reduction formula.

Examples

1. Evaluate $\int \tan^7 x \, dx$.

Solution. We know that

$$\int \tan^n x \, dx = \frac{\tan^{n-1} x}{n-1} - \int \tan^{n-2} x \, dx \tag{1}$$

Putting $n = 7, 5, 3$ successively in Eq. (1), we get

$$\int \tan^7 x \, dx = \frac{1}{6} \tan^6 x - \int \tan^5 x \, dx \tag{2}$$

$$\int \tan^5 x \, dx = \frac{1}{4} \tan^4 x - \int \tan^3 x \, dx \tag{3}$$

$$\int \tan^3 x \, dx = \frac{1}{2} \tan^2 x - \int \tan x = \frac{1}{2} \tan^2 x + \log |\cos x| \tag{4}$$

Hence, from Eqs. (2), (3) and (4), we get

$$\int \tan^7 x \, dx = \frac{1}{6} \tan^6 x - \left[\frac{1}{4} \tan^4 x - \frac{1}{2} \tan^2 x - \log |\cos x|\right]$$

$$= \frac{1}{6} \tan^6 x - \frac{1}{4} \tan^4 x + \frac{1}{2} \tan^2 x + \log |\cos x|$$

2. Evaluate $\int \cot^6 x \, dx$.

Solution. We know that

$$\int \cot^n x \, dx = -\frac{\cot^{n-1} x}{n-1} - \int \cot^{n-2} x \, dx \tag{1}$$

Putting $n = 6, 4, 2$ successively in Eq. (1), we get

$$\int \cot^6 x\, dx = -\frac{1}{5}\cot^5 x - \int \cot^4 x\, dx \tag{2}$$

$$\int \cot^4 x\, dx = -\frac{1}{3}\cot^3 x - \int \cot^2 x\, dx \tag{3}$$

$$\int \cot^2 x\, dx = -\cot x - \int dx = -\cot x - x \tag{4}$$

Hence, from Eqs. (2), (3) and (4), we get

$$\int \cot^6 x\, dx = -\frac{1}{5}\cot^5 x - \left[-\frac{1}{3}\cot^3 x - (-\cot x - x)\right]$$

$$= -\frac{1}{5}\cot^5 x + \frac{1}{3}\cot^3 x - \cot x - x$$

3. If $I_n = \int_0^{\pi/4} \tan^n x\, dx$, prove that

$$I_n + I_{n-2} = \frac{1}{n-1}$$

and deduce the value of I_5.

Solution. We have

$$I_n = \int_0^{\pi/4} \tan^n x\, dx = \int_0^{\pi/4} \tan^{n-2} x(\sec^2 x - 1)\, dx$$

$$= \int_0^{\pi/4} \tan^{n-2} x \sec^2 x\, dx - \int_0^{\pi/4} \tan^{n-2} x\, dx$$

$$= \left[\frac{\tan^{n-1} x}{n-1}\right]_0^{\frac{\pi}{4}} - I_{n-2} = \frac{1}{n-1} - I_{n-2}$$

Hence $$I_n + I_{n-2} = \frac{1}{n-1}$$

In the above formulae putting $n = 5, 3$, we get

$$I_5 = \frac{1}{4} - I_3$$

$$I_3 = \frac{1}{2} - I_1 = \frac{1}{2} - \int_0^{\pi/4} \tan x\, dx$$

$$= \frac{1}{2} + [\log \cos x]_0^{\pi/4} = \frac{1}{2} + \log\frac{1}{\sqrt{2}} = \frac{1}{2} - \frac{1}{2}\log 2$$

Therefore $\quad I_5 = \frac{1}{4} - \frac{1}{2} + \frac{1}{2}\log 2 = -\frac{1}{4} + \frac{1}{2}\log 2$

EXERCISES

Evaluate

1. $\int \tan^3 x\,dx$ 2. $\int \tan^6 x\,dx$ 3. $\int \cot^5 x\,dx$

4. $\int_0^{\pi/4} \tan^4 x\,dx$ 5. $\int_0^{\pi/4} \tan^5 x\,dx$ 6. $\int_{\pi/4}^{\pi/2} \cot^4 x\,dx$

7. If $I_n = \int_0^{\pi/4} \tan^n x\,dx$, prove that when n is a positive integer,

$$n(I_{n-1} + I_{n+1}) = 1$$

Hence prove that $\int_0^a \frac{x^5}{(2a^2 - x^2)^3}\,dx = \frac{1}{2}\left(\log 2 - \frac{1}{2}\right)$

6.5 Reduction Formulae for $\int \sec^n x dx$ and $\int \operatorname{cosec}^n x dx$

Writing $\sec^n x = \sec^{n-2} x \sec^2 x$ and integrating by parts, we get

$$\int \sec^n x\,dx = \int \sec^{n-2} x \sec^2 x\,dx$$

$$= \sec^{n-2} x \tan x - \int (n-2)\sec^{n-3} x \sec x \tan x \tan x\,dx$$

$$= \sec^{n-2} x \tan x - (n-2)\int \sec^{n-2} x\,(\sec^2 x - 1)\,dx$$

$$= \sec^{n-2} x \tan x - (n-2)\int \sec^n x\,dx + (n-2)\int \sec^{n-2} x\,dx$$

$$\Rightarrow \quad (n-1)\int \sec^n x\,dx = \sec^{n-2} x \tan x + (n-2)\int \sec^{n-2} x\,dx$$

$$\Rightarrow \quad \int \sec^n x\,dx = \frac{1}{n-1}\sec^{n-2} x \tan x + \frac{n-2}{n-1}\int \sec^{n-2} x\,dx$$

Similarly

$$\int \operatorname{cosec}^n x\,dx = -\frac{1}{n-1}\operatorname{cosec}^{n-2} x \cot x + \frac{n-2}{n-1}\int \operatorname{cosec}^{n-2} x\,dx$$

These are the required reduction formulae.

Examples

1. Evaluate $\int \sec^3 x\, dx$.

Solution. We know that

$$\int \sec^n x\, dx = \frac{1}{n-1} \sec^{n-2} x \tan x + \frac{n-2}{n-1} \int \sec^{n-2} x\, dx \qquad (1)$$

Putting $n = 3$ in Eq. (1), we get

$$\int \sec^3 x = \frac{1}{2} \sec x \tan x + \frac{1}{2} \int \sec x\, dx$$

$$= \frac{1}{2} \sec x \tan x + \frac{1}{2} \log |\sec x + \tan x|$$

2. Evaluate $\int (1 + x^2)^{3/2} dx$.

Solution. Let $x = \tan \theta$. Then $dx = \sec^2\theta\, d\theta$. Therefore

$$\int (1 + x^2)^{3/2} dx = \int (1 + \tan^2\theta)^{3/2} \sec^2\theta d\theta = \int \sec^5\theta\, d\theta$$

We know that

$$\int \sec^n \theta\, d\theta = \frac{1}{n-1} \sec^{n-2} \theta \tan \theta + \frac{n-2}{n-1} \int \sec^{n-2} \theta\, d\theta$$

Putting $n = 5, 3$ successively in the above equation, we get

$$\int \sec^5 \theta\, d\theta = \frac{1}{4} \sec^3 \theta \tan \theta + \frac{3}{4} \int \sec^3 \theta\, d\theta$$

$$\int \sec^3 \theta\, d\theta = \frac{1}{2} \sec \theta \tan \theta + \frac{1}{2} \int \sec \theta\, d\theta$$

$$= \frac{1}{2} \sec \theta \tan \theta + \frac{1}{2} \log |\sec \theta + \tan \theta|$$

Therefore

$$\int \sec^5 \theta\, d\theta = \frac{1}{4} \sec^3 \theta \tan \theta + \frac{3}{4}\left[\frac{1}{2} \sec \theta \tan \theta + \frac{1}{2} \log |\sec \theta + \tan \theta|\right]$$

$$= \frac{1}{4} \sec^3 \theta \tan \theta + \frac{3}{8} \sec \theta \tan \theta + \frac{3}{8} \log |\sec \theta + \tan \theta|$$

Thus

$$\int (1+x^2)^{3/2}\,dx = \frac{1}{4}x(1+x^2)^{3/2} + \frac{3}{8}x(1+x^2)^{1/2} + \frac{3}{8}\log|x+\sqrt{1+x^2}|$$

EXERCISES

Evaluate

1. $\int \sec^4 x\,dx$
2. $\int \operatorname{cosec}^3 x\,dx$
3. $\int \operatorname{cosec}^5 x\,dx$
4. $\int_0^a (a^2+x^2)^{5/2}\,dx$ [Hint. Put $x = a\tan\theta$]
5. Find a reduction formula for $\int \sec^n x\,dx$, and prove that

$$\int_0^{\pi/4} \sec^3 x\,dx = \frac{1}{2}\left(\sqrt{2} + \log|\sqrt{2}+1|\right)$$

6.6 Reduction Formula for $\int \sin^m x \cos^n x\,dx$

A reduction formula for this integral may be obtained by connecting it with any one of the following six integrals:

1. $\int \sin^{m-2} x\cos^n x\,dx$
2. $\int \sin^{m+2} x\cos^n x\,dx$
3. $\int \sin^{m} x\cos^{n-2} x\,dx$
4. $\int \sin^{m} x\cos^{n+2} x\,dx$
5. $\int \sin^{m-2} x\cos^{n+2} x\,dx$
6. $\int \sin^{m+2} x\cos^{n-2} x\,dx$

The method of connecting the original integral with any one of the six integrals is as follows:

I. Method of Connection

(i) Put $P = \sin^{\lambda+1} x\cos^{\mu+1} x$, where λ and μ are the smaller of the two indices of $\sin x$ and $\cos x$, respectively in the two expressions whose integrals are to be connected.

(ii) Find $\frac{dP}{dx}$ and resolve it into a linear function of the two expressions whose integrals are to be connected.

(iii) Integrate back and we get the required reduction formula.

As an example, we connect $\int \sin^m x\cos^n x\,dx$ with $\int \sin^{m-2} x\cos^n x\,dx$.

Here $\lambda = m - 2$, $\mu = n$. We, therefore, put $P = \sin^{m-1} x \cos^{n+1} x$.

Hence $\dfrac{dP}{dx} = (m-1)\sin^{m-2} x \cos^{n+2} x - (n+1)\sin^m x \cos^n x$

$$= (m-1)\sin^{m-2} x \cos^n x\,(1 - \sin^2 x) - (n+1)\sin^m x \cos^n x$$

$$= (m-1)\sin^{m-2} x \cos^n x - (m+n)\sin^m x \cos^n x$$

Integrating both sides, we get

$$P = (m-1)\int \sin^{m-2} x \cos^n x\,dx - (m+n)\int \sin^m x \cos^n x\,dx$$

Substituting the value of P, we get

$$\sin^{m-1} x \cos^{n+1} x = (m-1)\int \sin^{m-2} x \cos^n x\,dx - (m+n)\int \sin^m x \cos^n x\,dx$$

$$\Rightarrow \int \sin^m x \cos^n x\,dx = -\frac{1}{m+n}\sin^{m-1} x \cos^{n+1} x$$

$$+ \frac{m-1}{m+n}\int \sin^{m-2} x \cos^n x\,dx$$

II. Method of Integration by Parts

Let $I_{m,n} = \int \sin^m x \cos^n x\,dx$. We shall connect $I_{m,n}$ with $I_{m-2,n}$.

We have $\dfrac{d}{dx}(\sin^{m-1} x) = (m-1)\sin^{m-2} x \cos x$

and $\displaystyle\int \cos^m x \sin x\,dx = -\frac{\cos^{m+1} x}{m+1}$

Now

$$I_{m,n} = \int \sin^{m-1} x\,(\cos^n x \sin x)\,dx$$

$$= -\sin^{m-1} x \cdot \frac{\cos^{n+1} x}{n+1} + \int (m-1)\sin^{m-2} x \cos x \cdot \frac{\cos^{n+1} x}{n+1}\,dx$$

$$= -\frac{1}{n+1}\sin^{m-1} x \cos^{n+1} x + \frac{m-1}{n+1}\int \sin^{m-2} x \cos^n x\,(1 - \sin^2 x)\,dx$$

$$= -\frac{1}{n+1}\sin^{m-1} x \cos^{n+1} x + \frac{m-1}{n+1}\int \sin^{m-2} x \cos^n x\,dx$$

$$- \frac{m-1}{n+1}\int \sin^m x \cos^n x\,dx$$

$$= -\frac{1}{n+1}\sin^{m-1}x\cos^{n+1}x + \frac{m-1}{n+1}I_{m-2,n} - \frac{m-1}{n+1}I_{m,n}$$

$$\Rightarrow \quad \left(1 + \frac{m-1}{n+1}\right)I_{m,n} = -\frac{1}{n+1}\sin^{m-1}x\cos^{n+1}x + \frac{m-1}{n+1}I_{m-2,n}$$

$$\Rightarrow \quad I_{m,n} = -\frac{1}{m+n}\sin^{m-1}x\cos^{n+1}x + \frac{m-1}{m+n}I_{m-2,n}$$

6.7 Evaluation of $\int_0^{\pi/2} \sin^m x \cos^n x\, dx$

By reduction formula, we have

$$\int_0^{\pi/2} \sin^m x\cos^n x\,dx = \left[-\frac{1}{m+n}\sin^{m-1}x\cos^{n+1}x\right]_0^{\pi/2} + \frac{m-1}{m+n}\int_0^{\pi/2}\sin^{m-2}x\cos^n x\,dx$$

$$= \frac{m-1}{m+n}\int_0^{\pi/2}\sin^{m-2}x\cos^n x\,dx$$

Let $I_{m,n} = \int_0^{\pi/2}\sin^m x\cos^n x\,dx$. Then, we have

$$I_{m,n} = \frac{m-1}{m+n}I_{m-2,n}$$

By the repeated application of the recurrence formula we can evaluate $I_{m,n}$. There arise two cases.

Case I. When m is odd, we have

$$I_{m,n} = \frac{m-1}{m+n}I_{m-2,n}$$

$$= \frac{m-1}{m+n}\cdot\frac{m-3}{m+n-2}I_{m-4,n} = \frac{m-1}{m+n}\cdot\frac{m-3}{m+n-2}\cdots\frac{2}{n+3}I_{1,n}$$

But

$$I_{1,n} = \int_0^{\pi/2}\sin x\cos^n x\,dx = \left[-\frac{\cos^{n+1}x}{n+1}\right]_0^{\pi/2} = \frac{1}{n+1}$$

Therefore

$$I_{m,n} = \frac{m-1}{m+n}\cdot\frac{m-3}{m+n-2}\cdots\frac{2}{n+3}\cdot\frac{1}{n+1}$$

Case II. When m is even, we have

$$I_{m,n} = \frac{m-1}{m+n} I_{m-2,n} = \frac{m-1}{m+n} \cdot \frac{m-3}{m+n-2} I_{m-4,n}$$

$$= \frac{m-1}{m+n} \cdot \frac{m-3}{m+n-2} \cdots \frac{3}{n+4} \cdot \frac{1}{n+2} I_{0,n}$$

But $\quad I_{0,n} = \int_0^{\pi/2} \cos^n x \, dx$

$$= \begin{cases} \dfrac{n-1}{n} \cdot \dfrac{n-3}{n-2} \cdots \dfrac{4}{5} \cdot \dfrac{2}{3}, \text{ when } n \text{ is odd} \\ \dfrac{n-1}{n} \cdot \dfrac{n-3}{n-2} \cdots \dfrac{3}{4} \cdot \dfrac{1}{2} \cdot \dfrac{\pi}{2}, \text{ when } n \text{ is even} \end{cases}$$

Therefore

$$I_{m,n} = \begin{cases} \dfrac{m-1}{m+n} \cdot \dfrac{m-3}{m+n-2} \cdots \dfrac{3}{n+4} \cdot \dfrac{1}{n+2} \cdot \dfrac{n-1}{n} \cdot \dfrac{n-3}{n-2} \cdots \\ \qquad \dfrac{4}{5} \cdot \dfrac{2}{3}, \text{ when } n \text{ is odd} \\ \dfrac{m-1}{m+n} \cdot \dfrac{m-3}{m+n-2} \cdots \dfrac{3}{n+4} \cdot \dfrac{1}{n+2} \cdot \dfrac{n-1}{n} \cdot \dfrac{n-3}{n-2} \cdots \\ \qquad \dfrac{3}{4} \cdot \dfrac{1}{2} \cdot \dfrac{\pi}{2}, \text{ when } n \text{ is even} \end{cases}$$

Remark. The value of the integral $\int_0^{\pi/2} \sin^m x \cos^n x \, dx$ can be expressed very conveniently in terms of the Gamma function. For more details on Gamma function see the Appendix. Here the following properties of Gamma function will suffice our purpose.

For positive values of n

$$\Gamma(n+1) = n\,\Gamma(n) \tag{1}$$

Also

$$\Gamma(1) = 1 \text{ and } \Gamma\left(\frac{1}{2}\right) = \sqrt{\pi} \tag{2}$$

In case n is a positive integer, from Eqs. (1) and (2), we get

$$\Gamma(n+1) = n\Gamma(n) = n(n-1)\,\Gamma(n-1)$$

$$= n(n-1)(n-2) \cdots 3 \cdot 2 \cdot 1\, \Gamma(1)$$

$$= n(n-1)(n-2) \cdots 3 \cdot 2 \cdot 1 = \lfloor n$$

Hence $\Gamma(n+1) = \lfloor n$.

In terms of the Gamma function the value of the integral is given by

$$\int_0^{\pi/2} \sin^m x \cos^n x \, dx = \frac{\Gamma\left(\frac{m+1}{2}\right)\Gamma\left(\frac{n+1}{2}\right)}{2\Gamma\left(\frac{m+n+2}{2}\right)}$$

Examples

1. Evaluate $\int_0^{\pi/2} \sin^4 x \cos^2 x \, dx.$

Solution. We have

$$\int_0^{\pi/2} \sin^4 x \cos^2 x \, dx = \frac{\Gamma\left(\frac{4+1}{2}\right)\Gamma\left(\frac{2+1}{2}\right)}{2\Gamma\left(\frac{4+2+2}{2}\right)} = \frac{\Gamma\left(\frac{5}{2}\right)\Gamma\left(\frac{3}{2}\right)}{2\Gamma(4)}$$

$$= \frac{\frac{3}{2}\cdot\frac{1}{2}\sqrt{\pi}\cdot\frac{1}{2}\sqrt{\pi}}{2\cdot 3\cdot 2\cdot 1} = \frac{\pi}{32}$$

2. Evaluate $\int_0^1 x^4(1-x^2)^{5/2} \, dx.$

Solution. Let $x = \sin\theta$. Then $dx = \cos\theta \, d\theta$.

When $x = 1, \theta = \frac{\pi}{2}$

When $x = 0, \theta = 0$

Therefore

$$\int_0^1 x^4(1-x^2)^{5/2} \, dx = \int_0^{\pi/2} \sin^4\theta \cos^5\theta \cos\theta \, d\theta$$

$$= \int_0^{\pi/2} \sin^4\theta \cos^6\theta \, d\theta$$

$$= \frac{\Gamma\left(\frac{4+1}{2}\right)\Gamma\left(\frac{6+1}{2}\right)}{2\Gamma\left(\frac{4+6+2}{2}\right)} = \frac{\Gamma\left(\frac{5}{2}\right)\Gamma\left(\frac{7}{2}\right)}{2\Gamma(6)} = \frac{\frac{3}{2}\cdot\frac{1}{2}\sqrt{\pi}\cdot\frac{5}{2}\cdot\frac{3}{2}\cdot\frac{1}{2}\sqrt{\pi}}{2\cdot 5\cdot 4\cdot 3\cdot 2\cdot 1} = \frac{3\pi}{512}$$

3. Evaluate $\int_0^{\pi/2} \sin^5 x \cos^4 x \, dx.$

Solution. We have

$$\int_0^{\pi/2} \sin^5 x \cos^4 x \, dx = \frac{\Gamma\left(\frac{5+1}{2}\right)\Gamma\left(\frac{4+1}{2}\right)}{2\Gamma\left(\frac{5+4+2}{2}\right)} = \frac{\Gamma(3)\,\Gamma\left(\frac{5}{2}\right)}{2\Gamma\left(\frac{11}{2}\right)}$$

$$= \frac{2 \cdot 1 \cdot \frac{3}{2} \cdot \frac{1}{2}\sqrt{\pi}}{2 \cdot \frac{9}{2} \cdot \frac{7}{2} \cdot \frac{5}{2} \cdot \frac{3}{2} \cdot \frac{1}{2}\sqrt{\pi}} = \frac{8}{315}$$

4. Evaluate $\int_0^{\pi/2} \cos^8 x \, dx$.

Solution. We have

$$\int_0^{\pi/2} \cos^8 x \, dx = \int_0^{\pi/2} \sin^0 x \cos^8 x \, dx = \frac{\Gamma\left(\frac{0+1}{2}\right)\Gamma\left(\frac{8+1}{2}\right)}{2\Gamma\left(\frac{0+8+2}{2}\right)}$$

$$= \frac{\Gamma\left(\frac{1}{2}\right)\Gamma\left(\frac{9}{2}\right)}{2\Gamma(5)} = \frac{\sqrt{\pi} \cdot \frac{7}{2} \cdot \frac{5}{2} \cdot \frac{3}{2} \cdot \frac{1}{2}\sqrt{\pi}}{2 \cdot 4 \cdot 3 \cdot 2 \cdot 1} = \frac{35\pi}{256}$$

5. Evaluate $\int_0^{2a} x^m \sqrt{2ax - x^2} \, dx$, where m is a positive integer.

Solution. (***First Method***) The given integral can be written as

$$\int_0^{2a} x^m \sqrt{x(2a - x)} \, dx$$

Let $x = 2a \sin^2 \theta$. Then $dx = 4a \sin \theta \cos \theta \, d\theta$.

When $\quad x = 2a, \quad \theta = \frac{\pi}{2}$

When $\quad x = 0, \quad \theta = 0$

Therefore

$$\int_0^{2a} x^m \sqrt{x(2a - x)} \, dx = \int_0^{2a} x^{m+\frac{1}{2}} \sqrt{2a - x} \, dx$$

$$= \int_0^{\pi/2} (2a \sin^2 \theta)^{m+\frac{1}{2}} \sqrt{2a - 2a \sin^2 \theta} \cdot 4a \sin \theta \cos \theta \, d\theta$$

$$= 2 \cdot (2a)^{m+2} \int_0^{\pi/2} \sin^{2m+2} \theta \cos^2 \theta \, d\theta$$

$$= 2 \cdot (2a)^{m+2} \frac{\Gamma\left(\frac{2m+2+1}{2}\right)\Gamma\left(\frac{2+1}{2}\right)}{2\Gamma\left(\frac{2m+2+2}{2}\right)}$$

$$= (2a)^{m+2} \frac{\Gamma\left(m+\frac{3}{2}\right)\Gamma\left(\frac{3}{2}\right)}{\Gamma(m+2)}$$

$$= 2^{m+2} a^{m+2} \frac{\left(m+\frac{1}{2}\right)\left(m-\frac{1}{2}\right)\cdots\frac{3}{2}\cdot\frac{1}{2}\sqrt{\pi}\cdot\frac{1}{2}\sqrt{\pi}}{\Gamma(m+2)}$$

$$= \frac{2^{m+2} a^{m+2} \cdot \left(\frac{2m+1}{2}\right)\left(\frac{2m-1}{2}\right)\cdots\frac{3}{2}\cdot\frac{1}{2}\sqrt{\pi}\cdot\frac{1}{2}\sqrt{\pi}}{\Gamma(m+2)}$$

$$= \frac{2^{m+2} a^{m+2} \cdot \left(\frac{2m+1}{2}\right)\left(\frac{2m}{2}\right)\left(\frac{2m-1}{2}\right)\cdots\frac{4}{2}\cdot\frac{3}{2}\cdot\frac{2}{2}\cdot\frac{1}{2}\sqrt{\pi}\cdot\frac{1}{2}\sqrt{\pi}}{\left(\frac{2m}{2}\right)\left(\frac{2m-2}{2}\right)\cdots\frac{4}{2}\cdot\frac{2}{2}\Gamma(m+2)}$$

$$= \frac{2^{m+2} a^{m+2} (2m+1)(2m)(2m-1)\cdots 4\cdot 3\cdot 2\cdot 1\sqrt{\pi}\cdot\sqrt{\pi}}{2^{2m+2}\cdot m(m-1)\cdots 2\cdot 1\,\Gamma(m+2)}$$

$$= \frac{\pi a^{m+2} \lfloor 2m+1}{2^m \lfloor m \lfloor m+1}$$

Second Method. The given integral can be written as

$$\int_0^{2a} x^m \sqrt{a^2 - (x-a)^2} \, dx$$

Let $x - a = a \cos \theta$. Then $dx = -a \sin \theta \, d\theta$.

When $\quad x = 2a, \quad \theta = 0$

When $\quad x = 0, \quad \theta = \pi$

Therefore $\int_0^{2a} x^m \sqrt{a^2 - (x-a)^2}\, dx$

$$= \int_\pi^0 a^m (1 + \cos\theta)^m \sqrt{a^2 - a^2 \cos^2\theta}\, (-a\sin\theta) d\theta$$

$$= \int_0^\pi a^{m+2} (1 + \cos\theta)^m \sin^2\theta\, d\theta$$

$$= a^{m+2} \int_0^\pi \left(2\cos^2\frac{\theta}{2}\right)^m \left(2\sin\frac{\theta}{2}\cos\frac{\theta}{2}\right)^2 d\theta$$

$$= 2^{m+2} a^{m+2} \int_0^\pi \cos^{2m+2}\frac{\theta}{2} \sin^2\frac{\theta}{2}\, d\theta$$

$$= 2^{m+3} a^{m+2} \int_0^{\pi/2} \cos^{2m+2}\phi \sin^2\phi\, d\phi \left(\text{where } \phi = \frac{\theta}{2}\right)$$

$$= 2^{m+3} a^{m+2} \frac{\Gamma\left(\frac{2m+3}{2}\right)\Gamma\left(\frac{3}{2}\right)}{2\Gamma(m+2)} = \frac{\pi a^{m+2} \lfloor 2m+1}{\lfloor m \lfloor m+1} \qquad \text{(from above)}$$

EXERCISES

Evaluate

1. $\int \cos^4 x \sin^2 x\, dx$
2. $\int_0^{\pi/2} \sin^5 x \cos^6 x\, dx$
3. $\int_0^{\pi/2} \sin^3 x \cos^4 x\, dx$
4. $\int_0^{\pi/2} \sin^6 x \cos^8 x\, dx$
5. $\int_0^a x^2 (a^2 - x^2)^{3/2}\, dx$
6. $\int_0^{\pi/4} \sin^5 x \cos^2 x\, dx$
7. $\int_0^1 x^{3/2} (1-x)^{3/2}\, dx$
8. $\int_0^{\pi/2} \cos^3 2x \sin^4 4x\, dx$
9. $\int_0^2 x^3 \sqrt{2x - x^2}\, dx$
10. $\int_0^{2a} x^2 \sqrt{2ax - x^2}\, dx$
11. $\int_0^2 x^{5/2} \sqrt{2-x}\, dx$
12. $\int_0^\pi \sin^3 x (1 - \cos x)^2\, dx$
13. $\int_0^a x^4 \sqrt{a^2 - x^2}\, dx$
14. $\int_0^{\frac{\pi}{6}} \sin^2 6\theta \cos^4 3\theta\, d\theta$
15. Evaluate $\int_0^{\pi/2} \sin^m x \cos^n x\, dx$ for positive integral values of m and n and apply it to find the value of $\int_0^{\pi/2} \sin^5 x \cos^6 x\, dx$.

16. If $I_{m,n} = \int \sin^m x \cos^n x\, dx$, prove the following reduction formulae:

(i) $I_{m,n} = -\dfrac{1}{m+n} \sin^{m-1} x \cos^{n+1} x + \dfrac{m-1}{m+n} I_{m-2,n}$

(ii) $I_{m,n} = \dfrac{1}{m+1} \sin^{m+1} x \cos^{n+1} x + \dfrac{m+n+2}{m+1} I_{m+2,n}$

(iii) $I_{m,n} = \dfrac{1}{m+n} \sin^{m+1} x \cos^{n-1} x + \dfrac{n-1}{m+n} I_{m,n-2}$

(iv) $I_{m,n} = -\dfrac{1}{n+1} \sin^{m+1} x \cos^{n+1} x + \dfrac{m+n+2}{n+1} I_{m,n+2}$

(v) $I_{m,n} = -\dfrac{1}{n+1} \sin^{m-1} x \cos^{n+1} x + \dfrac{m-1}{n+1} I_{m-2,n+2}$

(vi) $I_{m,n} = \dfrac{1}{m+1} \sin^{m+1} x \cos^{n-1} x + \dfrac{n-1}{m+1} I_{m+2,n-2}$

6.8 Reduction Formulae for $\int \cos^m x \cos nx\, dx$ and $\int \cos^m x \sin nx\, dx$

Let $I_{m,n} = \int \cos^m x \cos nx\, dx$. Then, integrating $I_{m,n}$ by parts, we get

$$I_{m,n} = \frac{1}{n} \cos^m x \sin nx - \int m \cos^{m-1} x (-\sin x) \frac{\sin nx}{n} dx$$

$$= \frac{1}{n} \cos^m x \sin nx + \frac{m}{n} \int \cos^{m-1} x \sin x \sin nx\, dx$$

$$= \frac{1}{n} \cos^m x \sin nx + \frac{m}{n} \int \cos^{m-1} x [\cos(n-1)x - \cos nx \cos x] dx$$

$$= \frac{1}{n} \cos^m x \sin nx + \frac{m}{n} \int \cos^{m-1} x \cos(n-1) x\, dx$$

$$- \frac{m}{n} \int \cos^m x \cos nx\, dx$$

$$= \frac{1}{n} \cos^m x \sin nx + \frac{m}{n} I_{m-1,n-1} - \frac{m}{n} I_{m,n}$$

$$\Rightarrow \quad \left(1 + \frac{m}{n}\right) I_{m,n} = \frac{1}{n} \cos^m x \sin nx + \frac{m}{n} I_{m-1,n-1}$$

$$\Rightarrow \quad I_{m,n} = \frac{1}{m+n} \cos^m x \sin nx + \frac{m}{m+n} I_{m-1,n-1}$$

Similarly, if $I_{m,n} = \int \cos^m x \sin nx\, dx$, we can prove that

$$I_{m,n} = -\frac{1}{m+n}\cos^m x \cos nx + \frac{m}{m+n} I_{m-1,n-1}$$

6.9 Reduction Formulae for $\int \sin^m x \cos nx\, dx$ and $\int \sin^m x \sin nx\, dx$

Let $I_{m,n} = \int \sin^m x \cos nx\, dx$. Then, integrating $I_{m,n}$ by parts, we get

$$I_{m,n} = \frac{1}{n}\sin^m x \sin nx - \int m \sin^{m-1} x \cos x \frac{\sin nx}{n}\, dx$$

$$= \frac{1}{n}\sin^m x \sin nx - \frac{m}{n}\int \sin^{m-1} x \cos x \sin nx\, dx$$

$$= \frac{1}{n}\sin^m x \sin nx - \frac{m}{n}\left[\sin^{m-1} x \cos x \frac{(-\cos nx)}{n}\right.$$

$$\left. - \int \{(m-1)\sin^{m-2} x \cos^2 x - \sin^{m-1} x \sin x\}\left(\frac{-\cos nx}{n}\right) dx\right]$$

$$= \frac{1}{n}\sin^m x \sin nx + \frac{m}{n^2}\sin^{m-1} x \cos x \cos nx$$

$$- \frac{m}{n^2}\int \{(m-1)\sin^{m-2} x(1-\sin^2 x) - \sin^m x\}\cos nx\, dx$$

$$= \frac{1}{n}\sin^m x \sin nx + \frac{m}{n^2}\sin^{m-1} x \cos x \cos nx$$

$$- \frac{m}{n^2}\int \{(m-1)\sin^{m-2} x - m\sin^m x\}\cos nx\, dx$$

$$= \frac{1}{n}\sin^m x \sin nx + \frac{m}{n^2}\sin^{m-1} x \cos x \cos nx$$

$$- \frac{m(m-1)}{n^2}\int \sin^{m-2} x \cos nx\, dx + \frac{m^2}{n^2}\int \sin^m x \cos nx\, dx$$

$$\Rightarrow \quad I_{m,n} = \frac{1}{n}\sin^m x \sin nx + \frac{m}{n^2}\sin^{m-1} x \cos x \cos nx$$

$$- \frac{m(m-1)}{n^2} I_{m-2,n} + \frac{m^2}{n^2} I_{m,n}$$

$$\Rightarrow \quad \left(1-\frac{m^2}{n^2}\right)I_{m,n} = \frac{1}{n}\sin^m x \sin nx + \frac{m}{n^2}\sin^{m-1} x \cos x \cos nx$$

$$-\frac{m(m-1)}{n^2} I_{m-2,n}$$

$$\Rightarrow I_{m,n} = -\frac{(n\sin^m x \sin nx + m\sin^{m-1} x \cos x \cos nx)}{m^2-n^2} + \frac{m(m-1)}{m^2-n^2} I_{m-2,n}$$

Similarly, if $I_{m,n} = \int \sin^m x \sin nx\, dx$, we can prove that

$$I_{m,n} = \frac{n\sin^m x \cos nx - m\sin^{m-1} x \cos x \sin nx}{m^2-n^2} + \frac{m(m-1)}{m^2-n^2} I_{m-2,n}$$

6.10 Reduction Formula for $\int \frac{\sin nx}{\sin x}\, dx$

Let $$I_n = \int \frac{\sin nx}{\sin x}\, dx$$

Then $$I_{n-2} = \int \frac{\sin (n-2)x}{\sin x}\, dx$$

Now $$I_n - I_{n-2} = \int \frac{(\sin nx - \sin (n-2)x)}{\sin x}\, dx$$

$$= \int \frac{2\cos (n-1)x \sin x}{\sin x}\, dx$$

$$= 2\int \cos (n-1)x\, dx = \frac{2\sin (n-1)x}{n-1}$$

$$\Rightarrow \quad I_n = \frac{2}{n-1}\sin (n-1)x + I_{n-2}$$

Examples

1. If $I_{m,n} = \int_0^{\pi/2} \cos^m x \sin nx\, dx$, prove that

$$I_{m,n} = \frac{1}{m+n} + \frac{m}{m+n} I_{m-1,n-1}$$

Hence or otherwise evaluate $\int_0^{\pi/2} \cos^5 x \sin 3x\, dx$.

Solution. Integrating $I_{m,n}$ by parts, we get

$$I_{m,n} = \int_0^{\pi/2} \cos^m x \sin nx\, dx$$

$$= \left[-\frac{\cos^m x \cos nx}{n} \right]_0^{\pi/2} - \int_0^{\pi/2} m \cos^{m-1} x(-\sin x)\left(-\frac{\cos nx}{n} \right) dx$$

$$= \frac{1}{n} - \frac{m}{n} \int_0^{\pi/2} \cos^{m-1} x \cos nx \sin x\, dx$$

$$= \frac{1}{n} - \frac{m}{n} \int_0^{\pi/2} \cos^{m-1} x\, [\sin nx \cos x - \sin(n-1)x]\, dx$$

$$= \frac{1}{n} - \frac{m}{n} \int_0^{\pi/2} \cos^m x \sin nx\, dx + \frac{m}{n} \int_0^{\pi/2} \cos^{m-1} x \sin(n-1)x\, dx$$

$$= \frac{1}{n} - \frac{m}{n} I_{m,n} + \frac{m}{n} I_{m-1,n-1}$$

$$\Rightarrow \qquad \left(1 + \frac{m}{n}\right) I_{m,n} = \frac{1}{n} + \frac{m}{n} I_{m-1,n-1}$$

$$\Rightarrow \qquad I_{m,n} = \frac{1}{m+n} + \frac{m}{m+n} I_{m-1,n-1} \qquad (1)$$

Further, the given integral $\int_0^{\pi/2} \cos^5 x \sin 3x\, dx$ is denoted by $I_{5,3}$.

Therefore, putting $m = 5, 4$ and $n = 3, 2$ in Eq. (1), we get

$$I_{5,3} = \frac{1}{5+3} + \frac{5}{5+3} I_{4,2} = \frac{1}{8} + \frac{5}{8} I_{4,2}$$

$$I_{4,2} = \frac{1}{6} + \frac{2}{3} I_{3,1} = \frac{1}{6} + \frac{2}{3} \int_0^{\pi/2} \cos^3 x \sin x\, dx = \frac{1}{6} + \frac{2}{3} \cdot \frac{1}{4} = \frac{1}{3}$$

Hence $$I_{5,3} = \frac{1}{8} + \frac{5}{8} \cdot \frac{1}{3} = \frac{1}{3}$$

2. Prove that, if n is a positive integer,

$$\int_0^{\pi/2} \cos^n x \cos nx\, dx = \frac{\pi}{2^{n+1}}$$

Solution. Integrating $I_{n,n}$ by parts, we get

$$I_{n,n} = \int_0^{\pi/2} \cos^n x \cos nx\, dx$$

$$= \left[\frac{\cos^n x \sin nx}{n}\right]_0^{\pi/2} - \int_0^{\pi/2} n \cos^{n-1} x(-\sin x)\, \frac{\sin nx}{n}\, dx$$

$$= \int_0^{\pi/2} \cos^{n-1} x[\cos(n-1)x - \cos nx \cos x]dx$$

$$= \int_0^{\pi/2} \cos^{n-1} x \cos(n-1)x\, dx - \int_0^{\pi/2} \cos^n x \cos nx dx$$

Therefore, $$I_{n,n} = I_{n-1,n-1} - I_{n,n}$$

$\Rightarrow$ $$I_{n,n} = \frac{1}{2} I_{n-1,n-1} \qquad (1)$$

Replacing n by $n-1, n-2, \cdots, 3, 2, 1$ in Eq. (1), we get

$$I_{n-1,n-1} = \frac{1}{2} I_{n-2,n-2}$$

$$I_{n-2,n-2} = \frac{1}{2} I_{n-3,n-3}$$

$$\ldots \qquad \ldots$$

$$\ldots \qquad \ldots$$

$$I_{3,3} = \frac{1}{2} I_{2,2}$$

$$I_{2,2} = \frac{1}{2} I_{1,1}$$

$$I_{1,1} = \frac{1}{2} I_{0,0} = \frac{1}{2} \int_0^{\pi/2} dx = \frac{1}{2} \cdot \frac{\pi}{2}$$

Hence $$I_{n,n} = \frac{1}{2} \cdot \frac{1}{2} \cdots \frac{1}{2} \cdot \frac{\pi}{2} = \frac{\pi}{2^{n+1}}$$

3. If $I_{m,n} = \int_0^{\pi/2} \sin^m x \sin nx\, dx$, prove that $I_{m,n} = \dfrac{m(m-1)}{m^2 - n^2} I_{m-2,n}$

Hence or otherwise evaluate $\int_0^{\pi/2} \sin^4 x \sin 5x\, dx$.

Solution. Integrating $I_{m,n}$ by parts, we get

$$I_{m,n} = \int_0^{\pi/2} \sin^m x \cdot \sin nx\, dx$$

$$= \left[-\frac{\sin^m x \cos nx}{n} \right]_0^{\pi/2} - \int_0^{\pi/2} m \sin^{m-1} x \cos x \frac{(-\cos nx)}{n} dx$$

$$= \frac{m}{n} \int_0^{\pi/2} \sin^{m-1} x \cos x \cos nx \, dx$$

$$= \frac{m}{n} \left[\frac{\sin^{m-1} x \cos x \sin nx}{n} \right]_0^{\pi/2}$$

$$- \frac{m}{n} \int_0^{\pi/2} \{(m-1) \sin^{m-2} x \cos^2 x - \sin^{m-1} x \sin x\} \frac{\sin nx}{n} dx$$

$$= -\frac{m}{n^2} \int_0^{\pi/2} \{(m-1) \sin^{m-2} x (1 - \sin^2 x) - \sin^m x\} \sin nx \, dx$$

$$= -\frac{m(m-1)}{n^2} \int_0^{\pi/2} \sin^{m-2} x \sin nx \, dx + \frac{m^2}{n^2} \int_0^{\pi/2} \sin^m x \sin nx \, dx$$

Therefore, $I_{m,n} = -\dfrac{m(m-1)}{n^2} I_{m-2,n} + \dfrac{m^2}{n^2} I_{m,n}$

$$\Rightarrow \qquad \left(\frac{m^2}{n^2} - 1 \right) I_{m,n} = \frac{m(m-1)}{n^2} I_{m-2,n}$$

$$\Rightarrow \qquad I_{m,n} = \frac{m(m-1)}{m^2 - n^2} I_{m-2,n} \qquad (1)$$

Further, the given integral $\int_0^{\pi/2} \sin^4 x \sin 5x \, dx$ is denoted by $I_{4,5}$. Therefore, putting $m = 4, 2$ and $n = 5$ in Eq. (1), we get

$$I_{4,5} = \frac{4 \cdot (4-1)}{4^2 - 5^2} I_{2,5} = -\frac{4}{3} I_{2,5}$$

$$I_{2,5} = \frac{2 \cdot (2-1)}{2^2 - 5^2} I_{0,5} = -\frac{2}{21} \int_0^{\pi/2} \sin 5x \, dx = -\frac{2}{21} \left[-\frac{\cos 5x}{5} \right]_0^{\pi/2} = -\frac{2}{105}$$

Hence $\qquad I_{4,5} = -\dfrac{4}{3} \times \left(-\dfrac{2}{105} \right) = \dfrac{8}{315}$

EXERCISES

1. If $I_{m,n} = \int_0^{\pi/2} \cos^m x \cos nx \, dx$, prove that $I_{m,n} = \dfrac{m}{m-n} I_{m-1,n+1} = \dfrac{m}{m+n} I_{m-1,n-1}$.

2. If m and n are positive integers such that $m - n$ is even and positive, prove that

$$\int_0^{\pi/2} \cos^m x \cos nx\, dx = \frac{\lfloor m}{\lfloor \frac{m+n}{2} \lfloor \frac{m-n}{2}} \cdot \frac{\pi}{2^{m+1}}$$

3. Prove that $\displaystyle\int \frac{\cos nx}{\cos x} dx = \frac{2}{n-1} \sin (n-1)x - \int \frac{\cos(n-2)x}{\cos x} dx.$

4. Prove that $\displaystyle\int_0^{\pi/2} \cos^m x \sin mx\, dx = \frac{1}{2^{m+1}}\left(2 + \frac{2^2}{2} + \frac{2^3}{3} + \cdots + \frac{2^m}{m}\right)$

5. Find a reduction formula for $\displaystyle\int \frac{\sin nx}{\sin x} dx$, and prove that

$$\int_0^{\pi} \frac{\sin nx}{\sin x} dx = \pi \text{ or } 0$$

according as n an odd or even positive integer.

6. Evaluate $\displaystyle\int_0^{\pi/2} \cos^{m+1} x \sin mx\, dx.$

7. If $u_n = \displaystyle\int \frac{\sin nx}{\sin x} dx$, then prove that $u_n = \dfrac{2 \sin (n-1)x}{n-1} + u_{n-2}.$

Hence evaluate $\displaystyle\int_0^{\pi/2} \frac{\sin 7x}{\sin x} dx.$

6.11 Reduction Formulae for $\int x^m \cos nx\, dx$ and $\int x^m \sin nx\, dx$

Let $I_{m,n} = \int x^m \cos nx\, dx$. Then, integrating $I_{m,n}$ by parts, we get

$$I_{m,n} = \frac{1}{n} x^m \sin nx - \int m x^{m-1} \frac{\sin nx}{n} dx$$

$$= \frac{1}{n} x^m \sin nx - \frac{m}{n} \int x^{m-1} \sin nx\, dx$$

$$= \frac{1}{n} x^m \sin nx - \frac{m}{n}\left[x^{m-1}\left(-\frac{\cos nx}{n}\right) - \int (m-1)x^{m-2}\left(-\frac{\cos nx}{n}\right) dx\right]$$

$$= \frac{1}{n} x^m \sin nx + \frac{m}{n^2} x^{m-1} \cos nx - \frac{m(m-1)}{n^2} \int x^{m-2} \cos nx\, dx$$

Therefore, $I_{m,n} = \frac{1}{n} x^m \sin nx + \frac{m}{n^2} x^{m-1} \cos nx - \frac{m(m-1)}{n^2} I_{m-2,n}$

Similarly, integrating $x^m \sin nx$ by parts twice, we get

$$\int x^m \sin nx\, dx = -\frac{1}{n} x^m \cos nx + \frac{m}{n^2} x^{m-1} \sin nx - \frac{m(m-1)}{n^2} \int x^{m-2} \sin nx\, dx$$

6.12 Reduction Formula for $\int e^{mx} x^n dx$

If n is a positive integer, then integrating by parts, we get

$$\int e^{mx} x^n\, dx = \frac{1}{m} e^{mx} x^n - \frac{n}{m} \int e^{mx} x^{n-1} dx$$

By repeated application of this formula, the given integral is ultimately made to depend upon $\int e^{mx}\, dx$ whose integral is easily obtained as $\frac{1}{m} e^{mx}$

6.13 Reduction Formula for $\int x^m (\log x)^n\, dx$

Integrating by parts, we get

$$\int x^m (\log x)^n\, dx = \frac{1}{m+1} x^{m+1} (\log x)^n - \int \frac{x^{m+1}}{(m+1)} \cdot \frac{n(\log x)^{n-1}\, dx}{x}$$

$$= \frac{1}{m+1} x^{m+1} (\log x)^n - \frac{n}{m+1} \int x^m (\log x)^{n-1} dx$$

Examples

1. If $U_n = \int_0^{\pi/2} x^n \sin x\, dx$ and $n > 1$, prove that

$$U_n + n(n-1) U_{n-2} = \frac{n\pi^{n-1}}{2^{n-1}}$$

Hence evaluate U_3.

Solution. Integrating U_n by parts, we get

$$U_n = [x^n(-\cos x)]_0^{\pi/2} - \int_0^{\pi/2} nx^{n-1} (-\cos x)\, dx$$

$$= n \int_0^{\pi/2} x^{n-1} \cos x\, dx = n[x^{n-1} \sin x]_0^{\pi/2} - n \int_0^{\pi/2} (n-1) x^{n-2} \sin x\, dx$$

Therefore, $$U_n = n\left(\frac{\pi}{2}\right)^{n-1} - n(n-1) U_{n-2} \qquad (1)$$

$$\Rightarrow \qquad U_n + n(n-1)U_{n-2} = \frac{n\pi^{n-1}}{2^{n-1}}$$

Putting $n = 3$ in Eq. (1), we get

$$U_3 = 3 \cdot \left(\frac{\pi}{2}\right)^2 - 3 \cdot (3-1)U_1 = \frac{3}{4}\pi^2 - 6U_1$$

But $U_1 = \int_0^{\pi/2} x \sin x \, dx$. Therefore

$$U_1 = [x(-\cos x)]_0^{\pi/2} - \int_0^{\pi/2} 1 \cdot (-\cos x) dx = [\sin x]_0^{\pi/2} = 1$$

Hence $U_3 = \dfrac{3\pi^2}{4} - 6$.

2. If $U_n = \int_0^{\pi/2} \theta \sin^n \theta \, d\theta$ and $n > 1$, prove that

$$U_n = \frac{(n-1)}{n} U_{n-2} + \frac{1}{n^2}$$

Deduce that $$U_5 = \frac{149}{225}.$$

Solution. Integrating U_n by parts, we get

$$U_n = \int_0^{\pi/2} (\theta \sin^{n-1}\theta) \sin\theta \, d\theta = [\theta \sin^{n-1}\theta(-\cos\theta)]_0^{\pi/2} - \int_0^{\pi/2} [\sin^{n-1}\theta$$

$$+ (n-1)\,\theta \sin^{n-2}\theta \cos\theta]\,(-\cos\theta) d\theta$$

$$= \int_0^{\pi/2} \sin^{n-1}\theta \cos\theta d\theta + (n-1)\int_0^{\pi/2} \theta \sin^{n-2}\theta(1 - \sin^2\theta) d\theta$$

$$= \left[\frac{\sin^n\theta}{n}\right]_0^{\pi/2} + (n-1)\int_0^{\pi/2} \theta \sin^{n-2}\theta \, d\theta - (n-1)\int_0^{\pi/2} \theta \sin^n \theta \, d\theta$$

Therefore, $$U_n = \frac{1}{n} + (n-1)U_{n-2} - (n-1)U_n$$

$$\Rightarrow \qquad (1 + n - 1)U_n = \frac{1}{n} + (n-1)U_{n-2}$$

$$\Rightarrow \qquad U_n = \frac{1}{n^2} + \frac{(n-1)}{n}U_{n-2} \qquad (1)$$

Putting $n = 5, 3$ in Eq. (1), we get

$$U_5 = \frac{1}{25} + \frac{4}{5} U_3$$

$$U_3 = \frac{1}{9} + \frac{2}{3} U_1 = \frac{1}{9} + \frac{2}{3} \int_0^{\pi/2} \theta \sin \theta \, d\theta = \frac{1}{9} + \frac{2}{3} = \frac{7}{9}$$

Hence $\quad U_5 = \dfrac{1}{25} + \dfrac{4}{5} \cdot \dfrac{7}{9} = \dfrac{1}{25} + \dfrac{28}{45} = \dfrac{149}{225}$

EXERCISES

Evaluate

1. $\int x^3 \sin x \, dx$
2. $\int x^4 \cos nx \, dx$
3. $\int_0^1 x^4 (\log x)^3 \, dx$
4. $\int_0^{\pi/2} x^5 \sin x \, dx$
5. $\int_0^{\pi/2} x^3 \sin 3x \, dx$
6. If $U_n = \int_0^{\pi/2} x^n \sin mx \, dx$, then prove that

$$U_n = \frac{n}{m^2}\left(\frac{\pi}{2}\right)^{n-1} - \frac{n(n-1)}{m^2} U_{n-2}$$

where m is of the form $4r + 1$.

6.14 Reduction Formula for $\int x^m (a + bx^n)^p \, dx$

A reduction formula for $\int x^m (a + bx^n)^p$, where m, n, p are constants, positive or negative, integral or fractional, may be obtained by connecting it with any one of the following six integrals:

1. $\int x^{m-n} (a + bx^n)^p \, dx$
2. $\int x^{m+n} (a + bx^n)^p \, dx$
3. $\int x^m (a + bx^n)^{p-1} \, dx$
4. $\int x^m (a + bx^n)^{p+1} \, dx$
5. $\int x^{m-n} (a + bx^n)^{p+1} \, dx$
6. $\int x^{m+n} (a + bx^n)^{p-1} \, dx$

The method of connecting the original integral with any one of the six integrals is as follows:

I. Method of Connection

(i) Put $P = x^{\lambda+1} (a + bx^n)^{\mu+1}$, where λ is the smaller of the two indices of x, and μ is the smaller of the two indices of $(a + bx^n)$ in the two expressions whose integrals are to be connected.

(ii) Find $\frac{dP}{dx}$ and resolve it into a linear function of the two expressions whose integrals are to be connected.

(iii) Integrate back and we get the required reduction formulae.

As an example, we connect the integral

$$\int x^m (a + bx^n)^p \, dx \text{ with } \int x^{m-n} (a + bx^n)^p \, dx.$$

Here the smaller of the two indices of x is $m - n$ and therefore $\lambda = m - n$. The indices of $(a + bx^n)$ in the two integrals are each equal to p. Hence we put

$$P = x^{m-n+1} (a + bx^n)^{p+1}$$

so that

$$\frac{dP}{dx} = (m - n + 1)x^{m-n} (a + bx^n)^{p+1} + (p + 1) nbx^m (a + bx^n)^p$$

$$= (m - n + 1)x^{m-n} (a + bx^n) (a + bx^n)^p + (p + 1) \, nbx^m (a + bx^n)^p$$

$$= a(m - n + 1)x^{m-n}(a + bx^n)^p$$

$$+ [b(m - n + 1) + nb(p + 1)]x^m (a + bx^n)^p$$

$$= a(m - n + 1)x^{m-n}(a + bx^n)^p + b(np + m + 1)x^m (a + bx^n)^p$$

Therefore $\frac{dP}{dx}$ has been written as a linear function of the two integrands, we now integrate back and get

$$P = a(m - n + 1) \int x^{m-n}(a + bx^n)^p \, dx + b(np + m + 1) \int x^m (a + bx^n)^p \, dx$$

Hence $$\int x^m (a + bx^n)^p \, dx = \frac{x^{m-n+1}(a + bx^n)^{p+1}}{b(np + m + 1)}$$

$$- \frac{a(m - n + 1)}{b(np + m + 1)} \int x^{m-n} (a + bx^n)^p \, dx$$

II. Method of Integration by Parts

Let $I_{m,p} = \int x^m (a + bx^n)^p \, dx$. We shall connect $I_{m,p}$ with $I_{m-n,p}$. Since

$$\int (a + bx^n)^p x^{n-1} \, dx = \frac{(a + bx^n)^{p+1}}{nb(p + 1)}$$

We have

$$I_{m,p} = \int x^m (a + bx^n)^p dx = \int x^{m-n+1} (a + bx^n)^p x^{n-1} dx$$

$$\therefore\ I_{m,p} = x^{m-n+1} \frac{(a + bx^n)^{p+1}}{nb(p + 1)} - \int (m - n + 1) x^{m-n} \frac{(a + bx^n)^{p+1}}{nb(p + 1)} dx$$

$$\Rightarrow \quad nb(p + 1) I_{m,p} = x^{m-n+1} (a + bx^n)^{p+1} - (m - n + 1) I_{m-n,p+1} \qquad (1)$$

Now

$$I_{m-n,p+1} = \int x^{m-n}(a + bx^n)^{p+1} dx = \int x^{m-n}(a + bx^n)^p (a + bx^n) dx$$

$$= a\int x^{m-n}(a + bx^n)^p dx + b\int x^m (a + bx^n)^p dx = aI_{m-n,p} + bI_{m,p}$$

Substituting the value of $I_{m-n,p+1}$ in Eq. (1), we get

$$nb(p + 1) I_{m,p} = x^{m-n+1}(a + bx^n)^{p+1} - (m - n + 1)(aI_{m-n,p} + bI_{m,p})$$

$$\Rightarrow \quad [nb(p + 1) + (m - n + 1)b] I_{m,p} = x^{m-n+1}(a + bx^n)^{p+1}$$

$$- (m - n + 1) aI_{m-n,p}$$

$$\Rightarrow \quad I_{m,p} = \frac{x^{m-n+1}(a + bx^n)^{p+1}}{b(np + m + 1)} - \frac{a(m - n + 1)}{b(np + m + 1)} I_{m-n,p}$$

Examples

1. Find a reduction formula for $\int \frac{1}{(x^2 + a^2)^n} dx$.

Solution. The given integral can be written as

$$\int x^0 (a^2 + x^2)^{-n} dx$$

Here $m = 0$, $n = 2$ and $p = -n$. Therefore, it is convenient to increase p by unity and connect the given integral with

$$\int x^0 (a^2 + x^2)^{-n+1} dx, \text{ i.e. with } \int \frac{1}{(a^2 + x^2)^{n-1}} dx$$

Since in this case $\lambda = 0$ and $\mu = -n$, we put

$$P = x(a^2 + x^2)^{-n+1}$$

so that

$$\frac{dP}{dx} = (a^2 + x^2)^{-n+1} + 2(-n + 1) x^2 (a^2 + x^2)^{-n}$$

$$= (a^2 + x^2)^{-n+1} + 2(-n + 1)\,(a^2 + x^2 - a^2)\,(a^2 + x^2)^{-n}$$
$$= (a^2 + x^2)^{-n+1} + 2(-n + 1)\,(a^2 + x^2)^{-n+1} - 2a^2(-n + 1)\,(a^2 + x^2)^{-n}$$
$$= (3 - 2n)(a^2 + x^2)^{-n+1} + 2a^2(n - 1)\,(a^2 + x^2)^{-n}$$

Therefore, $\frac{dP}{dx}$ has been written as a linear function of the two integrands, we now integrate back and get

$$P = (3 - 2n)\int \frac{1}{(a^2 + x^2)^{n-1}}\,dx + 2a^2(n - 1)\int \frac{1}{(a^2 + x^2)^n}\,dx$$

$$\Rightarrow \int \frac{1}{(a^2 + x^2)^n}\,dx = \frac{x}{2a^2(n - 1)(a^2 + x^2)^{n-1}} + \frac{2n - 3}{2a^2(n - 1)}\int \frac{1}{(a^2 + x^2)^{n-1}}\,dx$$

2. If $U_n = \int x^n\sqrt{a^2 - x^2}\,dx$, prove that

$$U_n = -\frac{x^{n-1}(a^2 - x^2)^{3/2}}{n + 2} + \left(\frac{n - 1}{n + 2}\right)a^2 U_{n-2}$$

Hence or otherwise evaluate $\int_0^a x^4\sqrt{a^2 - x^2}\,dx$.

Solution. Here $m = n$, $n = 2$ and $p = \frac{1}{2}$. Therefore, it is convenient to decrease m by 2 and connect the given integral with

$$\int x^{n-2}\sqrt{a^2 - x^2}\,dx$$

Since in this case $\lambda = n$ and $\mu = \frac{1}{2}$, we put

$$P = x^{n-1}(c^2 - x^2)^{3/2}$$

so that

$$\frac{dP}{dx} = (n - 1)\,x^{n-2}(a^2 - x^2)^{3/2} - 3x^n(a^2 - x^2)^{1/2}$$
$$= (n - 1)x^{n-2}\,(a^2 - x^2)\,(a^2 - x^2)^{1/2} - 3x^n(a^2 - x^2)^{1/2}$$
$$= a^2(n - 1)\,x^{n-2}(a^2 - x^2)^{1/2} - (n - 1)x^n\,(a^2 - x^2)^{1/2} - 3x^n(a^2 - x^2)^{1/2}$$
$$= a^2(n - 1)x^{n-2}(a^2 - x^2)^{1/2} - (n + 2)x^n\,(a^2 - x^2)^{1/2}$$

Therefore, $\frac{dP}{dx}$ has been written as a linear function of the two integrands, we now integrate back and get

$$P = a^2(n-1)\int x^{n-2}(a^2 - x^2)^{1/2}\,dx - (n+2)\int x^n (a^2 - x^2)^{1/2}\,dx$$

$$\Rightarrow \quad (n+2)\int x^n (a^2 - x^2)^{1/2}\,dx = a^2(n-1)\int x^{n-2}(a^2 - x^2)^{1/2}\,dx$$

$$- x^{n-1}(a^2 - x^2)^{3/2}$$

$$\Rightarrow \quad (n+2)U_n = a^2(n-1)\,U_{n-2} - x^{n-1}(a^2 - x^2)^{3/2}$$

$$\Rightarrow \quad U_n = \left(\frac{n-1}{n+2}\right)a^2 U_{n-2} - \frac{x^{n-1}(a^2 - x^2)^{3/2}}{n+2}$$

Let

$$U_n = \int_0^a x^n \sqrt{a^2 - x^2}\,dx.$$

Then $$U_n = \left(\frac{n-1}{n+2}\right)a^2 U_{n-2} - \left[\frac{x^{n-1}(a^2 - x^2)^{3/2}}{n+2}\right]_0^a$$

$$\Rightarrow \quad U_n = \left(\frac{n-1}{n+2}\right)a^2 U_{n-2} \tag{1}$$

Putting $n = 4, 2$ in Eq. (1), we get

$$U_4 = \frac{3}{4}a^2 U_2$$

$$U_2 = \frac{1}{2}a^2 U_0 = \frac{a^2}{2}\int_0^a \sqrt{a^2 - x^2}\,dx$$

$$= \frac{a^2}{2}\left[\frac{x\sqrt{a^2 - x^2}}{2} + \frac{a^2}{2}\sin^{-1}\frac{x}{a}\right]_0^a = \frac{\pi a^4}{8}$$

Hence $$U_4 = \frac{3a^2}{4}\cdot\frac{\pi a^4}{8} = \frac{3\pi a^6}{32}$$

EXERCISES

1. If $I_n = \int x^n \sqrt{a - x}\,dx$, prove that $(2n+3)I_n = 2an\,I_{n-1} - 2x^n(a-x)^{3/2}$.

   Evaluate $\int_0^a x^2 \sqrt{ax - x^2}\,dx$.

2. If m be a positive integer, find a reduction formula for

$$\int x^m \sqrt{2ax - x^2}\, dx$$

Hence obtain the value of $\int_0^{2a} x^3 \sqrt{2ax - x^2}\, dx$.

3. If $I_n = \int_0^a (a^2 - x^2)^n dx$ and $n > 0$, prove that $I_n = \dfrac{2na^2}{2n+1} I_{n-1}$.

   Hence evaluate $\int_0^a (a^2 - x^2)^3 dx$.

4. If $U_n = \int_0^{2a} x^n \sqrt{2ax - x^2}\, dx$, prove that $U_n = \dfrac{(2n+1)}{n+2} aU_{n-1}$.

   Hence evaluate $\int_0^{2a} x^2 \sqrt{2ax - x^2}\, dx$.

5. If n bė a positive integer, find a reduction formula for

$$\int (a^2 + x^2)^{n/2}\, dx$$

   and apply it to evaluate $\int (a^2 + x^2)^{5/2}\, dx$.

6. If $U_n = \int x(1 + x^3)^n\, dx$, where n is positive, prove that

$$(3n + 2)U_n = x^2(1 + x^3)^n + 3nU_{n-1}$$

MISCELLANEOUS EXERCISES

1. Find a reduction formula for $\int \sin^n x\, dx$ and hence evaluate $\int \sin^4 x\, dx$.
2. Integrate $\sin^2 x \cos^6 x$.
3. Evaluate $\int \sin^4 x \cos^2 x\, dx$.
4. Evaluate $\int_0^{\pi/4} \sin^4 x\, dx$.
5. Evaluate $\int_0^{\pi/2} \sin^6 x\, dx$.
6. Prove that $\int_0^{\pi} \theta \sin^2 \theta \cos \theta\, d\theta = -\dfrac{4}{9}$.
7. Prove that $\int_0^{\pi/2} \cos^{n-2} x \sin nx\, dx = \dfrac{1}{n-1}$ ($n > 1$ and integral).
8. If $I_n = \int_0^{\pi/2} x^n \sin (2p + 1) x\, dx$, prove that

$$I_n + \frac{n(n-1)}{(2p+1)^2} I_{n-2} = \frac{(-1)^p n}{(2p+1)^2} \left(\frac{\pi}{2}\right)^{n-1}$$

   n and p being positive integers.

9. If m and n are positive integers and $I_{m,n} = \int_0^1 x^{n-1} (\log x)^m \, dx$, prove that $I_{m,n} = -\frac{m}{n} I_{m-1,n}$.

   Deduce that $I_{m,n} = \frac{(-1)^m \lfloor m}{n^{m+1}}$.

10. Evaluate

    (i) $\int_0^a x^3 (2ax - x^2)^{3/2} \, dx$ (ii) $\int_0^a x^2 (2ax - x^2)^{5/2} \, dx$

    (iii) $\int_0^1 x^2 (1 - x^2)^{3/2} \, dx$ (iv) $\int_0^1 x^2 \sqrt{1 - x} \, dx$

    (v) $\int_0^1 x^{3/2} (1 - x)^{3/2} \, dx$ (vi) $\int_0^1 x^m (1 - x)^n \, dx$

    (vii) $\int_0^a \frac{x^4}{(a^2 + x^2)^4} \, dx$

11. Connect $\int x^m (a + bx^n)^p \, dx$ with the following integrals:

    (i) $\int x^{m+n} (a + bx^n)^p \, dx$ (ii) $\int x^m (a + bx^n)^{p-1} \, dx$

    (iii) $\int x^m (a + bx^n)^{p+1} \, dx$ (iv) $\int x^{m-n} (a + bx^n)^{p+1} \, dx$

    (v) $\int x^{m+n} (a + bx^n)^{p-1} \, dx$

12. Connect $\int x^{m-1} (a + bx^n)^p \, dx$ with

    $\int x^{m-n-1} (a + bx^n)^p \, dx$ and evaluate $\int \frac{x^8}{(1 + x^3)^{1/3}} \, dx$.

13. If $I_n = \int_0^1 x^p (1 - x^q)^n \, dx$, where p, q and n are positive, then prove that $(p + nq + 1) I_n = nq I_{n-1}$.
    Hence evaluate I_n when n is a positive integer.

14. Find a reduction formulae for $\int x^m (1 + x^2)^{n/2} \, dx$, where m and n are positive integers and evaluate the integral when $m = 5$, $n = 7$.

15. Evaluate $\int \frac{x^4}{\sqrt{a^2 - x^2}} \, dx$.

16. If $U_n = \int x^n (1 + x^4)^{-1/2} \, dx$, prove that $U_n = \frac{x^{n-3} (1 + x^4)^{1/2}}{n - 1} - \left(\frac{n - 3}{n - 1} \right) U_{n-4}$.

7

Quadrature

7.1 Introduction

In Sec. 5.3 it was proved that the area bounded by a curve $y = f(x)$, the two ordinates $x = a$ and $x = b$, and the x-axis is given by

$$A = \int_a^b y\,dx = \int_a^b f(x)dx$$

It may similarly be proved that the area bounded by a curve $x = f(y)$, the two abscissae $y = c$ and $y = d$, and the y-axis is given by

$$A = \int_c^d x\,dy = \int_c^d f(y)dy$$

The process of determining the area of a plane region is known as *quadrature.*

7.2 Areas When Cartesian Equation of a Curve is Given

Examples

1. Find the area bounded by the parabola $y^2 = 4ax$, its latus rectum and the x-axis.

Solution. The required area is

$$\int_0^a y\,dx = \int_0^a 2a^{\frac{1}{2}}x^{\frac{1}{2}}\,dx$$

$$= 2a^{\frac{1}{2}}\left[\frac{2}{3}x^{\frac{3}{2}}\right]_0^a = \frac{4a^2}{3}$$

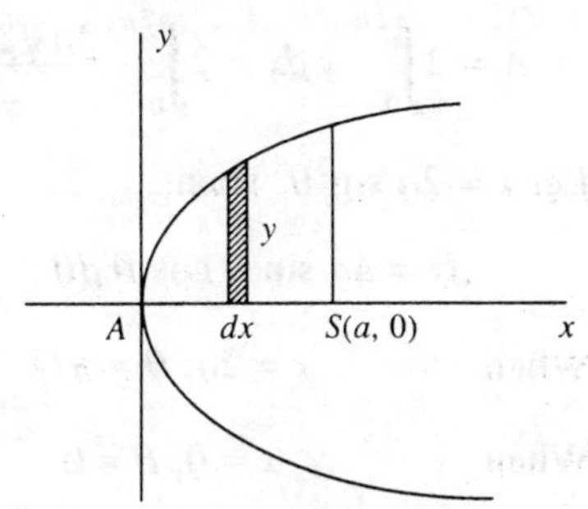

Fig. 7.1

2. Find the area bounded by the ellipse $\dfrac{x^2}{a^2} + \dfrac{y^2}{b^2} = 1.$

Solution. From the equation of the ellipse, we have

$$y = \frac{b}{a}\sqrt{a^2 - x^2}$$

Hence the required area is

$$4\int_0^a y\,dx = \frac{4b}{a}\int_0^a \sqrt{a^2 - x^2}\,dx = \frac{4b}{a}\left[\frac{x\sqrt{a^2 - x^2}}{2} + \frac{a^2}{2}\sin^{-1}\frac{x}{a}\right]_0^a$$

$$= \frac{4b}{a}\cdot\frac{a^2}{2}\cdot\frac{\pi}{2} = \pi ab$$

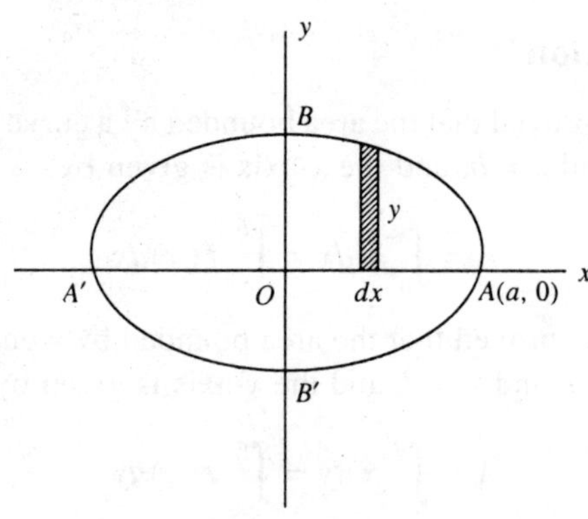

Fig. 7.2

3. Find the area included between the curve $x\,y^2 = 4a^2(2a - x)$ and its asymptote.

Solution. The curve is symmetrical about the x-axis. It cuts the x-axis at $x = 2a$ and has asymptote the y-axis.

Hence the required area is

$$A = 2\int_0^{2a} y\,dx = 2\int_0^{2a} \frac{2a\sqrt{2a - x}}{\sqrt{x}}\,dx$$

Let $x = 2a\sin^2\theta$. Then,

$$dx = 4a\sin\theta\cos\theta\,d\theta$$

When $x = 2a,\ \theta = \pi/2$

When $x = 0,\ \theta = 0$

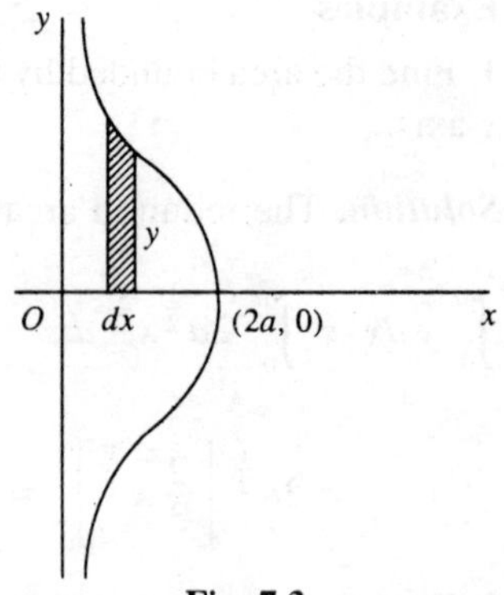

Fig. 7.3

Therefore $A = 2\displaystyle\int_0^{\pi/2} \frac{2a\sqrt{2a - 2a\sin^2\theta}}{\sqrt{2a\sin^2\theta}}\cdot 4a\sin\theta\cos\theta\,d\theta$

$$= 16a^2 \int_0^{\pi/2} \cos^2\theta\, d\theta = 16a^2 \cdot \frac{\Gamma\left(\frac{3}{2}\right)\Gamma\left(\frac{1}{2}\right)}{2\Gamma(2)}$$

$$= 16a^2 \cdot \frac{\frac{1}{2}\sqrt{\pi}\cdot\sqrt{\pi}}{2} = 4\pi a^2$$

4. Calculate the area of a loop of the curve $a^4y^2 = x^4(a^2 - x^2)$.

Solution. The curve is symmetrical about both the axes. It cuts x-axis at (0, 0), $(a, 0)$ and $(-a, 0)$.

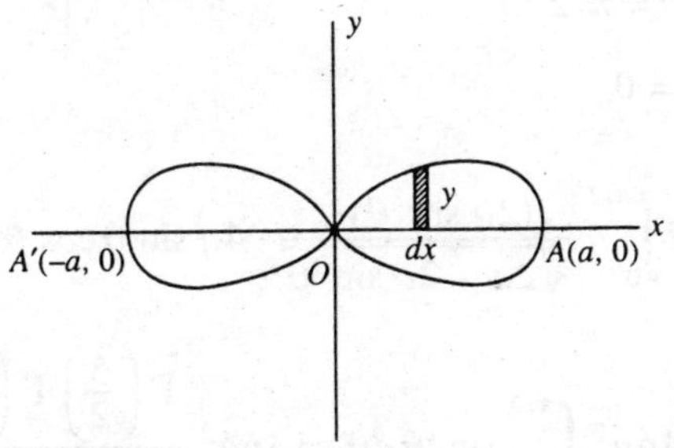

Fig. 7.4

Hence the area of a loop of the curve is

$$A = 2\int_0^a y\, dx = 2\int_0^a \frac{x^2\sqrt{a^2 - x^2}}{a^2}\, dx$$

Let $x = a \sin \theta$. Then, $dx = a \cos \theta\, d\theta$.

When $x = a$, $\theta = \frac{\pi}{2}$

When $x = 0$, $\theta = 0$

Therefore $A = \frac{2}{a^2}\int_0^{\pi/2} a^2\sin^2\theta\sqrt{a^2 - a^2\sin^2\theta}\; a\cos\theta\, d\theta$

$$= 2a^2\int_0^{\pi/2} \sin^2\theta\cos^2\theta\, d\theta$$

$$= 2a^2\,\frac{\Gamma\left(\frac{3}{2}\right)\Gamma\left(\frac{3}{2}\right)}{2\Gamma(3)} = 2a^2 \cdot \frac{\frac{1}{2}\sqrt{\pi}\cdot\frac{1}{2}\sqrt{\pi}}{2\cdot 2\cdot 1} = \frac{\pi a^2}{8}$$

5. Find the area of the infinite region between the curve $y^2(2a - x) = x^3$ and its asymptote.

Solution. The curve is symmetrical about the x-axis. It cuts the x-axis at $x = 2a$ and has asymptote $x = 2a$.

Hence the required area is

$$A = 2\int_0^{2a} y\,dx = 2\int_0^{2a} \frac{x^{\frac{3}{2}}}{\sqrt{2a - x}}\,dx$$

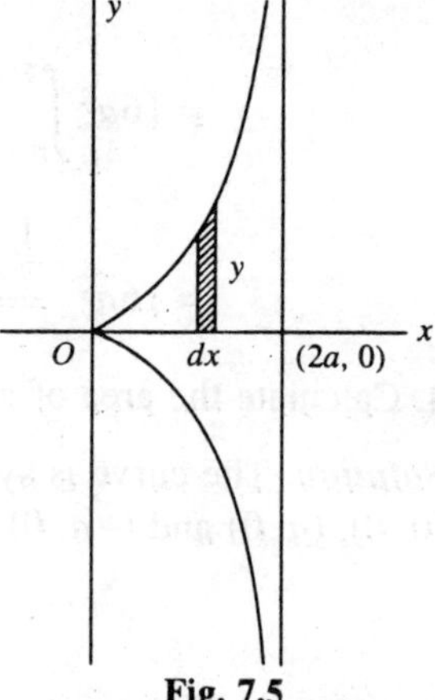

Fig. 7.5

Let $x = 2a\sin^2\theta$. Then,

$$dx = 4a\sin\theta\cos d\theta$$

When $x = 2a$, $\theta = \pi/2$

When $x = 0$, $\theta = 0$

$$\text{Therefore}\quad A = 2\int_0^{\pi/2} \frac{(2a\sin^2\theta)^{\frac{3}{2}}}{\sqrt{2a - 2a\sin^2\theta}} \cdot 4a\sin\theta\cos\theta\,d\theta$$

$$= 16a^2\int_0^{\pi/2} \sin^4\theta\,d\theta = 16a^2\,\frac{\Gamma\left(\frac{5}{2}\right)\Gamma\left(\frac{1}{2}\right)}{2\Gamma(3)}$$

$$= 16a^2\cdot\frac{\frac{3}{2}\cdot\frac{1}{2}\sqrt{\pi}\cdot\sqrt{\pi}}{2.2.1} = 3\pi a^2$$

6. Find the area between the parabolas $y^2 = 4ax$ and $x^2 = 4ay$.

Solution. The points of intersection of the parabolas are

$$(0, 0) \text{ and } (4a, 4a)$$

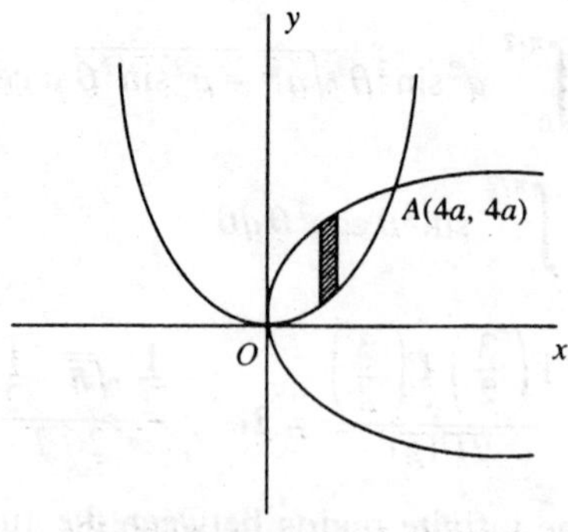

Fig. 7.6

Hence the required area is

$$\int_0^{4a} \left(\sqrt{4ax} - \frac{x^2}{4a}\right) dx = \left[2\sqrt{a} \cdot \frac{2}{3} x^{\frac{3}{2}} - \frac{1}{4a} \cdot \frac{x^3}{3}\right]_0^{4a}$$

$$= \frac{4\sqrt{a}}{3}(4a)^{\frac{3}{2}} - \frac{1}{12a}(4a)^3 = \frac{32a^2}{3} - \frac{16a^2}{3} = \frac{16a^2}{3}$$

EXERCISES

1. Find the area enclosed by the curve $xy^2 = 4(2 - x)$ and y-axis.
2. Trace the curve $ay^2 = x^2(a - x)$ and show that the area of its loop is $8a^2/15$.
3. Find the whole area of the curve $a^2y^2 = x^3\ (2a - x)$.
4. Show that the area of a loop of the curve $y^2 = x^2(4 - x^2)$ is 16/3.
5. Trace the curve $a^2y^2 = a^2x^2 - x^4$ and find its whole area.
6. Find the whole area included between the curve $x^2y^2 = a^2(y^2 - x^2)$ and its asymptote.
7. Show that the whole area of the curve $a^6y^2 = x^6(a^2 - x^2)$ is $8a^2/15$.
8. Prove that the area of the region bounded by the curve $a^4y^2 = x^5(2a - x)$ is to that of the circle whose radius a is 5 to 4.
9. Find the area between the curve $x(x^2 + y^2) = a(x^2 - y^2)$ and its asymptote. Also, find the area of its loop.
10. Find the whole area of the curve $x^2(x^2 + y^2) = a^2(x^2 - y^2)$.
11. Find the whole area of the curve. $a^2x^2 = y^3(2a - y)$.
12. Find the area of the loop of the curve $y^2(a + x) = x^2(3a - x)$.
13. Find the area enclosed by the curve $y^2(a - x) = x^3$.
14. Find the area enclosed by the curve $xy^2 = a^2(a - x)$ and its asymptote.
15. Trace the curve $y^2(a + x) = (a - x)^3$ and find the area enclosed by the curve and its asymptote.
16. Show that the area of the loop of the curve $ay^2 = (x - a)\ (x - 5a)^2$ is $256a^2/15$.
17. Find the area enclosed by the curves $x^2 = 4ay$ and $x^2 + 4a^2 = 8a^3/y$.
18. Find the area common to the two curves $y^2 = ax$, $x^2 + y^2 = 4ax$.
19. Find the whole area included between the curve $x^2y^2 = a^2(x^2 + y^2)$ and its asymptotes.
20. Find the area common to the circle $x^2 + y^2 = 4$ and the ellipse $x^2 + 4y^2 = 9$.

7.3 Areas When Parametric Equations of a Curve are Given

Examples

1. Find the area enclosed between one arch of the cycloid

$$x = a(\theta - \sin\theta),\ y = a(1 - \cos\theta)$$

and its base.

Solution. For tracing of the curve we consider the following:

When $\theta = 0$, $x = 0$ and $y = 0$

When $\theta = \pi$, $x = a\pi$ and $y = 2a$

When $\theta = 2\pi$, $x = 2a\pi$ and $y = 0$

Hence the required area is

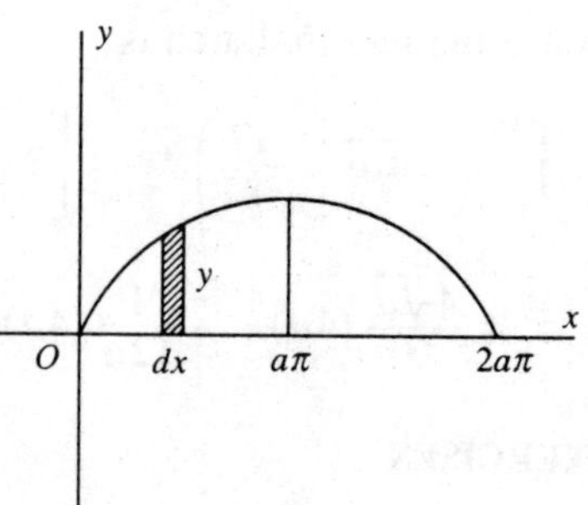

Fig. 7.7

$$A = 2\int_0^{a\pi} y\,dx = 2\int_0^{\pi} a(1 - \cos\theta)\,a(1 - \cos\theta)\,d\theta$$

$$= 2a^2 \int_0^{\pi} \left(2\sin^2\frac{\theta}{2}\right)^2 d\theta = 8a^2 \int_0^{\pi} \sin^4\frac{\theta}{2}\,d\theta$$

$$= 16a^2 \int_0^{\pi/2} \sin^4\phi\,d\phi, \quad \text{where } \frac{\theta}{2} = \phi$$

$$= 16a^2 \frac{\Gamma\left(\frac{5}{2}\right)\Gamma\left(\frac{1}{2}\right)}{2\Gamma(3)} = 16a^2 \cdot \frac{\frac{3}{2}\cdot\frac{1}{2}\sqrt{\pi}\cdot\sqrt{\pi}}{2\cdot 2\cdot 1} = 3\pi a^2$$

2. Find the area enclosed by the curve given by the equations

$$x = a\cos^3\theta,\ y = b\sin^3\theta$$

Solution. For tracing of the curve we consider the following:

When $\theta = 0$, $x = a$ and $y = 0$

When $\theta = \pi/2$, $x = 0$ and $y = b$

When $\theta = \pi$, $x = -a$ and $y = 0$

When $\theta = 3\pi/2$, $x = 0$ and $y = -b$

When $\theta = 2\pi$, $x = a$ and $y = 0$

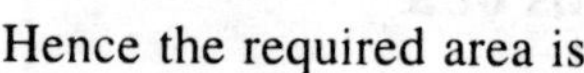
Hence the required area is

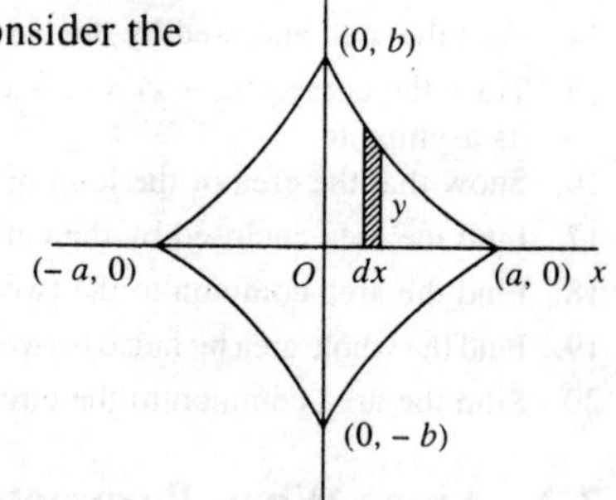

Fig. 7.8

$$A = 4\int_0^{a} y\,dx = 4\int_{\frac{\pi}{2}}^{0} b\sin^3\theta(-3a\cos^2\theta\sin\theta)\,d\theta$$

$$= 12ab\int_0^{\frac{\pi}{2}} \sin^4\theta\cos^2\theta\,d\theta = 12ab\,\frac{\Gamma\left(\frac{5}{2}\right)\Gamma\left(\frac{3}{2}\right)}{2\Gamma(4)}$$

$$= 12ab \cdot \frac{\frac{3}{2} \cdot \frac{1}{2}\sqrt{\pi} \cdot \frac{1}{2}\sqrt{\pi}}{2.3.2.1} = \frac{3\pi ab}{8}$$

EXERCISES

1. Show that the area of the loop of the curve
$$x = a(1 - t^2),\ y = at(1 - t^2),\ -1 \le t \le 1 \text{ is } 8a^2/15$$
2. Sketch the curve $x = t + t^2$, $y = t^2 + t^3$ and show that the area of its loop is 1/60.
3. Prove that the area bounded by the tractrix
$$x = a \cos t + \frac{a}{2} \log \tan^2 \frac{t}{2}, \quad y = a \sin t$$
the y-axis, the x-axis and any ordinate is $\frac{a^2}{4}(2t + \sin 2t - \pi)$.
Deduce that the area between the four infinite branches of the tractrix is πa^2.
4. For the cycloid $x = a(\theta + \sin \theta)$, $y = a(1 - \cos \theta)$, find the area included between the curve and its base.
5. Find the area of the ellipse $x = a \cos t$, $y = b \sin t$.
6. Trace the curve $x = t - t^3$, $y = 1 - t^4$ for all values of t and prove that it forms a loop of area 16/35.
7. Sketch the curve $x = a \sin 2t$, $y = a \sin t$ and find the area of one of its loops.
8. Show that the area bounded by the cissoid
$$x = a \sin^2 t, \quad y = \frac{a \sin^3 t}{\cos t}$$
and its asymptote is $\frac{3\pi a^2}{4}$.

7.4 Areas When Polar Equation of a Curve is Given

If $r = f(\theta)$ be the equation of a curve, then the area of the sector enclosed by the curve and the two radii vectors $\theta = \alpha$ and $\theta = \beta$ is

$$\frac{1}{2}\int_{\alpha}^{\beta} r^2 d\theta$$

Let AB be the curve $r = f(\theta)$, OA and OB be the radii vectors $\theta = \alpha$ and $\theta = \beta$, respectively. Let $P(r, \theta)$ be any point on the curve and let $Q(r + \delta r, \theta + \delta\theta)$ be another point on the curve close to P.

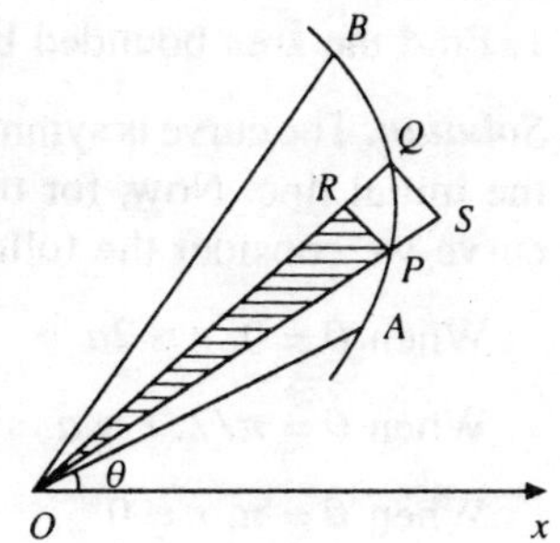

Fig. 7.9

With O as the centre and radii OP, OQ draw the circular arcs PR and QS. Then

$$PR = r\delta\theta \text{ and } QS = (r + \delta r)\delta\theta$$

Therefore, sectorial area OPR is

$$\frac{1}{2} r \cdot r\delta\theta = \frac{1}{2} r^2 \delta\theta$$

and the sectorial area OSQ is

$$\frac{1}{2} (r + \delta r) \cdot (r + \delta r)\delta\theta = \frac{1}{2} (r + \delta r)^2 \delta\theta$$

Let S and $S + \delta S$ denote the areas OAP and OAQ, respectively. Then

$$\delta S = \text{Area } OAQ - \text{Area } OAP = \text{Area } OPQ$$

Since the area OPQ lies between the areas OPR and OSQ, we have

$$\frac{1}{2} r^2 \delta\theta < \delta S < \frac{1}{2} (r + \delta r)^2 \delta\theta$$

$$\Rightarrow \qquad \frac{r^2}{2} < \frac{\delta S}{\delta \theta} < \frac{(r + \delta r)^2}{2}$$

As $\delta\theta \to 0$, $\delta r \to 0$ and hence from above, we get

$$\frac{dS}{d\theta} = \frac{r^2}{2}$$

Now

$$\int_\alpha^\beta \tfrac{1}{2} r^2 d\theta = \int_\alpha^\beta dS = [S]_\alpha^\beta$$

$$= (\text{Value of } S \text{ for } \theta = \beta) - (\text{Value of } S \text{ for } \theta = \alpha)$$

$$= \text{Area } OAB$$

Hence $\qquad \text{Area } OAB = \frac{1}{2} \int_\alpha^\beta r^2 d\theta$

Examples

1. Find the area bounded by the cardioide $r = a(1 + \cos\theta)$.

Solution. The curve is symmetrical about the initial line. Now, for tracing of the curve we consider the following:

When $\theta = 0$, $r = 2a$

When $\theta = \pi/2$, $r = a$

When $\theta = \pi$, $r = 0$

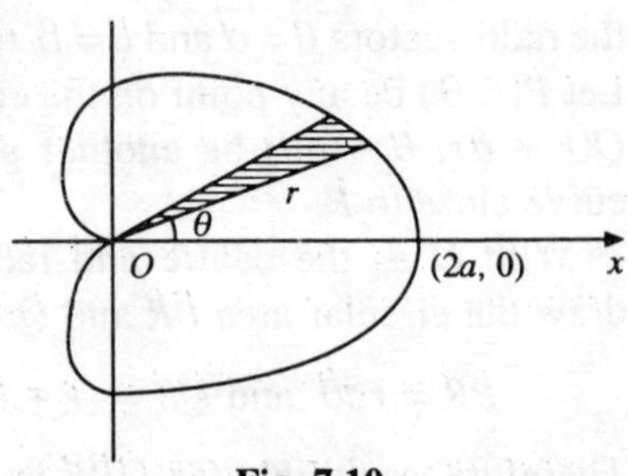

Fig. 7.10

Hence the required area is

$$2\int_0^{\pi} \frac{1}{2} r^2 d\theta = \int_0^{\pi} a^2 (1 + \cos\theta)^2 d\theta = a^2 \int_0^{\pi} 4\cos^4 \frac{\theta}{2} d\theta$$

$$= 8a^2 \int_0^{\pi/2} \cos^4\phi \, d\phi, \quad \text{where } \frac{\theta}{2} = \phi$$

$$= 8a^2 \frac{\Gamma\left(\frac{5}{2}\right)\Gamma\left(\frac{1}{2}\right)}{2\Gamma(3)} = 8a^2 \cdot \frac{\frac{3}{2}\cdot\frac{1}{2}\sqrt{\pi}\cdot\sqrt{\pi}}{2\cdot 2\cdot 1} = \frac{3\pi a^2}{2}$$

2. Find the sum of the areas of all the loops of the curve $r = a \sin 2\theta$.

Solution. For tracing of the curve we consider the following:

When $\theta = 0$, $r = 0$

When $\theta = \pi/4$, $r = a$

When $\theta = \pi/2$, $r = 0$

The curve consists of four equal loops. Hence the required area is

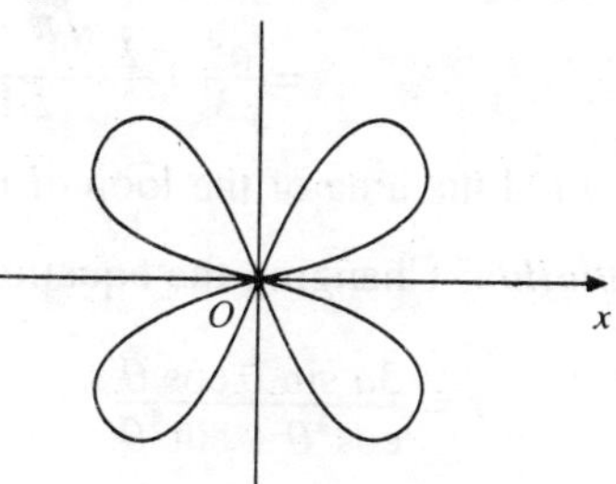

Fig. 7.11

$$4\int_0^{\pi/2} \frac{1}{2} r^2 \, d\theta = 2\int_0^{\pi/2} a^2 \sin^2 2\theta \, d\theta = 2a^2 \int_0^{\pi/2} (2\sin\theta\cos\theta)^2 d\theta$$

$$= 8a^2 \int_0^{\pi/2} \sin^2\theta\cos^2\theta \, d\theta = 8a^2 \frac{\Gamma\left(\frac{3}{2}\right)\Gamma\left(\frac{3}{2}\right)}{2\Gamma(3)}$$

$$= 8a^2 \cdot \frac{\frac{1}{2}\sqrt{\pi}\cdot\frac{1}{2}\sqrt{\pi}}{2.2.1} = \frac{\pi a^2}{2}$$

3. Find the area of a loop of the curve $r = a \sin 3\theta$.

Solution. For tracing of the curve we consider the following:

When $\theta = 0$, $r = 0$

When $\theta = \pi/6$, $r = a$

When $\theta = \pi/3$, $r = 0$

Hence the area of a loop is

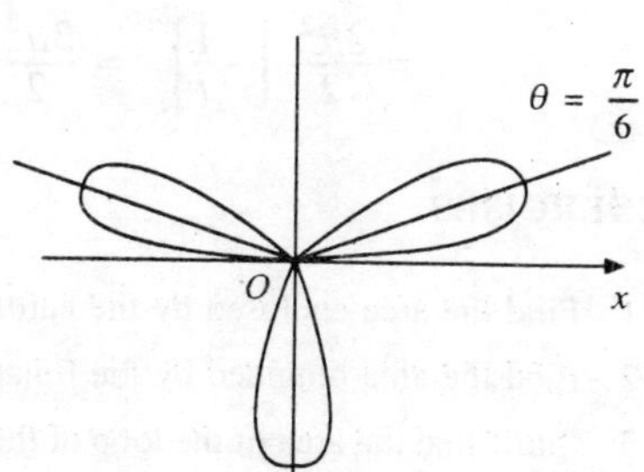

Fig. 7.12

$$\int_0^{\pi/3} \frac{1}{2} r^2 d\theta = \frac{1}{2} \int_0^{\pi/3} a^2 \sin^2 3\theta \, d\theta$$

$$= \frac{a^2}{2} \cdot \frac{1}{3} \int_0^{\pi} \sin^2\phi \, d\phi, \quad \text{where } 3\theta = \phi$$

$$= \frac{a^2}{6} \cdot 2 \int_0^{\pi/2} \sin^2\phi \, d\phi = \frac{a^2}{3} \frac{\Gamma\left(\frac{3}{2}\right)\Gamma\left(\frac{1}{2}\right)}{2\Gamma(2)}$$

$$= \frac{a^2}{3} \cdot \frac{\frac{1}{2}\sqrt{\pi} \cdot \sqrt{\pi}}{2.1} = \frac{\pi a^2}{12}$$

4. Find the area of the loop of the curve $x^3 + y^3 = 3axy$.

Solution. Changing the equation of the curve into polar form, we get

$$r = \frac{3a \sin\theta \cos\theta}{\cos^3\theta + \sin^3\theta}$$

When $\theta = 0$, $r = 0$

When $\theta = \frac{\pi}{4}$, $r = \frac{3a}{\sqrt{2}}$

When $\theta = \pi/2$, $r = 0$

Hence the area of the loop is

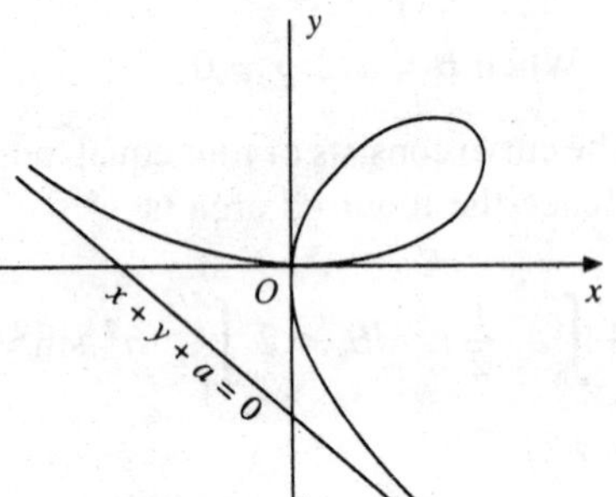

Fig. 7.13

$$\int_0^{\pi/2} \frac{1}{2} r^2 d\theta = \frac{1}{2} \int_0^{\pi/2} \frac{9a^2 \sin^2\theta \cos^2\theta}{(\cos^3\theta + \sin^3\theta)^2} d\theta = \frac{9a^2}{2} \int_0^{\pi/2} \frac{\tan^2\theta \sec^2\theta}{(1 + \tan^3\theta)^2} d\theta$$

$$= \frac{9a^2}{2} \int_1^{\infty} \frac{1}{t^2} \cdot \frac{1}{3} dt, \quad \text{where } 1 + \tan^3\theta = t$$

$$= \frac{3a^2}{2} \left[-\frac{1}{t}\right]_1^{\infty} = \frac{3a^2}{2}$$

EXERCISES

1. Find the area enclosed by the cardioide $r = a(1 - \cos\theta)$.
2. Find the area bounded by the limacon $r = a + b\cos\theta$, $a > b$.
3. Show that the area of the loop of the curve $r = a\theta\cos\theta$ lying in the first quadrant is $\frac{a^2\pi(\pi^2 - 6)}{96}$.

4. Prove that the area of the loop of the folium $x^3 + y^3 = 3axy$ is three times the area of one of the loops of the Lemniscate $r^2 = a^2 \cos 2\theta$.
5. Find the whole area of the curve $r = a \cos 2\theta$.
6. Show that the whole area of the curve $r^2 = a^2 \cos^2\theta + b^2 \sin^2\theta$ is $\frac{\pi}{2}(a^2 + b^2)$.
7. Find the area of a loop of the curve $r = \sqrt{3} \cos 3\theta + \sin 3\theta$.
8. Find the area of the loop of the curve $x^4 + y^4 = 2a^2xy$.
9. Find the area of the loop of the curve $x^5 + y^5 = 5a\, x^2y^2$.
10. Find the area common to the circles $r = a\sqrt{2}$ and $r = 2a \cos \theta$.
11. Show that the area of the region included between the cardioides $r = a(1 + \cos \theta)$ and $r = a(1 - \cos \theta)$ is $\frac{a^2}{2}(3\pi - 8)$.
12. Prove that the area of the loop of the curve $x^6 + y^6 = a^2x^2y^2$ is $\frac{\pi a^2}{12}$.

MISCELLANEOUS EXERCISES

1. Trace the curve $a^2y = x^2(x + a)$ and show that the curve includes with the axis of x an area $\frac{a^2}{12}$.
2. Find the area of the loop of the curve $a^3y^2 = x^4(b + x)$.
3. Find the area of the region lying above x-axis, and included between the circle $x^2 + y^2 = 2ax$ and the parabola $y^2 = ax$.
4. Show that the area enclosed between the parabolas $y^2 = 4a(x + a)$ and $y^2 = -4a(x - a)$ is $\frac{16a^2}{3}$.
5. Show that the area enclosed by the curves $xy^2 = a^2(a - x)$ and $(a - x)y^2 = a^2x$ is $(\pi - 2)a^2$.
6. Show that the area of the loop of the curve $a^2y^2 = x^2(2a - x)(x - a)$ is $\frac{3\pi a^2}{8}$.
7. Find the area between the curve $x^2(x^2 + y^2) = a^2(y^2 - x^2)$ and its asymptote.
8. Trace the curve $y^2(a - x) = x^2(a + x)$ and find the area of the loop.
9. Show that the ordinate $x = a$ divides the area between the curve

$$y^2(2a - x) = x^3$$

and its asymptote into two parts in the ratio $(3\pi - 8) : (3\pi + 8)$.
10. Find the area of the astroid $x^{2/3} + y^{2/3} = a^{2/3}$.
[Hint: Change the equation of the curve into parametric form]
11. Prove that the whole area between the four infinite branches of the tractrix

$$x = a \cos t + \frac{a}{2} \log \tan^2 \frac{t}{2},\ y = a \sin t$$

is equal to the area of the circle of radius a.
12. Find the area included between the curve $x^3 + y^3 = 3axy$ and its asymptote and prove that it is equal to the area of the loop of the curve.

13. Find the whole area of the curve

$$x = \frac{a(1 - t^2)}{1 + t^2}, y = \frac{2at}{1 + t^2}$$

14. Show that the area of the loop of the curve

$$x = \frac{a \sin 3t}{\sin t}, \quad y = \frac{a \sin 3t}{\cos t}$$

is $3\sqrt{3}a^2$.

15. Sketch the curve

$$x = \frac{1 - t^2}{1 + t^2}, y = \frac{t - t^3}{1 + t^2}, -1 \le t \le 1$$

and calculate the area enclosed by the curve.

16. Show that the area enclosed by the curve

$$x = a \sin 2\theta (1 + \cos 2\theta), y = a \cos 2\theta (1 - \cos 2\theta)$$

is $\dfrac{\pi a^2}{2}$.

17. Prove that the area of a loop of the curve $(x^2 + y^2)^2 = 4axy^2$ in the positive quadrant is $\dfrac{\pi a^2}{4}$.
18. Show that the area of the region lying in the second quadrant and bounded by the curve $x^5 + y^5 = 5ax^2y^2$ its asymptote and the y-axis is a^2.
19. Show that the area of the region enclosed between the two loops of the curve $r = a(1 + 2 \cos \theta)$ is $a^2(\pi + 3\sqrt{3})$.
20. Show that the area of the loop of the curve $r^2 \cos \theta = a^2 \sin 3\theta$ lying in the first quadrant is $\dfrac{a^2}{4} \log \dfrac{e^3}{4}$.
21. Show that the whole area of the curve $(x^2 + y^2)^2 = a^2x^2 + b^2y^2$ is $\frac{\pi}{2}(a^2 + b^2)$.
22. Trace the curve $r = a(\sec \theta + \cos \theta)$ and find the area between the curve and its asymptote.
23. Find the area lying between the cardioide $r = a(1 - \cos \theta)$ and its double tangent.
24. Find the larger of the areas into which the cardioide $r = 2a(1 + \cos \theta)$ is divided by the parabola $2a = r(1 + \cos \theta)$. Also, find the ratio of the areas of the two parts.
25. Show that the area of the loop of the curve

$$x^5 + y^5 = 5ax^2y^2$$

is five times the area of the loop of the curve $r^2 = a^2 \cos 2\theta$.

8

Rectification

8.1 Introduction

In this chapter, we determine the lengths of arcs of plane curves between two given points. This process is called *rectification*. Also, methods are given to find the intrinsic equations of curves.

8.2 Lengths of Curves

1. Let s denote the length of the arc AB of the curve $y = f(x)$, measured from a fixed point A upto to any point (x, y) on it. Then, from Differential Calculus, we know that

$$\frac{ds}{dx} = \sqrt{1 + \left(\frac{dy}{dx}\right)^2}$$

Therefore $$s = \int_a^b \sqrt{1 + \left(\frac{dy}{dx}\right)^2}\, dx$$

where a and b are abscissae of the points A and B, respectively.

2. If the equation of the curve is given in the form $x = f(y)$, then

$$\frac{ds}{dy} = \sqrt{1 + \left(\frac{dx}{dy}\right)^2}$$

Therefore $$s = \int_c^d \sqrt{1 + \left(\frac{dx}{dy}\right)^2}\, dy$$

where c and d are the ordinates of the points C and D, respectively, on the curve $x = f(y)$.

3. If the equation of the curve is given in the parametric form $x = f(t)$, $y = g(t)$, then

$$\frac{ds}{dt} = \sqrt{\left(\frac{dx}{dt}\right)^2 + \left(\frac{dy}{dt}\right)^2}$$

Therefore $$s = \int_{t_1}^{t_2} \sqrt{\left(\frac{dx}{dt}\right)^2 + \left(\frac{dy}{dt}\right)^2}\, dt$$

where t_1 and t_2 are the values of t for any two points A and B on the curve.

4. If the equation of the curve is given in the polar form $r = f(\theta)$, then

$$\frac{ds}{d\theta} = \sqrt{r^2 + \left(\frac{dr}{d\theta}\right)^2}$$

Therefore $$s = \int_{\theta_1}^{\theta_2} \sqrt{r^2 + \left(\frac{dr}{d\theta}\right)^2}\, d\theta$$

where θ_1 and θ_2 are vectorial angles of any two points A and B on the curve.

5. If the equation of the curve is given in the form $\theta = f(r)$, then

$$\frac{ds}{dr} = \sqrt{1 + r^2\left(\frac{d\theta}{dr}\right)^2}$$

Therefore $$s = \int_{r_1}^{r_2} \sqrt{1 + r^2\left(\frac{d\theta}{dr}\right)^2}\, dr$$

where r_1 and r_2 are the radii vectors of any two points A and B on the curve.

Examples

1. Find the length of the arc of the parabola $y^2 = 4ax$ measured from the vertex to an extremity of the latus rectum.

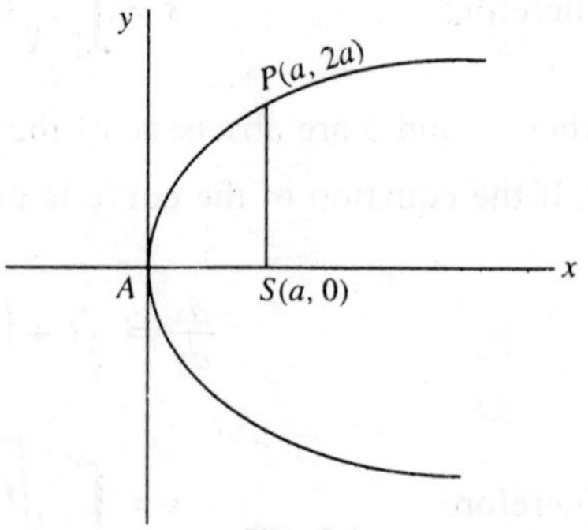

y

P(a, 2a)

A

S(a, 0)

x

Fig. 8.1

Solution. We have

$$\frac{dy}{dx} = \frac{2a}{y} = \frac{2a}{2\sqrt{ax}} = \sqrt{\frac{a}{x}}$$

Therefore $$\frac{ds}{dx} = \sqrt{1 + \left(\frac{dy}{dx}\right)^2} = \sqrt{1 + \frac{a}{x}} = \frac{\sqrt{a + x}}{\sqrt{x}}$$

Hence the required length of the arc is

$$s = \int_0^a \frac{\sqrt{a+x}}{\sqrt{x}}\,dx = 2a\int_0^1 \sqrt{t^2+1}\,dt, \quad \text{where } x = at^2$$

$$= 2a\left[\frac{t}{2}\sqrt{t^2+1} + \frac{1}{2}\log|t+\sqrt{t^2+1}|\right]_0^1$$

$$= 2a\left[\frac{\sqrt{2}}{2} + \frac{1}{2}\log(1+\sqrt{2})\right] = a\,[\sqrt{2} + \log(1+\sqrt{2})]$$

2. Show that the length of an arc of the cycloid

$$x = a(t - \sin t), \quad y = a(1 - \cos t) \text{ is } 8a$$

Solution. We have

$$\frac{dx}{dt} = a(1 - \cos t)$$

$$\frac{dy}{dt} = a\sin t$$

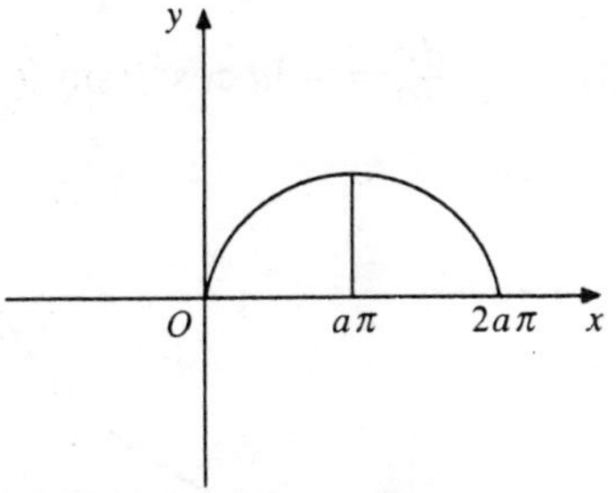

Fig. 8.2

Now

$$\frac{ds}{dt} = \sqrt{\left(\frac{dx}{dt}\right)^2 + \left(\frac{dy}{dt}\right)^2}$$

$$= \sqrt{a^2(1-\cos t)^2 + a^2\sin^2 t} = a\sqrt{1 - 2\cos t + \cos^2 t + \sin^2 t}$$

$$= a\sqrt{2(1-\cos t)} = a\sqrt{2.2\sin^2\frac{t}{2}} = 2a\sin\frac{t}{2}$$

Hence the required length of an arc of the cycloid is

$$\int_0^{2\pi}\frac{ds}{dt}\,dt = \int_0^{2\pi} 2a\sin\frac{t}{2}\,dt = 4a\left[-\cos\frac{t}{2}\right]_0^{2\pi} = 8a$$

3. Find the perimeter of the cardioide $r = a(1 + \cos\theta)$.

Solution. We have

$$\frac{dr}{d\theta} = -a\sin\theta$$

Therefore $\dfrac{ds}{d\theta} = \sqrt{r^2 + \left(\dfrac{dr}{d\theta}\right)^2}$

$$= \sqrt{a^2(1+\cos\theta)^2 + a^2\sin^2\theta}$$

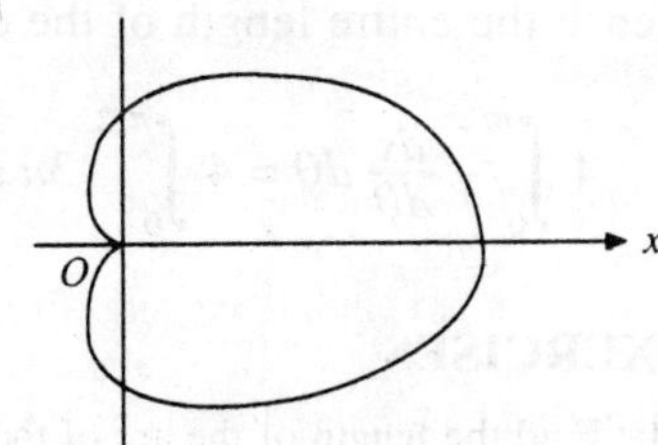

Fig. 8.3

$$= a\sqrt{1 + 2\cos\theta + \cos^2\theta + \sin^2\theta}$$

$$= a\sqrt{2(1 + \cos\theta)} = 2a\cos\frac{\theta}{2}$$

Hence the perimeter of the cardioide is

$$2\int_0^{\pi} \frac{ds}{d\theta}\, d\theta = 2\int_0^{\pi} 2a\cos\frac{\theta}{2}\, d\theta = 8a\left[\sin\frac{\theta}{2}\right]_0^{\pi} = 8a$$

4. Find the entire length of the astroid $x = a\cos^3\theta$, $y = a\sin^3\theta$.

Solution. We have

$$\frac{dx}{d\theta} = -3a\cos^2\theta\sin\theta \quad \text{and} \quad \frac{dy}{d\theta} = 3a\sin^2\theta\cos\theta$$

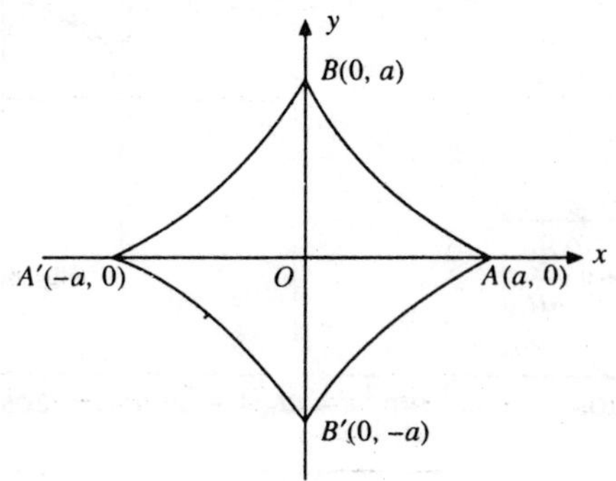

Fig. 8.4

Therefore $\dfrac{ds}{d\theta} = \sqrt{\left(\dfrac{dx}{d\theta}\right)^2 + \left(\dfrac{dy}{d\theta}\right)^2} = \sqrt{9a^2\cos^4\theta\sin^2\theta + 9a^2\sin^4\theta\cos^2\theta}$

$$= 3a\sin\theta\cos\theta\sqrt{\cos^2\theta + \sin^2\theta} = 3a\sin\theta\cos\theta$$

Hence the entire length of the cardioide is

$$4\int_0^{\pi/2} \frac{ds}{d\theta}\, d\theta = 4\int_0^{\pi/2} 3a\sin\theta\cos\theta\, d\theta = 12a\left[\frac{\sin^2\theta}{2}\right]_0^{\pi/2} = 6a$$

EXERCISES

1. Find the length of the arc of the parabola $x^2 = 4ay$ measured from the vertex to one extremity of the latus rectum.

2. Find the length of the arc of the catenary $y = c \cosh \frac{x}{c}$ measured from the vertex $(0, c)$ to any point (x, y).
3. Show that the length of the arc of the curve $y = \log \sec x$ from $x = 0$ to $x = \frac{\pi}{3}$ is $\log (2 + \sqrt{3})$.
4. Find the length of the arc of the curve
$$y = \log \frac{e^x - 1}{e^x + 1}$$
from $x = 1$ to $x = 2$.
5. Find the length of the arc of the semi-cubical parabola $ay^2 = x^3$ from the vertex to the point (a, a).
6. Find the arc length of the curve $y = \frac{1}{2}x^2 - \frac{1}{4}\log x$ from $x = 1$ to $x = 2$.
7. Find the perimeter of the loop of the curve $3ay^2 = x^2(a - x)$.
8. Prove that the length of an arc of the cycloid $x = a(\theta + \sin \theta)$, $y = a(1 - \cos \theta)$ from the vertex to any point on the curve is $\sqrt{8\,ay}$. Also, show that the whole length of an arch is $8a$.
9. Find the perimeter of the loop of the curve
$$x = t^2,\ y = t - \frac{1}{3}t^3$$
10. (a) Find the perimeter of the cardioide $r = a(1 - \cos \theta)$.
    (b) Show that the arc of the upper half of the above cardioide is bisected by the line $\theta = \frac{2\pi}{3}$.
11. Find the length of the arc of the parabola $\frac{2a}{r} = 1 + \cos \theta$ cut off by its latus rectum.
12. Find the length of an arc of the curve $r = ae^{\theta \cot \alpha}$ taking $s = 0$ when $\theta = 0$.
13. Find the length of the curve $x = e^{\theta}\sin \theta$, $y = e^{\theta}\cos \theta$ from $\theta = 0$ to $\theta = \frac{\pi}{2}$.
14. A curve is given by the equations $x = a(\cos \theta + \theta \sin \theta)$, $y = a(\sin \theta - \theta \cos \theta)$, find the length of the arc from $\theta = 0$ to $\theta = \alpha$.
15. Find the length of the loop of the curve $r = a(\theta^2 - 1)$
16. Prove that the whole length of the curve $x^{2/3} + y^{2/3} = a^{2/3}$ is $6a$.
17. Find the length of an arc of the cissoid $r = \frac{a \sin^2 \theta}{\cos \theta}$.

8.3 Intrinsic Equation

Let s be the length of the arc AP of a curve and ψ be the angle between the tangents at A and P (Fig. 8.5). Then, a relation between s and ψ of the form $f(s, \psi) = 0$ is called the *intrinsic equation* of the curve and s and ψ are called the *intrinsic coordinates*.

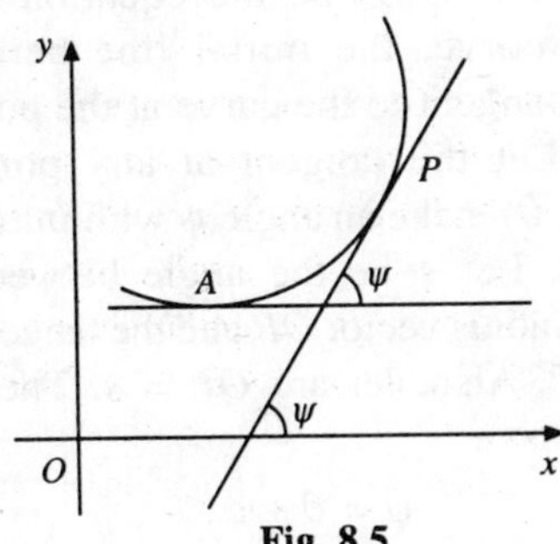

Fig. 8.5

8.3.1 Derivation of Intrinsic Equation from Cartesian Equation

Let the equation of the curve be $y = f(x)$, where the x-axis is a tangent to the curve at the origin O. Let the tangent at any point $P(x, y)$ on the curve make an angle ψ with the x-axis and let arc $OP = s$ (Fig. 8.6). Then, we have

$$\tan \psi = f'(x) \qquad (1)$$

and

$$s = \int_a^x \sqrt{1 + (f'(x))^2}\, dx \qquad (2)$$

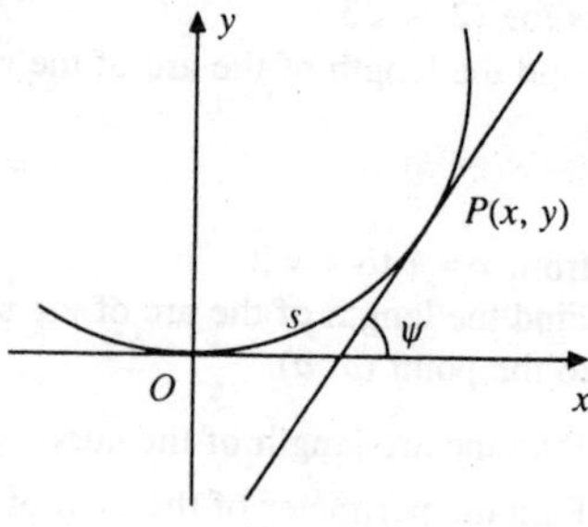

Fig. 8.6

Eliminating x from Eqs. (1) and (2), we get a relation between s and ψ, which is the required intrinsic equation.

8.3.2 Derivation of Intrinsic Equation from Parametric Equations

Let the parametric equations of the curve be

$$x = f(t), \; y = g(t)$$

Then

$$\tan \psi = \frac{dy}{dx} = \frac{\frac{dy}{dt}}{\frac{dx}{dt}} = \frac{g'(t)}{f'(t)} \qquad (1)$$

Also

$$s = \int_{t_1}^{t_2} \sqrt{(f'(t))^2 + (g'(t))^2}\, dt \qquad (2)$$

Eliminating t from Eqs. (1) and (2), we get the required intrinsic equation.

8.3.3 Derivation of Intrinsic Equation from Polar Equation

Let $r = f(\theta)$ be the equation of the curve, the initial line being the tangent to the curve at the pole O. Let the tangent at any point $P(r, \theta)$ make an angle ψ with initial line. Let ϕ be the angle between the radius vector OP and the tangent at P. Also, let arc $OP = s$. Then, we have

$$\psi = \theta + \phi \qquad (1)$$

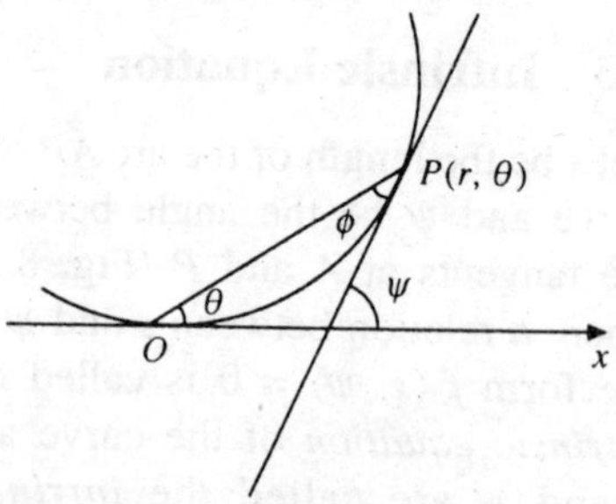

Fig. 8.7

where $$\tan \phi = \frac{r\,d\theta}{dr} = \frac{r}{f'(\theta)} \tag{2}$$

Also $$s = \int_0^\theta \sqrt{r^2 + \left(\frac{dr}{d\theta}\right)^2}\, d\theta \tag{3}$$

Eliminating θ and ϕ from Eqs. (1), (2) and (3), we get the required intrinsic equation.

Examples

1. Find the intrinsic equation of the catenary $y = c \cosh \dfrac{x}{c}$.

Solution. Let arc $AP = s$. Now

$$\tan \psi = \frac{dy}{dx} = \sinh \frac{x}{c} \qquad (1)$$

Therefore

$$s = \int_0^x \sqrt{1 + \left(\frac{dy}{dx}\right)^2}\, dx$$

$$= \int_0^x \sqrt{1 + \left(\sinh \frac{x}{c}\right)^2}\, dx$$

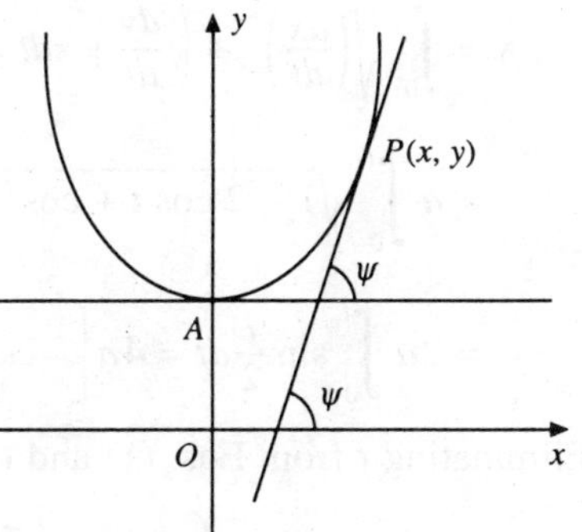

Fig. 8.8

$$= \int_0^x \cosh \frac{x}{c}\, dx = c \sinh \frac{x}{c} \qquad (2)$$

Eliminating x from Eqs. (1) and (2), we get

$$s = c \tan \psi$$

This is the required intrinsic equation.

2. Find the intrinsic equation of the cycloid

$$x = a(t - \sin t),\ y = a(1 - \cos t)$$

Solution. We have

$$\frac{dx}{dt} = a(1 - \cos t) \text{ and } \frac{dy}{dt} = a \sin t$$

Therefore

$$\tan \psi = \frac{dy}{dx} = \frac{a \sin t}{a(1 - \cos t)} = \frac{2 \sin \frac{t}{2} \cos \frac{t}{2}}{2 \sin^2 \frac{t}{2}} = \cot \frac{t}{2} = \tan\left(\frac{\pi}{2} - \frac{t}{2}\right)$$

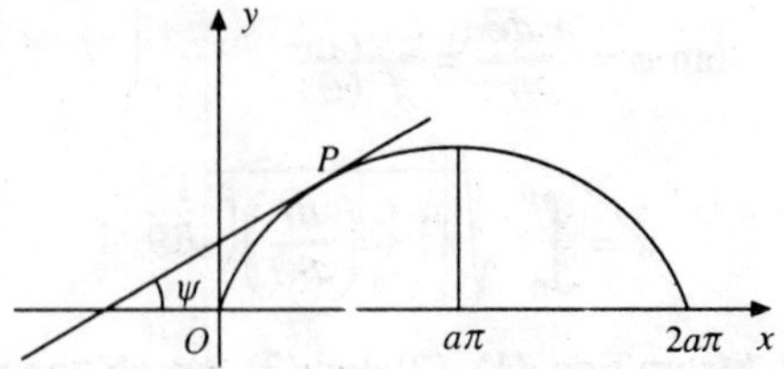

Fig. 8.9

$$\Rightarrow \qquad \psi = \frac{\pi}{2} - \frac{t}{2} \qquad (1)$$

Also

$$s = \int_0^t \sqrt{\left(\frac{dx}{dt}\right)^2 + \left(\frac{dy}{dt}\right)^2}\, dt = \int_0^t \sqrt{a^2(1-\cos t)^2 + a^2 \sin^2 t}\, dt$$

$$= a\int_0^t \sqrt{1 - 2\cos t + \cos^2 t + \sin^2 t}\, dt = a\int_0^t \sqrt{2(1-\cos t)}\, dt$$

$$= 2a\int_0^t \sin\frac{t}{2}\, dt = 4a\left[-\cos\frac{t}{2}\right]_0^t = 4a\left(1 - \cos\frac{t}{2}\right) \qquad (2)$$

Eliminating t from Eqs. (1) and (2), we get

$$s = 4a\left(1 - \cos\left(\frac{\pi}{2} - \psi\right)\right) = 4a(1 - \sin\psi)$$

This is the required intrinsic equation.

3. Show that the intrinsic equation of the semi-cubical parabola $ay^2 = x^3$ taking its cusp as the fixed point is $27s = 8a(\sec^3\psi - 1)$.

Solution. We have

$$\tan\psi = \frac{dy}{dx} = \frac{3x^2}{2ay}$$

$$\Rightarrow \qquad \tan^2\psi = \frac{9x^4}{4a^2y^2} = \frac{9x^4}{4a\cdot x^3} = \frac{9x}{4a} \qquad (1)$$

Since the cusp is at the origin, we have

$$s = \int_0^x \sqrt{1 + \left(\frac{dy}{dx}\right)^2}\, dx = \int_0^x \sqrt{1 + \frac{9x}{4a}}\, dx$$

$$= \frac{4a}{9}\left[\frac{2}{3}\left(1 + \frac{9x}{4a}\right)^{\frac{3}{2}}\right]_0^x = \frac{8a}{27}\left[\left(1 + \frac{9x}{4a}\right)^{\frac{3}{2}} - 1\right] \qquad (2)$$

Eliminating x from Eqs. (1) and (2), we get

$$s = \frac{8a}{27}[(1 + \tan^2 \psi)^{\frac{3}{2}} - 1]$$

$$\Rightarrow \qquad 27s = 8a(\sec^3 \psi - 1)$$

4. Find the intrinsic equation of the cardioide $r = a(1 + \cos\theta)$.

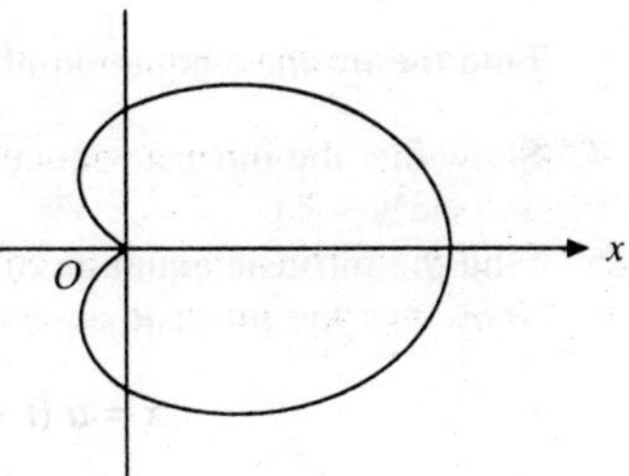

Fig. 8.10

Solution. We have

$$\frac{dr}{d\theta} = -a \sin\theta$$

Therefore $\tan\phi = r\dfrac{d\theta}{dr} = \dfrac{a(1+\cos\theta)}{-a\sin\theta}$

$$= \frac{2\cos^2\frac{\theta}{2}}{-2\sin\frac{\theta}{2}\cos\frac{\theta}{2}} = -\cot\frac{\theta}{2} = \tan\left(\frac{\pi}{2} + \frac{\theta}{2}\right)$$

$$\Rightarrow \qquad \phi = \frac{\pi}{2} + \frac{\theta}{2} \qquad (1)$$

Also $$\psi = \theta + \phi = \frac{\pi}{2} + \frac{3\theta}{2} \qquad (2)$$

Measuring s from the pole O (where $\theta = 0$), we have

$$s = \int_0^\theta \sqrt{r^2 + \left(\frac{dr}{d\theta}\right)^2}\, d\theta = \int_0^\theta \sqrt{a^2(1+\cos\theta)^2 + a^2\sin^2\theta}\, d\theta$$

$$= a\int_0^\theta \sqrt{1 + 2\cos\theta + \cos^2\theta + \sin^2\theta}\, d\theta = a\int_0^\theta \sqrt{2(1+\cos\theta)}\, d\theta$$

$$= 2a\int_0^\theta \cos\frac{\theta}{2}\, d\theta = 4a\left[\sin\frac{\theta}{2}\right]_0^\theta = 4a\sin\frac{\theta}{2} \qquad (3)$$

Eliminating θ from Eqs. (2) and (3), we get

$$s = 4a\sin\frac{\psi}{3}$$

EXERCISES

1. Find the intrinsic equation of the parabola $y^2 = 4ax$; origin being taken as the fixed point.

2. Show that the intrinsic equation of the astroid

$$x^{2/3} + y^{2/3} = a^{2/3}$$

taking $(a, 0)$ as the fixed point is $s = \frac{3}{2}\, a \sin^2\psi$.

3. Find the intrinsic equation of the curve $y = a \log \sec \dfrac{x}{a}$.
4. Show that the intrinsic equation of the semi-cubical parabola $3ay^2 = 2x^3$ is $9s = 4a\,(\sec^3\psi - 1)$.
5. Find the intrinsic equation of a circle of radius a.
6. Show that the intrinsic equation of the cycloid

$$x = a\,(t + \sin t),\ y = a\,(1 - \cos t)$$

is $s = 4a \sin\psi$.

Hence or otherwise find the length of an arch of the given cycloid.

7. Find the intrinsic equation of the curve whose parametric equations are

$$x = e^{\theta}\sin\theta,\ y = e^{\theta}\cos\theta$$

s being measured from the point for which $\theta = \pi/4$.

8. Find the intrinsic equation of the cardioide $r = a\,(1 - \cos\theta)$.
9. Find the intrinsic equation of the curve $x = a\,(1 + \sin t)$, $y = a\,(1 + \cos t)$.
10. Show that the intrinsic equation of the tractrix

$$x = a \cos t + \frac{a}{2} \log \tan^2 \frac{t}{2},\ y = a \sin t$$

taking $t = \pi/2$ as the fixed point is $s = a \log \cos\psi$.

MISCELLANEOUS EXERCISES

1. Find the length of the arc of the equiangular spiral $r = ae^{\theta \cot\alpha}$ between the points for which the radii vectors are r_1 and r_2.
2. Find the entire length of the cardioide $r = a\,(1 + \cos\theta)$, and show that the arc of the upper half is bisected by $\theta = \dfrac{\pi}{3}$.
3. Find the length of the curve $r^{1/3} = 8 \cos \dfrac{\theta}{3}$.
4. Show that the intrinsic equation of the parabola $y^2 = 4ax$ is

$$s = a \cot\psi \operatorname{cosec}\psi + a \log(\cot\psi + \operatorname{cosec}\psi)$$

ψ being the angle between the x-axis and the tangent at the point whose distance from the vertex is s.

5. Find the intrinsic equation of the cardioide

$$r = a\,(1 + \cos\theta)$$

and hence, or otherwise, prove that $s^2 + 9\rho^2 = 16a^2$,

where ρ is the radius of curvature at any point, and s is the length of the arc intercepted between the vertex and the point.

6. If s be the length of the arc of the catenary $y = c \cosh \frac{x}{c}$ from the vertex $(0, c)$ to the point (x, y), show that

$$s^2 = y^2 - c^2$$

7. Show that the length of the arc of the parabola $y^2 = 4ax$ cut off by the line $3y = 8x$ is $a \log\left(2 + \frac{15}{16}\right)$.
8. Find the length of one quadrant of the curve

$$\left(\frac{x}{a}\right)^{2/3} + \left(\frac{y}{b}\right)^{2/3} = 1$$

9. Trace the curve $8a^2y^2 = x^2(a^2 - 2x^2)$ and show that the whole length of the curve is $a\pi$.
10. Find the perimeter of the loop of the curve $9ay^2 = (x - 2a)(x - 5a)^2$.
11. Find the length of the arc of the parabola

$$\frac{l}{r} = 1 + \cos\theta$$

cut off by its latus rectum.
12. Find the length of the arc of the spiral $r = a\theta$ between the points whose radii vectors are r_1 and r_2.
13. In the four-cusped hypocycloid $x^{2/3} + y^{2/3} = a^{2/3}$ show that $s = \frac{3a}{4}\cos 2\psi$, s being measured from the vertex.
14. In the catenary $y = a \cosh \frac{x}{a}$, prove that the area between the curve, the axis of x and the ordinates of two points on the curve varies as the length of the intervening arc.
15. Prove that the cardioide $r = a(1 + \cos\theta)$ is divided by the line $4r\cos\theta = 3a$ into two parts such that the lengths of the arcs on either side of this line are equal.
16. Find the length of the arc of the curve

$$x = t^2 \cos t,\ y = t^2 \sin t$$

from the origin to the point t.
17. Trace the curve

$$x = a(\theta - \sin\theta),\ y = a(1 - \cos\theta)$$

as θ varies from 0 to 2π and show that the point $\theta = \frac{2\pi}{3}$ divides it in the ratio 1:3.
18. Show that the length of an arc of the curve

$$x \sin\theta + y\cos\theta = f'(\theta)$$
$$x\cos\theta - y\sin\theta = f''(\theta)$$

is given by $$s = f(\theta) + f''(\theta) + C$$
19. Find the length of the arc of the cissoid $y^2(a - x) = x^3$ measured from the vertex to the point (x, y).

9

Volumes and Surfaces of Solids of Revolution

9.1 Introduction

If a plane area rotates about an axis lying in its own plane, it generates a solid of revolution and its boundary generates a surface of revolution. Thus the volume of a solid of revolution is the space enclosed by a surface of revolution. For example, a right circular cylinder is generated by the revolution of a rectangle about one of its sides, a right circular cone by the revolution of right-angled triangle about one of the sides including the right angle and a sphere by the revolution of a semi-circle about its bounding diameter.

The section of a surface of revolution by a plane perpendicular to the axis of revolution is a circle having its centre on the axis. We shall now obtain expressions for the volumes and surfaces of solids of revolution.

9.2 Volume of a Solid of Revolution

Let the equation of the curve CD be $y = f(x)$ and let the abscissae of C and D be a and b, respectively. Let $P(x, y)$ be any point on the curve and let

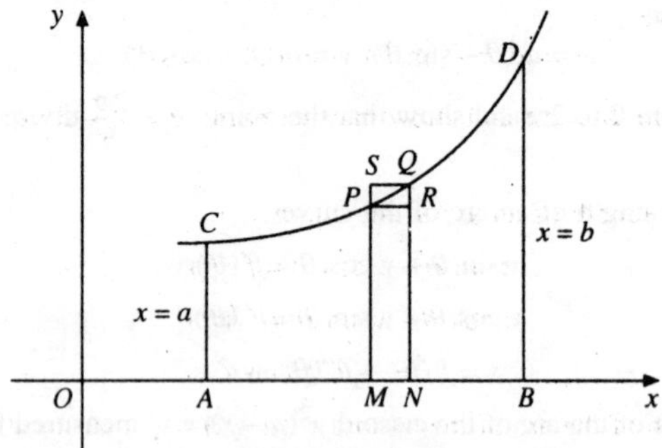

Fig. 9.1

PM be its ordinate. Take another point $Q(x + \delta x, y + \delta y)$ on the curve close to P and let QN be its ordinate. Let V be the volume of the solid of revolution obtained by revolving the arc CP about x-axis. Therefore, V is a function of x. Also, let $V + \delta V$ be the volume of the solid obtained by revolving the arc CQ about x-axis. Then, clearly δV is the volume of the solid of revolution obtained by revolving the arc PQ about x-axis and it lies between the volumes obtained by revolving the two rectangles $MNRP$ and $MNQS$ about x-axis. Therefore

$$\pi y^2 \delta x < \delta V < \pi(y + \delta y)^2 \, \delta x$$

$$\Rightarrow \qquad \pi y^2 < \frac{\delta V}{\delta x} < \pi(y + \delta y)^2$$

$$\Rightarrow \qquad \frac{dV}{dx} = \pi y^2 \quad (\text{when } \delta x \to 0, \delta y \to 0)$$

Hence $$\int_a^b \pi y^2 dx = \int_a^b \frac{dV}{dx} dx = [V]_a^b$$

= (Value of V when $x = b$) – (Value of V when $x = a$)
= (Value of V when $x = b$) – 0
= Volume of the solid obtained by revolving the arc CD about x-axis

Remark. The above theorem can be easily modified to give the volumes of the solids of revolution when the equation of the curve is known in other form or when the axis of revolution is other than the x-axis. Thus,

1. Volume of the solid of revolution obtained by revolving the area bounded by the curve $y = f(x)$, the y-axis and the abscissae $y = c$, $y = d$ about the y-axis is

$$\int_c^d \pi x^2 dy$$

2. If the equation of the curve be

$$x = f(t), \; y = g(t)$$

then the volume of the solid obtained by revolving the area bounded by the curve, the x-axis and the ordinates $x = a$, $x = b$ about the x-axis is

$$\int_a^b \pi y^2 dx = \int_{t_1}^{t_2} \pi(g(t))^2 \frac{dx}{dt} dt$$

where $t = t_1$ when $x = a$ and $t = t_2$ when $x = b$.

3. If the equation of the curve be $r = f(\theta)$, then the volume of the solid obtained by revolving the area bounded by the curve, the initial line and the ordinates $x = a$, $y = b$ about the initial line is

$$\int_a^b \pi y^2 dx = \int_\alpha^\beta \pi y^2 \frac{dx}{d\theta} d\theta$$

where $\theta = \alpha$ when $x = a$ and $\theta = \beta$ when $x = b$. But $x = r\cos\theta$, $y = r\sin\theta$. Therefore

$$\text{Volume} = \int_\alpha^\beta \pi(r\sin\theta)^2 \frac{d}{d\theta}(r\cos\theta)d\theta$$

4. The volume of the solid obtained by revolving the area bounded by the curve CD, the lines AC, AB and BD about the line AB is

$$\int_{OA}^{OB} \pi(MP)^2 \, d(OM)$$

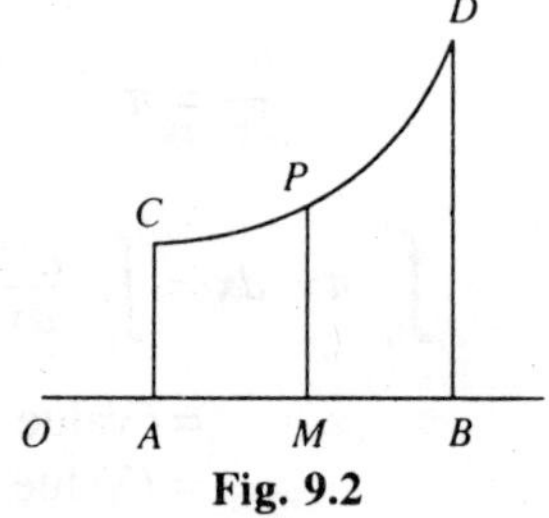

Fig. 9.2

Examples

1. Find the volume of a sphere of radius a.

Solution. The equation of the circle with centre at the origin and radius a is

$$x^2 + y^2 = a^2$$

The sphere is obtained by revolving the semi-circle about the x-axis. Hence, the volume of the sphere of radius a is

$$\int_{-a}^a \pi y^2 dx = \pi \int_{-a}^a (a^2 - x^2)dx$$

$$= 2\pi \int_0^a (a^2 - x^2)dx$$

$$= 2\pi \left[a^2 x - \frac{x^3}{3} \right]_0^a$$

$$= \frac{4\pi a^3}{3}$$

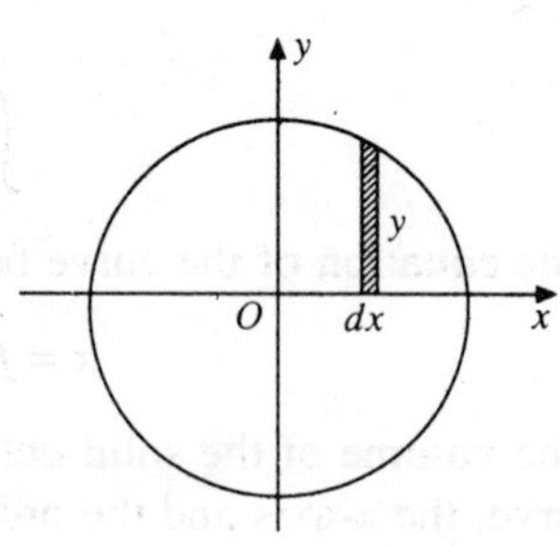

Fig. 9.3

2. The area bounded by the parabola $y^2 = 4ax$, the y-axis and the abscissae through the ends of the latus rectum revolves about the y-axis. Find the volume of the solid generated.

Solution. The required volume of the solid is

$$\int_{-2a}^{2a} \pi x^2 dy = \int_{-2a}^{2a} \pi \frac{y^4}{16a^2} dy$$

$$= \frac{\pi}{16a^2} \left[\frac{y^5}{5} \right]_{-2a}^{2a}$$

$$= \frac{4\pi a^3}{5}$$

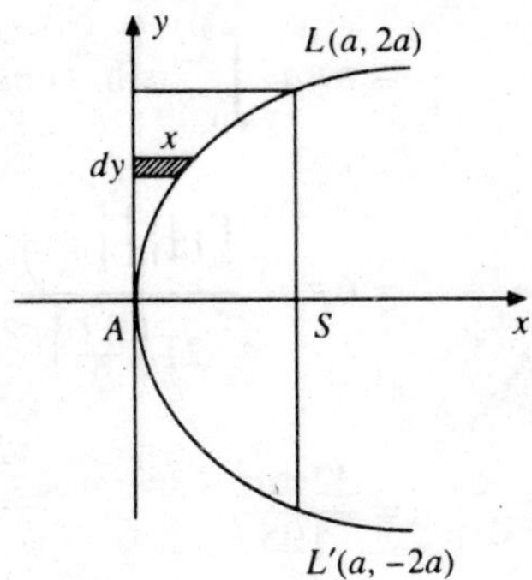

Fig. 9.4

3. The loop of the curve $2ay^2 = x(x - a)^2$ revolves about x-axis; find the volume of the solid so generated.

Solution. The required volume of the solid is

$$\int_0^a \pi y^2 dx = \pi \int_0^a \frac{x(x-a)^2}{2a} dx$$

$$= \frac{\pi}{2a} \int_0^a (x^3 - 2ax^2 + a^2x)\, dx$$

$$= \frac{\pi}{2a} \left[\frac{x^4}{4} - \frac{2ax^3}{3} + \frac{a^2x^2}{2} \right]_0^a$$

$$= \frac{\pi}{2a} \left[\frac{a^4}{4} - \frac{2a^4}{3} + \frac{a^4}{2} \right] = \frac{\pi a^3}{24}$$

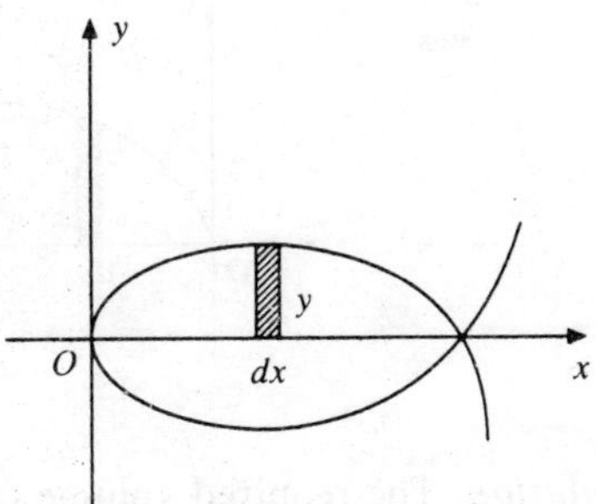

Fig. 9.5

4. Find the volume of the spindle-shaped solid generated by revolving the hypocycloid

$$x^{2/3} + y^{2/3} = a^{2/3}$$

about x-axis.

Solution. The equation of the curve in parametric form is given by

$$x = a\cos^3 t,\ y = a\sin^3 t$$

Hence, the required volume of the solid is

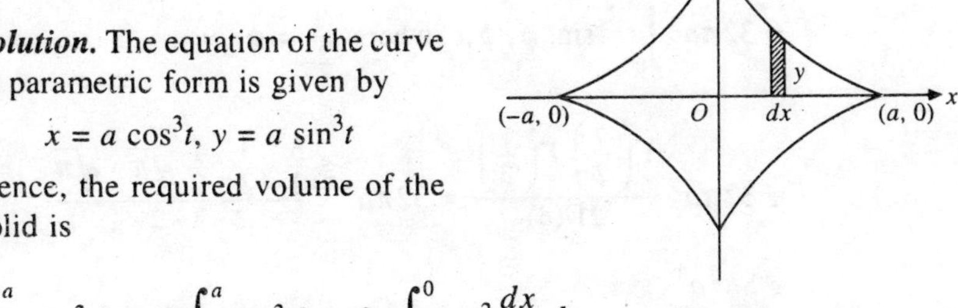

$$\int_{-a}^{a} \pi y^2 dx = 2.\int_0^a \pi y^2 dx = 2\pi \int_{\pi/2}^{0} y^2 \frac{dx}{dt} dt$$

Fig. 9.6

$$= 2\pi \int_{\pi/2}^{0} (a\sin^3 t)^2 (-3a\cos^2 t \sin t) dt$$

$$= 6\pi a^3 \int_0^{\pi/2} \sin^7 t \cos^2 t \, dt$$

$$= 6\pi a^3 \frac{\Gamma(4)\,\Gamma\left(\frac{3}{2}\right)}{2\Gamma\left(\frac{11}{2}\right)} = 6\pi a^3 \cdot \frac{3 \cdot 2 \cdot 1 \cdot \frac{1}{2}\sqrt{\pi}}{2 \cdot \frac{9}{2} \cdot \frac{7}{2} \cdot \frac{5}{2} \cdot \frac{3}{2} \cdot \frac{1}{2}\sqrt{\pi}}$$

$$= \frac{32\pi a^3}{105}$$

5. Find the volume of the solid generated by the revolution of the cycloid

$$x = a(\theta - \sin\theta),\ y = a(1 - \cos\theta)$$

about its base.

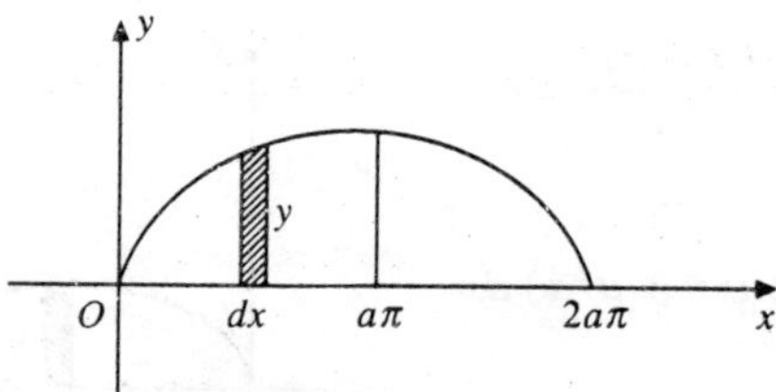

Fig. 9.7

Solution. The required volume of the solid is

$$2\int_0^{a\pi} \pi y^2 dx = 2\int_0^{\pi} \pi y^2 \frac{dx}{d\theta} d\theta = 2\int_0^{\pi} \pi a^2 (1 - \cos\theta)^2 a(1 - \cos\theta) d\theta$$

$$= 2\pi a^3 \int_0^{\pi} \left(2\sin^2\frac{\theta}{2}\right)^3 d\theta = 16\pi a^3 \int_0^{\pi} \sin^6\frac{\theta}{2}\, d\theta$$

$$= 32\pi a^3 \int_0^{\pi/2} \sin^6\phi \, d\phi, \quad \text{where } \frac{\theta}{2} = \phi$$

$$= 32\pi a^3 \frac{\Gamma\left(\frac{7}{2}\right)\Gamma\left(\frac{1}{2}\right)}{2\Gamma(4)} = 32\pi a^3 \cdot \frac{\frac{5}{2} \cdot \frac{3}{2} \cdot \frac{1}{2}\sqrt{\pi} \cdot \sqrt{\pi}}{2 \cdot 3 \cdot 2 \cdot 1}$$

$$= 5\pi^2 a^3$$

6. Prove that the volume of the reel formed by the revolution of the cycloid

$$x = a(\theta + \sin\theta),\ y = a(1 - \cos\theta)$$

about the tangent at the vertex is $\pi^2 a^3$.

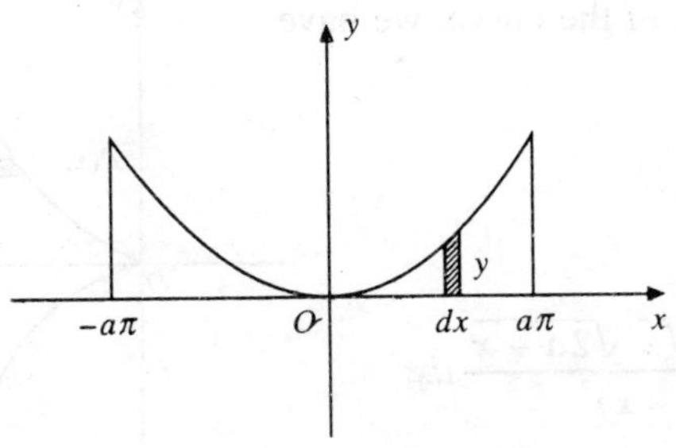

Fig. 9.8

Solution. The tangent at the vertex of the cycloid is x-axis. Hence the required volume is

$$2\int_0^{a\pi} \pi y^2 dx = 2\pi \int_0^{\pi} y^2 \frac{dx}{d\theta}\, d\theta = 2\pi \int_0^{\pi} a^2(1-\cos\theta)^2\, a(1+\cos\theta) d\theta$$

$$= 2\pi a^3 \int_0^{\pi} \left(2\sin^2\frac{\theta}{2}\right)^2 \left(2\cos^2\frac{\theta}{2}\right) d\theta$$

$$= 16\pi a^3 \int_0^{\pi} \sin^4\frac{\theta}{2}\cos^2\frac{\theta}{2}\, d\theta$$

$$= 32\pi a^3 \int_0^{\pi/2} \sin^4\phi \cos^2\phi\, d\phi, \quad \text{where } \frac{\theta}{2} = \phi$$

$$= 32\pi a^3 \frac{\Gamma\left(\frac{5}{2}\right)\Gamma\left(\frac{3}{2}\right)}{2\Gamma(4)} = 32\pi a^3 \cdot \frac{\frac{3}{2}\cdot\frac{1}{2}\sqrt{\pi}\cdot\frac{1}{2}\sqrt{\pi}}{2\cdot 3\cdot 2\cdot 1}$$

$$= \pi^2 a^3$$

7. Find the volume of the solid obtained by the revolution of the cissoid

$$y^2(2a - x) = x^3$$

about its asymptote.

Solution. The asymptote of the curve is $x = 2a$. Let $P(x, y)$ be a point on the curve. Draw perpendicular PM on the asymptote. Then, the required volume of the solid is

$$2\int_0^{\infty} \pi(PM)^2\, d(AM) = 2\pi \int_0^{\infty} \pi(2a-x)^2\, dy$$

From the equation of the curve, we have

$$y = \frac{x^{3/2}}{\sqrt{2a-x}}$$

Therefore

$$dy = \frac{(3a-x)\sqrt{x}\,\sqrt{2a-x}}{(2a-x)^2}\, dx$$

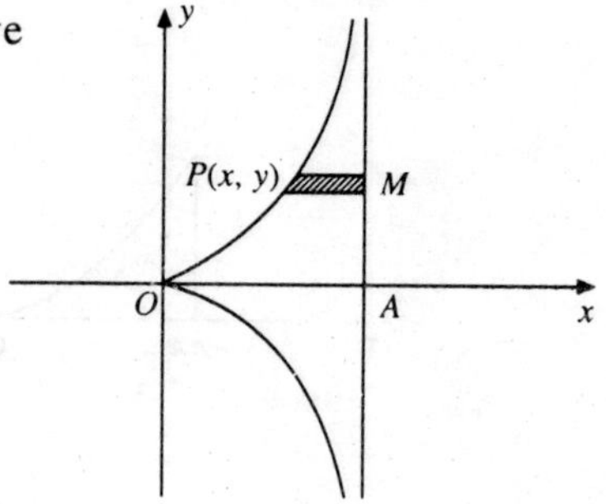

Thus the volume is

Fig. 9.9

$$2\pi\int_0^{2a} (2a-x)^2 \frac{(3a-x)\sqrt{x}\sqrt{2a-x}}{(2a-x)^2}\, dx = 2\pi\int_0^{2a} (3a-x)\sqrt{2a-x}\sqrt{x}\, dx$$

$$= 2\pi\int_0^{\pi/2} (3a-2a\sin^2\theta)\sqrt{2a-2a\sin^2\theta}\sqrt{2a\sin^2\theta}$$

$$\times\, 4a\sin\theta\cos\theta\, d\theta, \text{ where } x = 2a\sin^2\theta$$

$$= 16\pi a^3\int_0^{\pi/2} (3-2\sin^2\theta)\cos\theta\sin\theta\,\sin\theta\cos\theta\, d\theta$$

$$= 16\pi a^3\left[3\int_0^{\pi/2}\sin^2\theta\cos^2\theta\, d\theta - 2\int_0^{\pi/2}\sin^4\theta\cos^2\theta\, d\theta\right]$$

$$= 16\pi a^3\left[3\cdot\frac{\Gamma\left(\frac{3}{2}\right)\Gamma\left(\frac{3}{2}\right)}{2\Gamma(3)} - 2\cdot\frac{\Gamma\left(\frac{5}{2}\right)\Gamma\left(\frac{3}{2}\right)}{2\Gamma(4)}\right]$$

$$= 16\pi a^3\left[3\cdot\frac{\frac{1}{2}\sqrt{\pi}\cdot\frac{1}{2}\sqrt{\pi}}{2\cdot 2} - 2\cdot\frac{\frac{3}{2}\cdot\frac{1}{2}\sqrt{\pi}\cdot\frac{1}{2}\sqrt{\pi}}{2\cdot 3\cdot 2\cdot 1}\right]$$

$$= 16\pi a^3\left[\frac{3\pi}{16} - \frac{\pi}{16}\right] = 2\pi^2 a^3$$

8. The cardioide $r = a(1+\cos\theta)$ revolves about the initial line. Find the volume of the solid thus generated.

Solution. We have

$$x = r\cos\theta,\ y = r\sin\theta$$

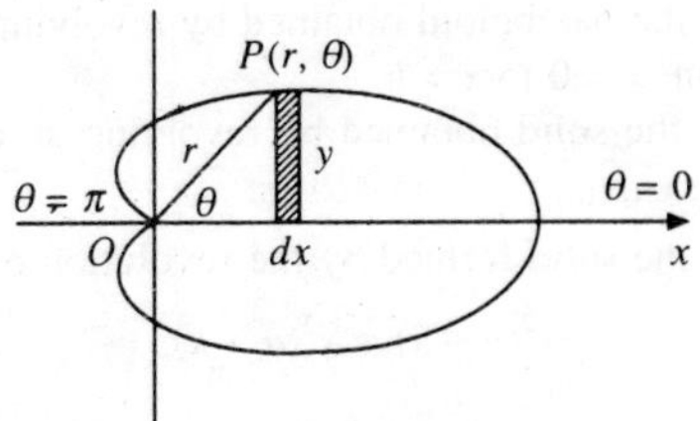

Fig. 9.10

The required volume is $\int_0^{2a} \pi y^2 dx = \int_\pi^0 \pi (r \sin \theta)^2 \, d(r \cos \theta)$

$$= \pi \int_\pi^0 a^2 (1 + \cos \theta)^2 \sin^2\theta \, d(a(1 + \cos \theta) \cos \theta)$$

$$= \pi a^3 \int_\pi^0 (1 + \cos \theta)^2 \sin^2\theta \, (-\sin \theta - 2 \sin \theta \cos \theta) d\theta$$

$$= \pi a^3 \int_0^\pi (1 + \cos \theta)^2 \sin^3\theta \, (1 + 2 \cos \theta) d\theta$$

$$= \pi a^3 \int_0^\pi \left(2 \cos^2 \frac{\theta}{2} \right)^2 \left(2 \sin \frac{\theta}{2} \cos \frac{\theta}{2} \right)^3 \left(4 \cos^2 \frac{\theta}{2} - 1 \right) d\theta$$

$$= 128 \pi a^3 \int_0^\pi \cos^9 \frac{\theta}{2} \sin^3 \frac{\theta}{2} \, d\theta - 32 \pi a^3 \int_0^\pi \cos^7 \frac{\theta}{2} \sin^3 \frac{\theta}{2} \, d\theta$$

$$= 256 \pi a^3 \int_0^{\pi/2} \cos^9 \phi \sin^3 \phi \, d\phi - 64 \pi a^3 \int_0^{\pi/2} \cos^7 \phi \sin^3 \phi \, d\phi, \text{ where } \frac{\theta}{2} = \phi$$

$$= 256 \pi a^3 \cdot \frac{\Gamma(5)\,\Gamma(2)}{2\Gamma(7)} - 64 \pi a^3 \cdot \frac{\Gamma(4)\,\Gamma(2)}{2\Gamma(6)}$$

$$= 256 \pi a^3 \cdot \frac{4 \cdot 3 \cdot 2 \cdot 1 \cdot 1}{2 \cdot 6 \cdot 5 \cdot 4 \cdot 3 \cdot 2 \cdot 1} - 64 \pi a^3 \cdot \frac{3 \cdot 2 \cdot 1 \cdot 1}{2 \cdot 5 \cdot 4 \cdot 3 \cdot 2 \cdot 1}$$

$$= \frac{64 \pi a^3}{15} - \frac{8 \pi a^3}{5} = \frac{8 \pi a^3}{3}$$

EXERCISES

1. Find the volume of the solid obtained by revolving the ellipse $\frac{x^2}{a^2} + \frac{y^2}{b^2} = 1$ about the x-axis.

2. Find the volume of the paraboloid obtained by revolving the parabola $y^2 = 4ax$ about the x-axis from $x = 0$ to $x = h$.
3. Find the volume of the solid obtained by revolving an arc of the catenary $y = c \cosh \frac{x}{c}$ about the x-axis.
4. Find the volume of the solid formed by the revolution of the loop of the curve

$$y^2(a - x) = x^2(a + x)$$

about the x-axis.
5. A basin is formed by the revolution of the curve

$$x^3 = 64y,\ y > 0$$

about the axis of y. If the depth of the basin is 8 cm, how many cubic cm of water would it hold?
6. Find the volume of the solid generated by the revolution of cissoid $y^2(2 - x) = x^3$ about its asymptote.
7. The curve $a^2x^2 = y^3(2a - y)$ is revolved about the y-axis, find the volume of the figure so formed.
8. Find the volume of the solid obtained by revolving the area enclosed by one arch of the cycloid

$$x = a(\theta + \sin \theta),\ y = a(1 + \cos \theta)$$

about the x-axis.
9. Find the volume of the solid formed by the revolution, about the straight line $x = a$, of that part of the parabola $y^2 = 4ax$, which is cut off by this line.
10. Show that the volume of the solid generated by the revolution of the cycloid

$$x = a(\theta + \sin \theta),\ y = a(1 - \cos \theta)$$

about the y-axis is $\pi a^3\left(\frac{3\pi^2}{2} - \frac{8}{3}\right)$.
11. Find the volume of the solid generated by revolving the area between the curve

$$x(a^2 + y^2) = ay^2$$

and its asymptote about the asymptote.
12. The part of the ellipse $\frac{x^2}{a^2} + \frac{y^2}{b^2} = 1$ cut off by a latus rectum revolves about the tangent at the nearer vertex. Find the volume of the reel thus formed.
13. Prove that the volume of the solid generated by the revolution of the curve $y(a^2 + x^2) = a^3$ about its asymptote is $\frac{1}{2}\pi^2 a^3$.
14. The part of the curve $y^2 = x^2(1 - x^2)$ between $x = 0$ and $x = 1$ rotates about the x-axis. Find the volume of the solid thus generated.
15. Show that the volume of the solid generated by the revolution of the curve $(a - x)y^2 = a^2x$ about its asymptote is $\frac{1}{2}\pi^2 a^3$.
16. Find the volume of the solid generated by the revolution of the loop of the curve $y^2 = x^4(x + 2)$ about the axis of x.
17. Find the volume of the solid generated by the revolution of the cardioide $r = a(1 - \cos \theta)$ about the initial line.

18. Find the volume of the solid generated by the revolution of the lemniscate $r^2 = a^2 \cos 2\theta$ about the

    (i) initial line and (ii) line $\theta = \frac{\pi}{2}$.

19. Find the volume of the solid generated by the revolution of the loop of the curve

$$x = t^2, \quad y = t - \frac{t^3}{3}$$

    about the x-axis.

20. Find the volume of the solid generated by the revolution of the cissoid

$$x = 2a \sin^2 t, \quad y = \frac{2a \sin^3 t}{\cos t}$$

    about its asymptote.

21. Find the volume of the solid generated by the revolution of the loop of the curve $y^2 = x^2(a - x)$ about the x-axis.
22. The curve $y^2(a + x) = x^2(3a - x)$ revolves about the x-axis. Find the volume generated by the loop.
23. Find the volume of the solid generated by revolving the tractrix

$$x = a \cos t + \frac{a}{2} \log \tan^2 \frac{t}{2}, \; y = a \sin t$$

    about its asymptote.

24. Find the area included between the curves $y^2 = x^3$ and $x^2 = y^3$, and find the volume of the solid of revolution obtained by rotating this area about the x-axis.

9.3 Surface of a Solid of Revolution

Let $y = f(x)$ be the equation of the curve CD and let the abscissae of C and D be a and b, respectively.

Let $P(x, y)$ be any point on the curve and let PM be its ordinate. Take another point $Q(x+\delta x, y+\delta y)$ on the curve close to P and let QN be its ordinate. Let δS be the length of the arc PQ and let δs be the area of the surface obtained by the revolution of the arc PQ about x-axis. Then

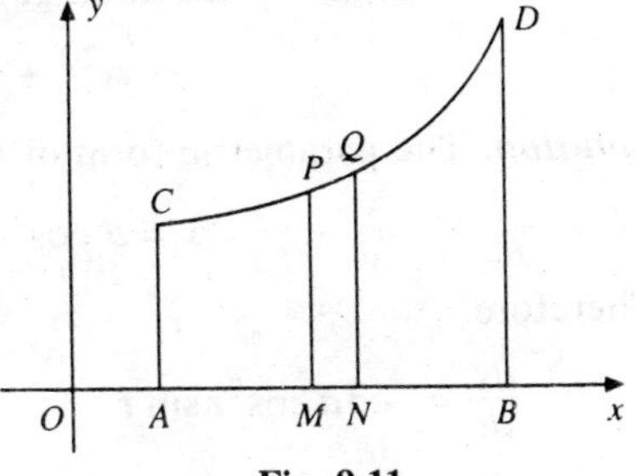

Fig. 9.11

$$2\pi y \, \delta s < \delta S < 2\pi(y + \delta y) \, \delta s$$

$$\Rightarrow \qquad 2\pi y < \frac{\delta S}{\delta s} < 2\pi(y + \delta y) \qquad (1)$$

When $\delta s \to 0$, δy also tends to zero. Then from (1), we get

$$\frac{dS}{ds} = 2\pi y$$

Hence $\int_a^b 2\pi y ds = \int_a^b \frac{dS}{ds} ds = [S]_a^b$

$= $ (Value of S when $x = b$) – (Value of S when $x = a$)
$= $ (Value of S when $x = b$) – 0
$= $ Surface of the solid obtained by revolving the arc CD about x-axis

Examples

1. Find the surface area of a sphere of radius a.

Solution. The equation of the circle with centre at origin and radius a is

$$x^2 + y^2 = a^2$$

The surface of the sphere is obtained by revolving the semi-circle about the x-axis. Hence the surface area of the sphere of radius a is

$$\int_{-a}^{a} 2\pi y\, ds = 2\pi \int_{-a}^{a} y \frac{ds}{dx} dx = 2\pi \int_{-a}^{a} y\sqrt{1 + \left(\frac{dy}{dx}\right)^2}\, dx$$

$$= 2\pi \int_{-a}^{a} y\sqrt{1 + \frac{x^2}{y^2}}\, dx = 2\pi \int_{-a}^{a} y \frac{\sqrt{x^2 + y^2}}{y} dx$$

$$= 2\pi \int_{-a}^{a} a\, dx = 2\pi a\, [x]_{-a}^{a} = 4\pi a^2$$

2. Find the surface of the solid generated by the revolution of the curve

$$x^{2/3} + y^{2/3} = a^{2/3}$$

Solution. The parametric form of the curve is

$$x = a\cos^3 t, \quad y = a\sin^3 t$$

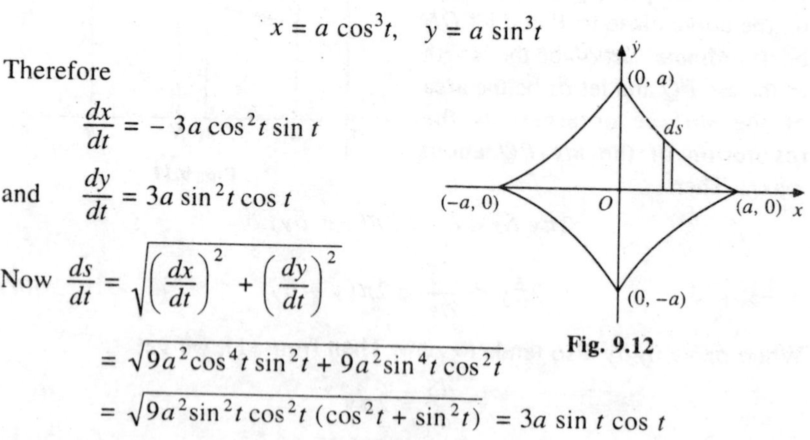

Fig. 9.12

Therefore

$$\frac{dx}{dt} = -3a\cos^2 t \sin t$$

and $$\frac{dy}{dt} = 3a\sin^2 t\cos t$$

Now $$\frac{ds}{dt} = \sqrt{\left(\frac{dx}{dt}\right)^2 + \left(\frac{dy}{dt}\right)^2}$$

$$= \sqrt{9a^2\cos^4 t\sin^2 t + 9a^2\sin^4 t\cos^2 t}$$

$$= \sqrt{9a^2\sin^2 t\cos^2 t\,(\cos^2 t + \sin^2 t)} = 3a\sin t\cos t$$

Hence, the required surface area is

$$2\int 2\pi y\, ds = 4\pi \int_0^{\pi/2} y\frac{ds}{dt}\, dt = 4\pi \int_0^{\pi/2} a \sin^3 t \cdot 3a \sin t \cos t\, dt$$

$$= 12\pi a^2 \int_0^{\pi/2} \sin^4 t \cos t\, dt = 12\pi a^2 \left[\frac{\sin^5 t}{5}\right]_0^{\pi/2} = \frac{12\,\pi a^2}{5}$$

3. Find the surface of the solid generated by the revolution of the cycloid

$$x = a(\theta + \sin\theta),\ y = a(1 - \cos\theta)$$

about the tangent at the vertex.

Solution. We have

$$\frac{dx}{d\theta} = a(1 + \cos\theta)$$

and $$\frac{dy}{d\theta} = a\sin\theta$$

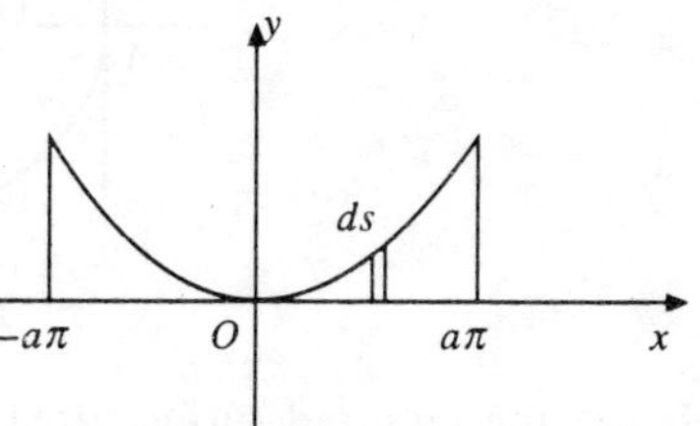

Fig. 9.13

Now

$$\frac{ds}{d\theta} = \sqrt{\left(\frac{dx}{d\theta}\right)^2 + \left(\frac{dy}{d\theta}\right)^2} = \sqrt{a^2(1+\cos\theta)^2 + a^2\sin^2\theta}$$

$$= a\sqrt{1 + 2\cos\theta + \cos^2\theta + \sin^2\theta} = a\sqrt{2(1+\cos\theta)} = 2a\cos\frac{\theta}{2}$$

Hence, the required surface area is

$$\int 2\pi y ds = 2\int_0^{\pi} 2\pi y\frac{ds}{d\theta}\, d\theta = 4\pi\int_0^{\pi} a(1-\cos\theta)\cdot 2a\cos\frac{\theta}{2}\, d\theta$$

$$= 8\pi a^2\int_0^{\pi} 2\sin^2\frac{\theta}{2}\cos\frac{\theta}{2}\, d\theta = 16\pi a^2\left[\frac{2\sin^3\frac{\theta}{2}}{3}\right]_0^{\pi} = \frac{32\pi a^2}{3}$$

4. Find the surface of the solid generated by revolving the arc of the parabola $y^2 = 4ax$ bounded by its latus rectum about x-axis.

Solution. We have

$$\frac{dy}{dx} = \frac{2a}{y}$$

Therefore

$$\frac{ds}{dx} = \sqrt{1 + \left(\frac{dy}{dx}\right)^2} = \sqrt{1 + \frac{4a^2}{y^2}} = \frac{\sqrt{y^2 + 4a^2}}{y}$$

$$= \frac{\sqrt{4ax + 4a^2}}{y} = \frac{2\sqrt{a}\sqrt{x + a}}{y}$$

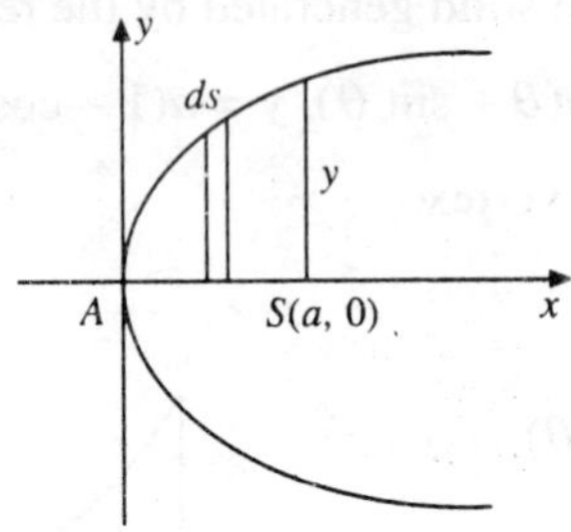

Fig. 9.14

Hence, the required surface area is

$$\int 2\pi y\, ds = \int_0^a 2\pi y \frac{ds}{dx}\, dx = 2\pi \int_0^a y \cdot \frac{2\sqrt{a}\sqrt{x + a}}{y}\, dx$$

$$= 4\pi\sqrt{a} \int_0^a \sqrt{x + a}\, dx = 4\pi\sqrt{a} \left[\frac{2(x + a)^{3/2}}{3}\right]_0^a$$

$$= \frac{8\pi\sqrt{a}}{3}\left[(2a)^{3/2} - a^{3/2}\right] = \frac{8\pi a^2}{3}(2\sqrt{2} - 1)$$

5. Prove that the surface generated by the revolution of the tractrix

$$x = a\cos t + \frac{a}{2}\log\tan^2\frac{t}{2},\ y = a\sin t$$

about its asymptote is eqúal to the surface of a sphere of radius a.

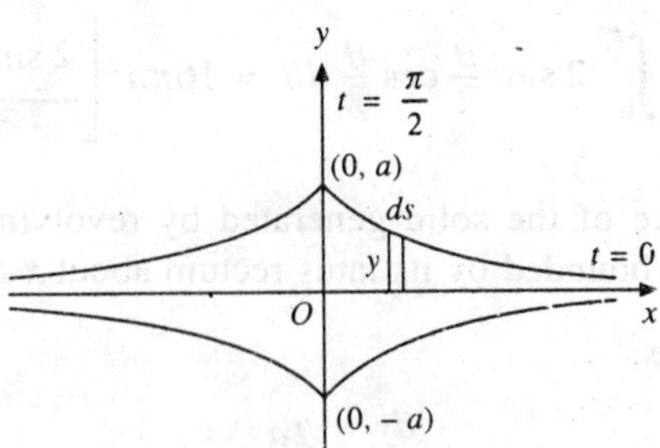

Fig. 9.15

Solution. We have

$$\frac{dx}{dt} = \frac{a\cos^2 t}{\sin t} \text{ and } \frac{dy}{dt} = a\cos t$$

Now
$$\frac{ds}{dt} = \sqrt{\left(\frac{dx}{dt}\right)^2 + \left(\frac{dy}{dt}\right)^2} = \sqrt{\frac{a^2\cos^4 t}{\sin^2 t} + a^2\cos^2 t}$$

$$= \frac{a\cos t}{\sin t}\sqrt{\cos^2 t + \sin^2 t} = \frac{a\cos t}{\sin t}$$

Hence, the required surface area is

$$2\int 2\pi y\, ds = 2\int_0^{\pi/2} 2\pi y \frac{ds}{dt}\, dt$$

$$= 4\pi\int_0^{\pi/2} a\sin t \cdot \frac{a\cos t}{\sin t}\, dt$$

$$= 4\pi a^2\int_0^{\pi/2} \cos t\, dt = 4\pi a^2[\sin t]_0^{\pi/2}$$

$$= 4\pi a^2 = \text{Surface area of a sphere of radius } a$$

6. Find the surface of the solid obtained by revolving the cardioide $r = a(1 - \cos\theta)$ about the initial line.

Solution. We have

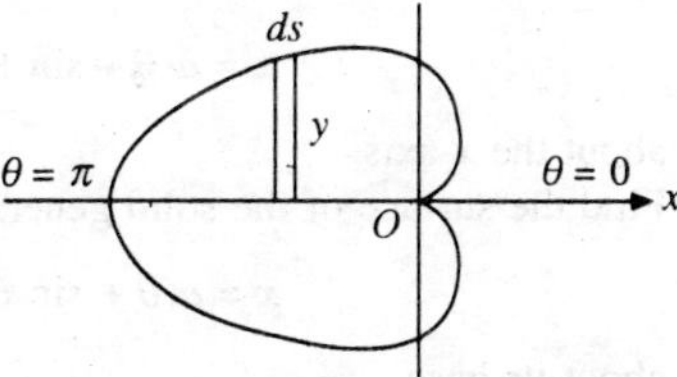

Fig. 9.16

$$\frac{dr}{d\theta} = a\sin\theta$$

Therefore

$$\frac{ds}{d\theta} = \sqrt{r^2 + \left(\frac{dr}{d\theta}\right)^2} = \sqrt{a^2(1 - \cos\theta)^2 + a^2\sin^2\theta}$$

$$= a\sqrt{1 - 2\cos\theta + \cos^2\theta + \sin^2\theta} = a\sqrt{2(1 - \cos\theta)} = 2a\sin\frac{\theta}{2}$$

Hence, the required surface area is

$$\int 2\pi y\, ds = \int_0^{\pi} 2\pi y \frac{ds}{d\theta}\, d\theta$$

$$= 2\pi \int_0^{\pi} r \sin \theta \cdot 2a \sin \frac{\theta}{2}\, d\theta \qquad (\because y = r \sin \theta)$$

$$= 4\pi a \int_0^{\pi} a(1 - \cos \theta) \sin \theta \sin \frac{\theta}{2}\, d\theta$$

$$= 4\pi a^2 \int_0^{\pi} 2 \sin^2 \frac{\theta}{2} \cdot 2 \sin \frac{\theta}{2} \cos \frac{\theta}{2} \sin \frac{\theta}{2}\, d\theta$$

$$= 16\pi a^2 \int_0^{\pi} \sin^4 \frac{\theta}{2} \cos \frac{\theta}{2}\, d\theta = 16\pi a^2 \left[\frac{2 \sin^5 \frac{\theta}{2}}{5} \right]_0^{\pi} = \frac{32\pi a^2}{5}$$

EXERCISES

1. Find the surface of the hemi-sphere of radius a.
2. The part of the parabola $y^2 = 4ax$ cut off by the latus rectum revolves about the tangent at the vertex. Find the curved surface of the reel thus generated.
3. Find the surface generated by the revolution of an arc of catenary $y = c \cosh \frac{x}{c}$ about the x-axis.
4. Find the surface of the solid generated by the revolution of ellipse $x^2 + 4y^2 = 16$ (i) about its minor axis and (ii) about its major axis.
5. Find the surface area of the solid generated by revolving one complete arch of the cycloid

$$x = a(\theta - \sin \theta),\ y = a(1 - \cos \theta)$$

about the x-axis.
6. Find the surface of the solid generated by the revolution of the cycloid

$$x = a(\theta + \sin \theta),\ y = a(1 + \cos \theta)$$

about its base.
7. Find the surface of the solid generated by the revolution of curve $r = 2a \cos \theta$ about the initial line.
8. The cardioide $r = a(1 + \cos \theta)$ revolves about the initial line. Find the surface area of the solid thus generated.
9. The arc of the cardioide $r = a(1 + \cos \theta)$ included between $-\frac{\pi}{2} \le \theta \le \frac{\pi}{2}$ is rotated about the line $\theta = \frac{\pi}{2}$. Find the area of the surface generated.
10. Find the area of the surface generated by the revolution of the lemniscate $r^2 = a^2 \cos 2\theta$ about the initial line.
11. Find the surface of the solid generated by the revolution of the loop of the curve

$$x = t^2,\ y = t - \frac{t^3}{3}$$

about the x-axis.

12. If an ellipse of eccentricity e and semi-major axis a revolves about its minor axis, show that the surface of the spheroid thus generated is

$$2\pi a^2 \left[1 + \frac{1 - e^2}{2e} \log \left| \frac{1 + e}{1 - e} \right| \right]$$

13. Find the area of the surface generated by revolving the arc of the curve

$$9y^2 = (2x - 1)^3$$

about the y-axis from $x = \frac{1}{2}$ to $x = 2$.

14. Find the surface area of the solid generated by the revolution of the loop of the curve

$$3ay^2 = x(x - a)^2$$

about the x-axis.

MISCELLANEOUS EXERCISES

1. Find the volume of the solid of revolution formed by revolving the area included between the curves $y^2 = ax^3$ and $x^2 = ay^3$ about the x-axis.
2. Find the volume of the solid formed by the revolution about the x-axis, of the area enclosed between the curve $27ay^2 = 4(x - 2a)^3$, the x-axis and the parabola $y^2 = 4ax$.
3. The area cut off from the parabola $y^2 = 4ax$ by the chord joining the vertex to an end of the latus rectum is revolved through four right angles about the chord. Find the volume of the solid generated.
4. Find the volume of the paraboloid generated by the revolution about the x-axis, of the parabola $y^2 = 4ax$ from $x = 0$ to $x = h$. If R is the area of cross-section at $x = h$, show that the volume is half that of cylinder of base R and length h.
5. The smaller segment of the ellipse $\frac{x^2}{a^2} + \frac{y^2}{b^2} = 1$ cut off by the chord $\frac{x}{a} + \frac{y}{b} = 1$ revolves completely about this chord. Show that the volume generated is $\frac{\pi a^2 b^2}{\sqrt{a^2 + b^2}} \left(\frac{5}{3} - \frac{\pi}{2} \right)$.
6. Sketch the curves $xy^2 = a^2(a - x)$, $(a - x)y^2 = a^2 x$. Prove that the volume obtained by revolution about $x = a/2$ of the area enclosed between the curves is $\frac{\pi a^3}{4}(4 - \pi)$.
7. Show that if the area lying within the cardioide $r = 2a(1 + \cos\theta)$ and without the parabola $r(1 + \cos\theta) = 2a$ revolves about the initial line, the volume generated is $18\pi a^3$.
8. The figure bounded by a quadrant of a circle of radius a and the tangents at its extremities revolves about one of the tangents. Prove that the volume of the solid generated is $\pi a^3 \left(\frac{5}{3} - \frac{\pi}{2} \right)$.
9. A circular arc revolves about its chord. Prove that the area of the surface generated is

$$4\pi a^2(\sin\alpha - \alpha\cos\alpha)$$

where a is the radius and 2α the angle subtended by the arc at the centre.

10. The area between a parabola and its latus rectum revolves about the directrix. Find the ratio of the volume of the ring thus generated to the volume of the sphere whose diameter is the latus rectum.
11. Trace roughly the curve $xy^2 = 4(2 - x)$. Find the area enclosed by the curve and y-axis, and also the volume of the solid formed by the revolution through four right angles about the axis.
12. Prove that the volume of the solid generated by the revolution of an ellipse about its minor axis is a mean proportional between those generated by the revolution of the ellipse and of the auxiliary circle about the major axis.
13. Find the surface of the solid formed by the revolution, about the axis of y, of the part of the curve $ay^2 = x^3$ from $x = 0$ to $x = 4a$ which is above the x-axis.
14. Prove that the surface of the prolate spheroid formed by the revolution of an ellipse of eccentricity e about its major axis is equal to

$$2 \cdot \text{Area of ellipse} \cdot \left[\sqrt{1 - e^2} + \frac{\sin^{-1} e}{e}\right]$$

15. The coordinates of a point on a cycloid are

$$x = a(\theta + \sin \theta),\ y = a(1 + \cos \theta)$$

Show that the ratio of the area of the surface formed by the rotation of the arc of the cycloid between two consecutive cusps about the axis of x, to the area enclosed by the cycloid and the axis of x is 64 : 9.
16. A quadrant of a circle of radius a revolves about its chord. Show that the volume of the spindle generated is $\dfrac{\pi a^3}{6\sqrt{2}}(10 - 3\pi)$.
17. Find the volumes of the oblate and prolate spheroids generated by an ellipse whose major and minor axes are $(24\pi)^{1/3}$ and $(3\pi)^{1/3}$.
(If an ellipse is revolved about the major axis, the solid obtained is known as prolate spheroid. In case the ellipse is revolved about minor axis, the solid obtained is called oblate spheroid.)
18. Prove that the volume generated by the revolution of limacon $r = a + b \cos \theta$, $a > b$, about the initial line, is $\dfrac{4\pi a}{3}(a^2 + b^2)$.
19. Find the area of the surface wept out by the arc of the rectangular hyperbola $x^2 - y^2 = a^2$, extending from the vertex to the end of the latus rectum, when rotated through four right angles about the x-axis.
20. The lemniscate $r^2 = a^2 \cos 2\theta$ revolves about a tangent at the pole. Show that the surface of the solid generated is $4\pi a^2$.
21. Prove that the surface and volume generated by the revolution of the tractrix

$$x = a \cos t + \frac{a}{2} \log \tan^2 \frac{t}{2},\ y = a \sin t$$

about its asymptote are, respectively, equal to the surface and half the volume of a sphere of radius a.
22. Find the volume of the solid obtained by revolving the loop of the curve $a^2y^2 = x^2(2a - x)(x - a)$ about x-axis.

Solution. Differentiating successively twice, we get

$$\frac{dy}{dx} = m, \quad \frac{d^2y}{dx^2} = 0$$

Hence, the required differential equation is $\frac{d^2y}{dx^2} = 0$.

2. Eliminate A and B from $y = A \sin mx + B \cos mx$ and form the differential equation of second order.

Solution. Differentiating successively twice, we get

$$\frac{dy}{dx} = Am \cos mx - Bm \sin mx$$

$$\frac{d^2y}{dx^2} = -Am^2 \sin mx - Bm^2 \cos mx$$

$$\Rightarrow \quad \frac{d^2y}{dx^2} = -m^2 (A \sin mx + B \cos mx) = -m^2y$$

Hence, the required differential equation is $\frac{d^2y}{dx^2} + m^2y = 0$.

EXERCISES

1. Form the differential equation by eliminating m from $y = e^{mx}$.
2. Form the differential equation by eliminating h and k from the equation
$$(x-h)^2 + (y-k)^2 = a^2$$
3. Form the differential equation by eliminating A and B from the equation
$$y = Ae^{3x} + Be^{5x}$$
4. By the elimination of the constants a, b, obtain the differential equation of which
$$xy = ce^x + be^{-x} + x^2$$
is a solution.
5. By the elimination of the constant a, obtain the differential equation of which $y^2 = 4a(x + a)$ is the solution.
6. Find the differential equation of the family of curves $y = e^x (A \cos x + B \sin x)$, where A and B are constants.
7. Form the differential equation by eliminating A and B from the equation
$$Ax^2 + By^2 = 1.$$

The differential equations of the first order and the first degree can be written in the form

$$M + N\frac{dy}{dx} = 0$$

10 Differential Equations of the First Order and the First Degree

10.1 Introduction

An equation which involves independent and dependent variables and the derivatives of the dependent variables or their differentials is called the *differential equation*.

A differential equation is said to be *ordinary differential equation* if it involves a single independent variable and it is said to be *partial differential equation* if it involves two or more independent variables.

The following are a few examples of differential equations:

$$y\,dx + x\,dy = 0 \tag{1}$$

$$\frac{dy}{dx} = \sin x \tag{2}$$

$$\frac{d^2y}{dx^2} = k^2y \tag{3}$$

$$\frac{dy}{dx} + \sin\frac{dy}{dx} = 0 \tag{4}$$

$$\frac{d^2y}{dx^2} + (x+2)\frac{dy}{dx} - 2y = 0 \tag{5}$$

$$\left(\frac{d^3y}{dx^3}\right)^2 + \left(\frac{d^2y}{dx^2}\right)^3 + \left(\frac{dy}{dx}\right)^4 + y^5 = 0 \tag{6}$$

$$x\frac{\partial z}{\partial x} + y\frac{\partial z}{\partial y} = z \tag{7}$$

$$\frac{\partial^2 z}{\partial x \partial y} = \frac{\partial z}{\partial x} \tag{8}$$

Note. The differential Eqs. (1) to (6) are ordinary and (7) and (8) are partial.

Remark. In this book we shall be concerned with ordinary differential equations only.

The *order* of a differential equation is defined to be the order of the highest order derivative occurring in the differential equation.

The *degree* of a differential equation whose terms are polynomials in the derivatives is defined to be the highest power (positive integral index) of the highest order derivative in it.

Hence, of the above differential Eqs. (1), (2) and (7) are of the first order and the first degree; (3), (5) and (8) are of the second order and the first degree; (6) is of the third order and the second degree; (4) is of the first order but the degree is not defined since the left hand side is not a polynomial in dy/dx.

A *solution* or *integral* of a differential equation is a relation between the variables which is free from the differential coefficients or the differentials such that this relation and the derivatives or differentials obtained from it satisfy the given differential equation.

10.2 Formation of Differential Equations

Ordinary differential equations are associated with families of curves and are obtained by the elimination of arbitrary constants, called parameters, from the given equation.

10.2.1 Differential Equation of Single Parameter Family of Curves

Let the equation of the family of curves be

$$f(x, y, c) = 0 \tag{1}$$

where c is a parameter. Differentiating Eq. (1) w.r.t. x, we get

$$\frac{\partial f}{\partial x} + \frac{\partial f}{\partial y} \cdot \frac{dy}{dx} = 0$$

where $\dfrac{\partial f}{\partial x}$ and $\dfrac{\partial f}{\partial y}$ are generally functions of x, y and c. Therefore, this relation may be written as

$$\phi\left(x, y, \frac{dy}{dx}, c\right) = 0 \tag{2}$$

Eliminating c from Eqs. (1) and (2), we get a relation in x, y and dy/dx. Let this relation be

$$F\left(x, y, \frac{dy}{dx}\right) = 0$$

This is the required differential equation.

10.2.2 Differential Equation of a Two-parameter Family of Curves

Let the equation of a two-parameter family of curves be

$$f(x, y, c_1, c_2) = 0 \tag{1}$$

To eliminate c_1 and c_2, we require three equations in general. Besides the given Eq. (1), we obtain two more equations by differentiating Eq. (1) twice successively w.r.t. x. Let these equations be

$$\left.\begin{aligned} \phi_1\left(x, y, c_1, c_2, \frac{dy}{dx}\right) &= 0 \\ \phi_2\left(x, y, c_1, c_2, \frac{dy}{dx}, \frac{d^2y}{dx^2}\right) &= 0 \end{aligned}\right\} \tag{2}$$

Eliminating c_1 and c_2 from Eqs. (1) and (2), we get a relation among x, y, $\dfrac{dy}{dx}$ and $\dfrac{d^2y}{dx^2}$. Let this relation be

$$F\left(x, y, \frac{dy}{dx}, \frac{d^2y}{dx^2}\right) = 0$$

This is the required differential equation.

Remark. A differential equation of the first order possesses a solution containing one arbitrary constant and a differential equation of the second order possesses a solution containing two arbitrary constants. In general, a differential equation of the nth order will possess a solution containing n arbitrary constants.

Note. A solution of a differential equation which contains the maximum number n of parameters or arbitrary constants is called the *general solution* or *primitive* of the differential equation. Any solution obtained from the general solution by assigning particular values to one or more of th arbitrary constants is called a *particular solution*. Certain differenti equations possess solutions which can not be derived from the gene solution by assigning particular values to the arbitrary constants. S solutions are called *singular solutions*.

Examples

1. Form the differential equation by eliminating m and c from the ec $y = mx + c$.

where M and N are functions of x and y. For convenience the above equation is generally written as

$$M\,dx + N\,dy = 0$$

For finding the solution of such equations we employ the following methods.

10.3 Variables Separable

Sometimes the equation of the form $M\,dx + N\,dy = 0$, where M and N are functions of x and y, can be put in the form

$$f(x)\,dx + \phi(y)\,dy = 0$$

A solution of such equations can be written by direct integration, i.e.

$$\int f(x)dx + \int \phi(y)dy = c$$

where c is an arbitrary constant.

Examples

1. Solve the equation

$$\frac{dy}{dx} = \frac{x(2\log x + 1)}{\sin y + y\cos y}, \quad (x > 0)$$

Solution. The given equation can be written as

$$(\sin y + y\cos y)\,dy = x\,(2\log x + 1)\,dx$$

Since the variables are separable, on integrating, we get

$$y\sin y = x^2\log x + c,\ (x > 0)$$

2. Solve the equation $3e^x\tan y\,dx + (1 - e^x)\sec^2 y\,dy = 0$.

Solution. On dividing by $(1 - e^x)\tan y$, the given equation can be written as

$$\frac{3e^x}{1 - e^x}\,dx + \frac{\sec^2 y}{\tan y}\,dy = 0$$

Since the variables are separable, on integrating, we get

$$-3\log|1 - e^x| + \log|\tan y| = \log|c|$$

$$\Rightarrow \quad \log|\tan y| = \log|c| + 3\log|1 - e^x|$$

$$\Rightarrow \quad \log|\tan y| = \log|c(1 - e^x)^3|$$

$$\Rightarrow \quad |\tan y| = |c(1 - e^x)^3|$$

3. Solve the equation

$$\frac{dy}{dx} = e^{x-y} + x^2 e^{-y}$$

Solution. Multiplying by e^y, the given equation can be written as

$$e^y dy = (e^x + x^2)\, dx$$

Since the variables are separable, on integrating, we get

$$e^y = e^x + \frac{1}{3}x^3 + c$$

4. Solve the equation $\sec^2 x \tan y\, dx + \sec^2 y \tan x\, dy = 0$.

Solution. Dividing by $\tan x \tan y$, the given equation can be written as

$$\frac{\sec^2 x}{\tan x}\, dx + \frac{\sec^2 y}{\tan y}\, dy = 0$$

Since the variables are separable, on integrating, we get

$$\log|\tan x| + \log|\tan y| = \log|c|$$

$$\Rightarrow \qquad |\tan x \tan y| = |c|$$

5. Solve the initial value problem which consists of the differential equation

$$x \sin y\, dx + (x^2 + 1)\cos y\, dy = 0$$

and the initial condition $y(1) = \dfrac{\pi}{2}$.

Solution. First of all we obtain the general solution of the differential equation. Separating the variables dividing by $(x^2 + 1)\sin y$, we get

$$\frac{x}{x^2+1}\, dx + \frac{\cos y}{\sin y}\, dy = 0$$

which on integration gives

$$\frac{1}{2}\log|x^2 + 1| + \log|\sin y| = \log|c_1|$$

$$\Rightarrow \qquad \log|x^2 + 1| + 2\log|\sin y| = 2\log|c_1|$$

$$\Rightarrow \qquad \log(x^2 + 1)\sin^2 y = \log c_1^2$$

$$\Rightarrow \qquad (x^2 + 1)\sin^2 y = c, \quad \text{where } c = c_1^2$$

Initial condition is given that when $x = 1$, $y = \dfrac{\pi}{2}$. Substituting these values in the general solution, we get

$$(1 + 1)\sin^2\frac{\pi}{2} = c \quad \Rightarrow c = 2$$

Thus, the solution of the initial value problem is $(x^2 + 1)\sin^2 y = 2$.

EXERCISES

Solve the following differential equations:

1. $\dfrac{dy}{dx} = xy$
2. $y\dfrac{dy}{dx} = ax$
3. $(1 + x)\,y\,dx + (1 - y)\,x\,dy = 0$
4. $\dfrac{dy}{dx} = \sin x \sin y$
5. $\dfrac{dy}{dx} = ye^x$
6. $\dfrac{dy}{dx} = \dfrac{1 + y^2}{1 + x^2}$
7. $(x^2 - yx^2)\dfrac{dy}{dx} + y^2 + xy^2 = 0$
8. $(1 - x^2)(1 - y)\,dx = xy\,(1 + y)\,dy$
9. $x^2(y + 1)\,dx + y^2(x - 1)\,dy = 0$
10. $y - x\dfrac{dy}{dx} = a\left(y^2 + \dfrac{dy}{dx}\right)$
11. $(e^x + 1)\,y\,dy = (y + 1)\,e^x dx$
12. $xy\dfrac{dy}{dx} + \sqrt{1 + x^2 + y^2 + x^2y^2} = 0$
13. $(e^y + 1)\cos x\,dx + e^y \sin x\,dy = 0$
14. $e^y(1 + x^2)\,dy - 2x\,(1 + e^y)\,dx = 0$
15. $x\sqrt{1 + y^2} + y\sqrt{1 + x^2}\dfrac{dy}{dx} = 0$
16. $xy^3\,dx + e^{x^2}\,dy = 0$
17. $x\cos^2 y\,dx + \tan y\,dy = 0$
18. $xy^3dx + (y + 1)\,e^{-x}dy = 0$
19. $y \log x \log y\,dx + dy = 0$
20. $(xy + x)\,dx = (x^2y^2 + x^2 + y^2 + 1)\,dy$
21. Find a curve for which the tangent at each point makes a constant angle α with the radius vector.
22. Show that the curve in which the cartesian subnormal is constant is a parabola.
23. Show that the curve in which the polar subnormal is constant is $r = a\theta + c$.
24. Find the equations of the curves for which: (i) the cartesian subtangent is constant and (ii) the polar subtangent is constant.
25. Show that the curve in which the angle between the tangent and the radius vector at every point is one half of the vectorial angle is a cardioide.
26. Find the curves for which the sum of the reciprocals of the radius vector and the polar subtangent is constant.
27. Show that the curve in which the square of the normal is equal to the radius vector is either a circle or a hyperbola.
28. Show that the curve in which the perpendicular from the foot of the ordinate upon the tangent is constant is a catenary.
29. Find the curve for which the angle between the tangent and the radius vector at any point is twice the vectorial angle.
30. Find the curve for which the angle between the tangent and the radius vector at any point is supplement of half the vectorial angle.
31. Show that the curve in which the angle between the radius vector and the tangent is m times the vectorial angle is $r^m = a^m \sin m\theta$.

10.4 Linear Equations

An equation of the form

$$\frac{dy}{dx} + Py = Q$$

where P and Q are any functions of x only or constants, is called a *linear differential equation.*

To find the solution of such equations, multiplying by $e^{\int P dx}$ (called the *integrating factor* and denoted by I.F.) in both the sides of the equation, we get

$$e^{\int P dx} \frac{dy}{dx} + (P e^{\int P dx}) y = Q e^{\int P dx}$$

This on integration gives

$$y e^{\int P dx} = \int (Q e^{\int P dx})\, dx + c$$

Working Rule

(i) Write the equation in the standard form $\frac{dy}{dx} + Py = Q$.

(ii) Obtain the integrating factor $e^{\int P dx}$

(iii) Multiply throughout by $e^{\int P dx}$ to the equation in its standard form.

(iv) Integrate the resultant equation.

Note. Sometimes it is easy to find the solution of the equation if we write the given equation in the form

$$\frac{dx}{dy} + Px = Q$$

where P and Q are any functions of y alone or constants. An integrating factor in this case is $e^{\int P dy}$.

Examples

1. Solve the equation

$$x(x-1)\frac{dy}{dx} - y = x^2(x-1)^2, \quad (x > 1)$$

Solution. Writing the equation in the standard form

$$\frac{dy}{dx} - \frac{1}{x(x-1)} y = x(x-1)$$

Now $\quad \text{I.F.} = e^{\int\left[-\frac{1}{x(x-1)}\right]dx} = e^{\int\left(\frac{1}{x}-\frac{1}{x-1}\right)dx}$

$$= e^{\log x - \log(x-1)} = e^{\log\frac{x}{x-1}} = \frac{x}{x-1}, \quad (x > 1)$$

Multiplying the above equation by the I.F. $\dfrac{x}{x-1}$, we get

$$\frac{x}{x-1}\frac{dy}{dx} - \frac{1}{(x-1)^2}y = x^2$$

On integrating, we get $\dfrac{xy}{x-1} = \dfrac{1}{3}x^3 + c, \quad (x > 1)$

2. Solve the equation

$$\frac{dy}{dx} + \frac{y}{x} = x^2, \ (x > 0)$$

given $y = 1$ when $x = 1$.

Solution. The given equation is in standard form.

Now $\quad \text{I.F.} = e^{\int\frac{1}{x}dx} = e^{\log x} = x, \quad (x > 0)$

Multiplying the given equation by I.F. x, we get

$$x\frac{dy}{dx} + y = x^3$$

On integrating, we get $\quad xy = \dfrac{x^4}{4} + c \quad$ (1)

Putting $x = 1$, $y = 1$ in Eq. (1), we get $c = 3/4$.
Substituting this value of c in Eq. (1), we obtain

$$4xy = x^4 + 3, \quad (x > 0)$$

3. Solve the equation

$$\frac{dy}{dx} + xe^x y = e^{(1-x)e^x}$$

Solution. The given equation is in standard form.

Now $\quad \text{I.F.} = e^{\int xe^x dx} = e^{(x-1)e^x}$

Multiplying the given equation by I.F. $e^{(x-1)e^x}$, we get

$$e^{(x-1)e^x}\frac{dy}{dx} + xe^x y e^{(x-1)e^x} = 1$$

On integrating, we get $ye^{(x-1)e^x} = x + c$

$\Rightarrow$ $$y = (x+c)e^{(1-x)e^x}$$

4. Solve the equation

$$x \log x \frac{dy}{dx} - y = 3x^3(\log x)^2, \quad (x > 0)$$

Solution. Writing the equation in standard form

$$\frac{dy}{dx} - \frac{1}{x\log x}y = 3x^2 \log x$$

Now $$\text{I.F.} = e^{\int\left(-\frac{1}{x\log x}\right)dx} = e^{-\log(\log x)} = \frac{1}{\log x}, \quad (x > 0)$$

Multiplying the above equation by I.F. $\frac{1}{\log x}$, we get

$$\frac{1}{\log x}\frac{dy}{dx} - \frac{1}{x(\log x)^2}y = 3x^2$$

On integrating, we get $\frac{1}{\log x}y = x^3 + c$

$\Rightarrow$ $$y = (x^3 + c)\log x, \quad (x > 0)$$

5. Solve the equation

$$\left(e^{-\frac{y^2}{2}} - xy\right)dy - dx = 0$$

Solution. Writing the equation in standard form

$$\frac{dx}{dy} + yx = e^{-\frac{y^2}{2}}$$

Now $$\text{I.F.} = e^{\int y\,dy} = e^{\frac{y^2}{2}}$$

Multiplying the above equation by I.F. $e^{\frac{y^2}{2}}$, we get

$$e^{\frac{y^2}{2}}\frac{dy}{dx} + yx\,e^{\frac{y^2}{2}} = 1$$

10

Differential Equations of the First Order and the First Degree

10.1 Introduction

An equation which involves independent and dependent variables and the derivatives of the dependent variables or their differentials is called the *differential equation.*

A differential equation is said to be *ordinary differential equation* if it involves a single independent variable and it is said to be *partial differential equation* if it involves two or more independent variables.

The following are a few examples of differential equations:

$$y\,dx + x\,dy = 0 \tag{1}$$

$$\frac{dy}{dx} = \sin x \tag{2}$$

$$\frac{d^2y}{dx^2} = k^2 y \tag{3}$$

$$\frac{dy}{dx} + \sin\frac{dy}{dx} = 0 \tag{4}$$

$$\frac{d^2y}{dx^2} + (x+2)\frac{dy}{dx} - 2y = 0 \tag{5}$$

$$\left(\frac{d^3y}{dx^3}\right)^2 + \left(\frac{d^2y}{dx^2}\right)^3 + \left(\frac{dy}{dx}\right)^4 + y^5 = 0 \tag{6}$$

$$x\frac{\partial z}{\partial x} + y\frac{\partial z}{\partial y} = z \tag{7}$$

$$\frac{\partial^2 z}{\partial x \partial y} = \frac{\partial z}{\partial x} \tag{8}$$

Note. The differential Eqs. (1) to (6) are ordinary and (7) and (8) are partial.

Remark. In this book we shall be concerned with ordinary differential equations only.

The *order* of a differential equation is defined to be the order of the highest order derivative occurring in the differential equation.

The *degree* of a differential equation whose terms are polynomials in the derivatives is defined to be the highest power (positive integral index) of the highest order derivative in it.

Hence, of the above differential Eqs. (1), (2) and (7) are of the first order and the first degree; (3), (5) and (8) are of the second order and the first degree; (6) is of the third order and the second degree; (4) is of the first order but the degree is not defined since the left hand side is not a polynomial in dy/dx.

A *solution* or *integral* of a differential equation is a relation between the variables which is free from the differential coefficients or the differentials such that this relation and the derivatives or differentials obtained from it satisfy the given differential equation.

10.2 Formation of Differential Equations

Ordinary differential equations are associated with families of curves and are obtained by the elimination of arbitrary constants, called parameters, from the given equation.

10.2.1 Differential Equation of Single Parameter Family of Curves

Let the equation of the family of curves be

$$f(x, y, c) = 0 \tag{1}$$

where c is a parameter. Differentiating Eq. (1) w.r.t. x, we get

$$\frac{\partial f}{\partial x} + \frac{\partial f}{\partial y} \cdot \frac{dy}{dx} = 0$$

where $\frac{\partial f}{\partial x}$ and $\frac{\partial f}{\partial y}$ are generally functions of x, y and c. Therefore, this relation may be written as

$$\phi\left(x, y, \frac{dy}{dx}, c\right) = 0 \tag{2}$$

Eliminating c from Eqs. (1) and (2), we get a relation in x, y and dy/dx. Let this relation be

$$F\left(x, y, \frac{dy}{dx}\right) = 0$$

This is the required differential equation.

10.2.2 Differential Equation of a Two-parameter Family of Curves

Let the equation of a two-parameter family of curves be

$$f(x, y, c_1, c_2) = 0 \tag{1}$$

To eliminate c_1 and c_2, we require three equations in general. Besides the given Eq. (1), we obtain two more equations by differentiating Eq. (1) twice successively w.r.t. x. Let these equations be

$$\left.\begin{aligned} \phi_1\left(x, y, c_1, c_2, \frac{dy}{dx}\right) &= 0 \\ \phi_2\left(x, y, c_1, c_2, \frac{dy}{dx}, \frac{d^2y}{dx^2}\right) &= 0 \end{aligned}\right\} \tag{2}$$

Eliminating c_1 and c_2 from Eqs. (1) and (2), we get a relation among x, y, $\frac{dy}{dx}$ and $\frac{d^2y}{dx^2}$. Let this relation be

$$F\left(x, y, \frac{dy}{dx}, \frac{d^2y}{dx^2}\right) = 0$$

This is the required differential equation.

Remark. A differential equation of the first order possesses a solution containing one arbitrary constant and a differential equation of the second order possesses a solution containing two arbitrary constants. In general, a differential equation of the nth order will possess a solution containing n arbitrary constants.

Note. A solution of a differential equation which contains the maximum number n of parameters or arbitrary constants is called the *general solution* or *primitive* of the differential equation. Any solution obtained from the general solution by assigning particular values to one or more of the arbitrary constants is called a *particular solution.* Certain differential equations possess solutions which can not be derived from the general solution by assigning particular values to the arbitrary constants. Such solutions are called *singular solutions.*

Examples

1. Form the differential equation by eliminating m and c from the equation $y = mx + c$.

Solution. Differentiating successively twice, we get

$$\frac{dy}{dx} = m, \quad \frac{d^2y}{dx^2} = 0$$

Hence, the required differential equation is $\frac{d^2y}{dx^2} = 0$.

2. Eliminate A and B from $y = A \sin mx + B \cos mx$ and form the differential equation of second order.

Solution. Differentiating successively twice, we get

$$\frac{dy}{dx} = Am \cos mx - Bm \sin mx$$

$$\frac{d^2y}{dx^2} = -Am^2 \sin mx - Bm^2 \cos mx$$

$$\Rightarrow \quad \frac{d^2y}{dx^2} = -m^2 (A \sin mx + B \cos mx) = -m^2 y$$

Hence, the required differential equation is $\frac{d^2y}{dx^2} + m^2y = 0$.

EXERCISES

1. Form the differential equation by eliminating m from $y = e^{mx}$.
2. Form the differential equation by eliminating h and k from the equation
$$(x - h)^2 + (y - k)^2 = a^2$$
3. Form the differential equation by eliminating A and B from the equation
$$y = Ae^{3x} + Be^{5x}$$
4. By the elimination of the constants a, b, obtain the differential equation of which
$$xy = ce^x + be^{-x} + x^2$$
is a solution.
5. By the elimination of the constant a, obtain the differential equation of which $y^2 = 4a(x + a)$ is the solution.
6. Find the differential equation of the family of curves $y = e^x (A \cos x + B \sin x)$, where A and B are constants.
7. Form the differential equation by eliminating A and B from the equation
$$Ax^2 + By^2 = 1.$$

The differential equations of the first order and the first degree can be written in the form

$$M + N\frac{dy}{dx} = 0$$

where M and N are functions of x and y. For convenience the above equation is generally written as

$$M\,dx + N\,dy = 0$$

For finding the solution of such equations we employ the following methods.

10.3 Variables Separable

Sometimes the equation of the form $M\,dx + N\,dy = 0$, where M and N are functions of x and y, can be put in the form

$$f(x)\,dx + \phi(y)\,dy = 0$$

A solution of such equations can be written by direct integration, i.e.

$$\int f(x)dx + \int \phi(y)dy = c$$

where c is an arbitrary constant.

Examples

1. Solve the equation

$$\frac{dy}{dx} = \frac{x(2\log x + 1)}{\sin y + y\cos y}, \quad (x > 0)$$

Solution. The given equation can be written as

$$(\sin y + y\cos y)\,dy = x\,(2\log x + 1)\,dx$$

Since the variables are separable, on integrating, we get

$$y\sin y = x^2 \log x + c, \; (x > 0)$$

2. Solve the equation $3e^x \tan y\,dx + (1 - e^x)\sec^2 y\,dy = 0$.

Solution. On dividing by $(1 - e^x)\tan y$, the given equation can be written as

$$\frac{3e^x}{1 - e^x}\,dx + \frac{\sec^2 y}{\tan y}\,dy = 0$$

Since the variables are separable, on integrating, we get

$$-3\log|1 - e^x| + \log|\tan y| = \log|c|$$

$$\Rightarrow \quad \log|\tan y| = \log|c| + 3\log|1 - e^x|$$

$$\Rightarrow \quad \log|\tan y| = \log|c\,(1 - e^x)^3|$$

$$\Rightarrow \quad |\tan y| = |c\,(1 - e^x)^3|$$

3. Solve the equation

$$\frac{dy}{dx} = e^{x-y} + x^2 e^{-y}$$

Solution. Multiplying by e^y, the given equation can be written as

$$e^y dy = (e^x + x^2)\, dx$$

Since the variables are separable, on integrating, we get

$$e^y = e^x + \frac{1}{3}x^3 + c$$

4. Solve the equation $\sec^2 x \tan y\, dx + \sec^2 y \tan x\, dy = 0$.

Solution. Dividing by $\tan x \tan y$, the given equation can be written as

$$\frac{\sec^2 x}{\tan x}\, dx + \frac{\sec^2 y}{\tan y}\, dy = 0$$

Since the variables are separable, on integrating, we get

$$\log|\tan x| + \log|\tan y| = \log|c|$$

$$\Rightarrow \qquad |\tan x \tan y| = |c|$$

5. Solve the initial value problem which consists of the differential equation

$$x \sin y\, dx + (x^2 + 1)\cos y\, dy = 0$$

and the initial condition $y(1) = \dfrac{\pi}{2}$.

Solution. First of all we obtain the general solution of the differential equation. Separating the variables dividing by $(x^2 + 1)\sin y$, we get

$$\frac{x}{x^2+1}\, dx + \frac{\cos y}{\sin y}\, dy = 0$$

which on integration gives

$$\frac{1}{2}\log|x^2 + 1| + \log|\sin y| = \log|c_1|$$

$$\Rightarrow \qquad \log|x^2 + 1| + 2\log|\sin y| = 2\log|c_1|$$

$$\Rightarrow \qquad \log(x^2 + 1)\sin^2 y = \log c_1^2$$

$$\Rightarrow \qquad (x^2 + 1)\sin^2 y = c, \quad \text{where } c = c_1^2$$

Initial condition is given that when $x = 1$, $y = \dfrac{\pi}{2}$. Substituting these values in the general solution, we get

$$(1 + 1)\sin^2\frac{\pi}{2} = c \quad \Rightarrow \quad c = 2$$

Thus, the solution of the initial value problem is $(x^2 + 1)\sin^2 y = 2$.

EXERCISES

Solve the following differential equations:

1. $\dfrac{dy}{dx} = xy$ 2. $y\dfrac{dy}{dx} = ax$ 3. $(1 + x)\,y\,dx + (1 - y)\,x\,dy = 0$

4. $\dfrac{dy}{dx} = \sin x \sin y$ 5. $\dfrac{dy}{dx} = ye^x$ 6. $\dfrac{dy}{dx} = \dfrac{1 + y^2}{1 + x^2}$

7. $(x^2 - yx^2)\dfrac{dy}{dx} + y^2 + xy^2 = 0$ 8. $(1 - x^2)(1 - y)\,dx = xy\,(1 + y)\,dy$

9. $x^2(y + 1)\,dx + y^2(x - 1)\,dy = 0$ 10. $y - x\dfrac{dy}{dx} = a\left(y^2 + \dfrac{dy}{dx}\right)$

11. $(e^x + 1)\,y\,dy = (y + 1)\,e^x dx$ 12. $xy\dfrac{dy}{dx} + \sqrt{1 + x^2 + y^2 + x^2y^2} = 0$

13. $(e^y + 1)\cos x\,dx + e^y \sin x\,dy = 0$

14. $e^y(1 + x^2)\,dy - 2x\,(1 + e^y)\,dx = 0$ 15. $x\sqrt{1 + y^2} + y\sqrt{1 + x^2}\,\dfrac{dy}{dx} = 0$

16. $xy^3\,dx + e^{x^2}\,dy = 0$ 17. $x\cos^2 y\,dx + \tan y\,dy = 0$

18. $xy^3dx + (y + 1)\,e^{-x}dy = 0$ 19. $y\log x \log y\,dx + dy = 0$

20. $(xy + x)\,dx = (x^2y^2 + x^2 + y^2 + 1)\,dy$

21. Find a curve for which the tangent at each point makes a constant angle α with the radius vector.
22. Show that the curve in which the cartesian subnormal is constant is a parabola.
23. Show that the curve in which the polar subnormal is constant is $r = a\theta + c$.
24. Find the equations of the curves for which: (i) the cartesian subtangent is constant and (ii) the polar subtangent is constant.
25. Show that the curve in which the angle between the tangent and the radius vector at every point is one half of the vectorial angle is a cardioide.
26. Find the curves for which the sum of the reciprocals of the radius vector and the polar subtangent is constant.
27. Show that the curve in which the square of the normal is equal to the radius vector is either a circle or a hyperbola.
28. Show that the curve in which the perpendicular from the foot of the ordinate upon the tangent is constant is a catenary.
29. Find the curve for which the angle between the tangent and the radius vector at any point is twice the vectorial angle.
30. Find the curve for which the angle between the tangent and the radius vector at any point is supplement of half the vectorial angle.
31. Show that the curve in which the angle between the radius vector and the tangent is m times the vectorial angle is $r^m = a^m \sin m\theta$.

10.4 Linear Equations

An equation of the form

$$\frac{dy}{dx} + Py = Q$$

where P and Q are any functions of x only or constants, is called a *linear differential equation.*

To find the solution of such equations, multiplying by $e^{\int P dx}$ (called the *integrating factor* and denoted by I.F.) in both the sides of the equation, we get

$$e^{\int P dx}\frac{dy}{dx} + (Pe^{\int P dx})y = Qe^{\int P dx}$$

This on integration gives

$$ye^{\int P dx} = \int (Qe^{\int P dx})\,dx + c$$

Working Rule

(i) Write the equation in the standard form $\frac{dy}{dx} + Py = Q$.

(ii) Obtain the integrating factor $e^{\int P dx}$

(iii) Multiply throughout by $e^{\int P dx}$ to the equation in its standard form.

(iv) Integrate the resultant equation.

Note. Sometimes it is easy to find the solution of the equation if we write the given equation in the form

$$\frac{dx}{dy} + Px = Q$$

where P and Q are any functions of y alone or constants. An integrating factor in this case is $e^{\int P dy}$.

Examples

1. Solve the equation

$$x(x-1)\frac{dy}{dx} - y = x^2(x-1)^2, \quad (x > 1)$$

Solution. Writing the equation in the standard form

$$\frac{dy}{dx} - \frac{1}{x(x-1)}y = x(x-1)$$

Now $\qquad \text{I.F.} = e^{\int\left[-\frac{1}{x(x-1)}\right]dx} = e^{\int\left(\frac{1}{x}-\frac{1}{x-1}\right)dx}$

$$= e^{\log x - \log(x-1)} = e^{\log\frac{x}{x-1}} = \frac{x}{x-1}, \quad (x > 1)$$

Multiplying the above equation by the I.F. $\frac{x}{x-1}$, we get

$$\frac{x}{x-1}\frac{dy}{dx} - \frac{1}{(x-1)^2}y = x^2$$

On integrating, we get $\frac{xy}{x-1} = \frac{1}{3}x^3 + c, \quad (x > 1)$

2. Solve the equation

$$\frac{dy}{dx} + \frac{y}{x} = x^2, \ (x > 0)$$

given $y = 1$ when $x = 1$.

Solution. The given equation is in standard form.

Now $\qquad \text{I.F.} = e^{\int\frac{1}{x}dx} = e^{\log x} = x, \quad (x > 0)$

Multiplying the given equation by I.F. x, we get

$$x\frac{dy}{dx} + y = x^3$$

On integrating, we get $\qquad xy = \frac{x^4}{4} + c \qquad (1)$

Putting $x = 1$, $y = 1$ in Eq. (1), we get $c = 3/4$.

Substituting this value of c in Eq. (1), we obtain

$$4xy = x^4 + 3, \quad (x > 0)$$

3. Solve the equation

$$\frac{dy}{dx} + xe^x y = e^{(1-x)e^x}$$

Solution. The given equation is in standard form.

Now $\qquad \text{I.F.} = e^{\int xe^x dx} = e^{(x-1)e^x}$

Multiplying the given equation by I.F. $e^{(x-1)e^x}$, we get

$$e^{(x-1)e^x}\frac{dy}{dx} + xe^x y e^{(x-1)e^x} = 1$$

On integrating, we get $ye^{(x-1)e^x} = x + c$

$\Rightarrow$ $$y = (x + c)e^{(1-x)e^x}$$

4. Solve the equation

$$x \log x \frac{dy}{dx} - y = 3x^3(\log x)^2, \quad (x > 0)$$

Solution. Writing the equation in standard form

$$\frac{dy}{dx} - \frac{1}{x \log x} y = 3x^2 \log x$$

Now $$\text{I.F.} = e^{\int\left(-\frac{1}{x\log x}\right)dx} = e^{-\log(\log x)} = \frac{1}{\log x}, \quad (x > 0)$$

Multiplying the above equation by I.F. $\frac{1}{\log x}$, we get

$$\frac{1}{\log x}\frac{dy}{dx} - \frac{1}{x(\log x)^2} y = 3x^2$$

On integrating, we get $\frac{1}{\log x} y = x^3 + c$

$\Rightarrow$ $$y = (x^3 + c) \log x, \quad (x > 0)$$

5. Solve the equation

$$\left(e^{-\frac{y^2}{2}} - xy\right) dy - dx = 0$$

Solution. Writing the equation in standard form

$$\frac{dx}{dy} + yx = e^{-\frac{y^2}{2}}$$

Now $$\text{I.F.} = e^{\int y\,dy} = e^{\frac{y^2}{2}}$$

Multiplying the above equation by I.F. $e^{\frac{y^2}{2}}$, we get

$$e^{\frac{y^2}{2}}\frac{dy}{dx} + yx\,e^{\frac{y^2}{2}} = 1$$

On integrating, we get $\quad x e^{\frac{y^2}{2}} = y + c$

6. Solve the equation

$$\frac{dx}{dy} + \frac{x}{1+y^2} = \frac{e^{-\tan^{-1} y}}{1+y^2}$$

Solution. The given equation is in standard form.

Now $\quad \text{I.F.} = e^{\int \frac{1}{1+y^2} dy} = e^{\tan^{-1} y}$

Multiplying the given equation by I.F. $e^{\tan^{-1} y}$, we get

$$e^{\tan^{-1} y} \frac{dx}{dy} + \frac{x}{1+y^2} e^{\tan^{-1} y} = \frac{1}{1+y^2}$$

On integrating, we get

$$x e^{\tan^{-1} y} = \tan^{-1} y + c$$

EXERCISES

Solve the following differential equations:

1. $\frac{dy}{dx} + ay = e^{mx}$
2. $\frac{dy}{dx} + \frac{y}{x} = x + \frac{2}{x} + 3$
3. $\frac{dy}{dx} = (1-x) + (1-y)$
4. $\frac{dy}{dx} + \frac{2xy}{1+x^2} = \frac{\cos x}{1+x^2}$
5. $(1-x^2)\frac{dy}{dx} + 2xy = x\sqrt{1-x^2}$
6. $\frac{dy}{dx} - \frac{y}{x} = 2x \operatorname{cosec} 2x$
7. $\frac{dy}{dx} + \frac{2}{x} y = \sin x$
8. $\frac{dy}{dx} + y\left(\tan x + \frac{1}{x}\right) = \frac{1}{x} \sec x$
9. $\frac{dy}{dx} + y \cos x = \sin x \cos x$
10. $\frac{dy}{dx} + y \tan x - \sec x = 0$
11. $(1+y^2)\, dx = (\tan^{-1} y - x)\, dy$
12. $\cos^2 x \frac{dy}{dx} + y = \tan x$
13. $\frac{dy}{dx} + \frac{2}{x} y = x \log x, (x > 0)$
14. $(1+x^2)\frac{dy}{dx} + 2xy = 4x^2$
15. $x \log x \frac{dy}{dx} + y = 2 \log x, (x > 0)$
16. $\frac{dy}{dx} + y \cot x = 2 \cos x$
17. $(x + 2y^3) \frac{dy}{dx} = y$
18. $\frac{dx}{dy} + x = e^{-y} \sec^2 y$
19. $\frac{dy}{dx} + \frac{y}{1+x^2} = \frac{e^{\tan^{-1} x}}{1+x^2}$
20. $x(x-1) \frac{dy}{dx} - (x-2)y = x^3(2x-1)$

21. $(y - x)\dfrac{dy}{dx} = a^2$

22. $\dfrac{dx}{dy} + \dfrac{2x}{y} = \dfrac{2}{y^2}$

23. $\dfrac{dy}{dx} + \dfrac{3x^2y}{1+x^3} = \dfrac{\sin^2 x}{1+x^3}$

24. $\dfrac{dy}{dx} + 2y\tan x = \sin x$, given that $y = 0$ when $x = \dfrac{\pi}{3}$. Hence find the maximum value of y.

10.5 The Bernoulli's Equation

An equation of the form

$$\frac{dy}{dx} + Py = Qy^n$$

where P and Q are functions of x alone and $n \neq 0$, is called the *Bernoulli equation.*

To find the solution of such equation dividing it by y^n, we get

$$y^{-n}\frac{dy}{dx} + Py^{-n+1} = Q \tag{1}$$

Let $y^{-n+1} = v$. Then $(-n+1)\,y^{-n}\dfrac{dy}{dx} = \dfrac{dv}{dx}$. Now, Eq. (1) transforms into the form

$$\frac{dv}{dx} + (1-n)\,Pv = (1-n)Q$$

This equation is linear in v and hence the solution can be obtained by using the technique for solving linear equations.

Examples

1. Solve the Bernoulli equation

$$\frac{dy}{dx} + 2xy = 2xy^2$$

Solution. Dividing by y^2 the given equation reduces to

$$\frac{1}{y^2}\frac{dy}{dx} + \frac{2x}{y} = 2x \tag{1}$$

Let $\dfrac{1}{y} = v$. Then $-\dfrac{1}{y^2}\dfrac{dy}{dx} = \dfrac{dv}{dx}$. Therefore, Eq. (1) transforms into the form

$$\frac{dv}{dx} - 2x\,v = -2x \tag{2}$$

This equation is linear in v. Now

$$\text{I.F.} = e^{\int(-2x)dx} = e^{-x^2}$$

Multiplying Eq. (2) by I.F. e^{-x^2}, we get

$$e^{-x^2}\frac{dv}{dx} - 2xve^{-x^2} = -2xe^{-x^2}$$

On integrating, we get $\quad ve^{-x^2} = e^{-x^2} + c$

$$\Rightarrow \qquad v = 1 + ce^{x^2}$$

$$\Rightarrow \qquad y(1 + ce^{x^2}) = 1$$

2. Solve the equation

$$(2\log x)\frac{dy}{dx} + \frac{y}{x} = \frac{1}{y}\cos x, \quad (x > 0)$$

Solution. Dividing by $\dfrac{\log x}{y}$ the given equation reduces to

$$2y\frac{dy}{dx} + \frac{1}{x\log x}y^2 = \frac{\cos x}{\log x} \tag{1}$$

Let $y^2 = v$. Then $2y\dfrac{dy}{dx} = \dfrac{dv}{dx}$. Therefore, Eq. (1) transforms into the form

$$\frac{dv}{dx} + \frac{1}{x\log x}v = \frac{\cos x}{\log x} \tag{2}$$

This equation is linear in v. Now

$$\text{I.F.} = e^{\int \frac{1}{x\log x}dx} = e^{\log(\log x)} = \log x, \ (x > 0)$$

Multiplying Eq. (2) by I.F. $\log x$, we get

$$(\log x)\frac{dv}{dx} + \frac{1}{x}v = \cos x$$

On integrating, we get $\quad v\log x = \sin x + c$

$$\Rightarrow \qquad y^2\log x = \sin x + c, \quad (x > 0)$$

3. Solve the equation

$$(2\sin x)\frac{dy}{dx} + y\cos x = y^3\sin^2 x$$

Solution. Dividing by $y^3 \sin x$ the given equation reduces to

$$\frac{2}{y^3}\frac{dy}{dx}+\frac{1}{y^2}\cot x=\sin x \tag{1}$$

Let $\frac{1}{y^2}=v$. Then $-\frac{2}{y^3}\frac{dy}{dx}=\frac{dv}{dx}$. Therefore, Eq. (1) transforms into the form

$$\frac{dv}{dx}-v\cot x=-\sin x \tag{2}$$

This equation is linear in v. Now

$$\text{I.F.}=e^{\int(-\cot x)dx}=e^{\log\frac{1}{\sin x}}=\frac{1}{\sin x}$$

Multiplying Eq. (2) by I.F. $\frac{1}{\sin x}$, we get

$$\frac{1}{\sin x}\frac{dv}{dx}-\frac{v\cos x}{\sin^2 x}=-1$$

On integrating, we get $\qquad \frac{v}{\sin x}=-x+c$

$\Rightarrow \qquad \frac{1}{y^2\sin x}=-x+c$

$\Rightarrow \qquad y^2(c-x)\sin x=1$

4. Solve the equation

$$\frac{dy}{dx}+x\sin 2y=2xe^{-x^2}\cos^2 y$$

Solution. The given equation can be written as

$$\frac{dy}{dx}+2x\sin y\cos y=2xe^{-x^2}\cos^2 y$$

Dividing by $\cos^2 y$ the above equation reduces to

$$\sec^2 y\frac{dy}{dx}+2x\tan y=2xe^{-x^2} \tag{1}$$

Let $\tan y=v$. Then $\sec^2 y\frac{dy}{dx}=\frac{dv}{dx}$. Therefore, Eq. (1) transforms into the form

$$\frac{dv}{dx}+2xv=2xe^{-x^2} \tag{2}$$

This equation is linear in v. Now

$$\text{I.F.} = e^{\int 2x\,dx} = e^{x^2}$$

Multiplying Eq. (2) by I.F. e^{x^2}, we get

$$e^{x^2}\frac{dv}{dx} + 2xve^{x^2} = 2x$$

On integrating, we get $\quad ve^{x^2} = x^2 + c$

$$\Rightarrow \qquad e^{x^2}\tan y = x^2 + c$$

$$\Rightarrow \qquad \tan y = (c + x^2)e^{-x^2}$$

5. Solve the equation

$$\frac{dx}{dy} - xy = x^2y^3$$

Solution. Dividing by x^2 the given equation reduces to

$$\frac{1}{x^2}\frac{dx}{dy} - \frac{1}{x}y = y^3 \qquad (1)$$

Let $\frac{1}{x} = v$. Then $-\frac{1}{x^2}\frac{dx}{dy} = \frac{dv}{dy}$. Therefore, Eq. (1) transforms into the form

$$\frac{dv}{dy} + vy = -y^3 \qquad (2)$$

This equation is linear in v. Now

$$\text{I.F.} = e^{\int y\,dy} = e^{\frac{y^2}{2}}$$

Multiplying Eq. (2) by I.F. $e^{\frac{y^2}{2}}$, we get

$$e^{\frac{y^2}{2}}\frac{dv}{dy} + vye^{\frac{y^2}{2}} = -y^3e^{\frac{y^2}{2}}$$

On integrating, we get $\quad ve^{\frac{y^2}{2}} = (2 - y^2)e^{\frac{y^2}{2}} + c$

$$\Rightarrow \qquad v = (2 - y^2) + ce^{-\frac{y^2}{2}}$$

$$\Rightarrow \qquad \frac{1}{x} = (2 - y^2) + ce^{-\frac{y^2}{2}}$$

EXERCISES

Solve the following differential equations:

1. $x\dfrac{dy}{dx} + y = xy^3$
2. $\dfrac{dy}{dx} + \dfrac{y}{x} = y^2$
3. $x\dfrac{dy}{dx} + y = x^3y^6$
4. $2\dfrac{dy}{dx} = \dfrac{y}{x} + \dfrac{y^2}{x^2}$
5. $x\dfrac{dy}{dx} + y = y^2 \log x$, $(x > 0)$
6. $(1 - x^2)\dfrac{dy}{dx} + xy = xy^2$
7. $\dfrac{dy}{dx} - 2y\tan x = y^2 \tan^2 x$
8. $xy - \dfrac{dy}{dx} = y^3 e^{-x^2}$
9. $\dfrac{dy}{dx} = x^3y^3 - xy$
10. $3\dfrac{dy}{dx} + \dfrac{2}{x+1}y = \dfrac{x^3}{y^2}$
11. $\dfrac{dy}{dx} + xy = y^2 e^{\frac{x^2}{2}} \sin x$
12. $y(2xy + e^x)\,dx - e^x dy = 0$
13. $\dfrac{dy}{dx} - \dfrac{1}{1+x}\tan y = (1 + x)e^x \sec y$
14. $\dfrac{dy}{dx} - y\cot x = -y^2 \operatorname{cosec} x$
15. $\dfrac{dy}{dx} + \dfrac{y}{x}\log y = \dfrac{y}{x^2}(\log y)^2$, $(y > 0)$
16. $\sec^2 y\dfrac{dy}{dx} + 2x\tan y = x^3$
17. $(1 + x^2)\dfrac{dy}{dx} = xy - y^2$
18. $\dfrac{dy}{dx} - \dfrac{x}{1 - x^2}y = \dfrac{x^3y^3}{1 - x^2}$
19. $\cos x\,dy = y(\sin x - y)\,dx$
20. $\dfrac{dy}{dx} + xy = xy^n$
21. $\dfrac{dy}{dx} + y\cos x + y^3 \cos x \sin^2 x = 0$
22. $\dfrac{dx}{dy} + x = -\dfrac{x^2}{y}$

10.6 Homogeneous Equations

An equation of the form

$$\frac{dy}{dx} = \phi(x, y)$$

is said to be *homogeneous* in x and y if $\phi(x, y)$ is a homogeneous function of its variables of degree zero. A homogeneous equation can always be represented by

$$\frac{dy}{dx} = f\left(\frac{y}{x}\right)$$

To find the solution of such equations substitute $y = vx$. Then $\dfrac{dy}{dx} = v + x\dfrac{dv}{dx}$. Therefore, the given equation reduces to

$$v + x\frac{dv}{dx} = f(v)$$

Now, the variables are separable. Hence the solution is

$$\int \frac{1}{x}\,dx = \int \frac{1}{v - f(v)}\,dv + c$$

After integration replacing v by $\frac{y}{x}$, we get the required solution.

Examples

1. Solve the equation

$$x\frac{dy}{dx} = y + x\cos^2\frac{y}{x}$$

Solution. The given equation can be written as

$$\frac{dy}{dx} = \frac{y}{x} + \cos^2\frac{y}{x} \qquad (1)$$

This is a homogeneous equation. Let $y = vx$. Then $\frac{dy}{dx} = v + x\frac{dv}{dx}$. Substituting these values in Eq. (1), we get

$$v + x\frac{dv}{dx} = v + \cos^2 v$$

$$\Rightarrow \qquad x\frac{dv}{dx} = \cos^2 v$$

Now, the variables are separable. Hence the solution is

$$\int \sec^2 v\,dv = \int \frac{1}{x}\,dx + c$$

$$\Rightarrow \qquad \tan v = \log|x| + c$$

Replacing v by $\frac{y}{x}$ the required solution is given by

$$\tan\frac{y}{x} = \log|x| + c$$

2. Solve the equation $(x^3 + y^3)\,dx - 3xy^2dy = 0$.

Solution. The given equation can be written as

$$\frac{dy}{dx} = \frac{x^3 + y^3}{3xy^2} \qquad (1)$$

This is a homogeneous equation. Let $y = vx$. Then $\frac{dy}{dx} = v + x\frac{dv}{dx}$. Substituting these values in Eq. (1), we get

$$v + x\frac{dy}{dx} = \frac{1 + v^3}{3v^2}$$

$$\Rightarrow \qquad x\frac{dv}{dx} = \frac{1 - 2v^3}{3v^2}$$

Now, the variables are separable. Hence the solution is

$$\int \frac{1}{x}\,dx - \int \frac{3v^2}{1 - 2v^3}\,dv = c'$$

$$\Rightarrow \qquad \log|x| + \frac{1}{2}\log|1 - 2v^3| = c'$$

$$\Rightarrow \qquad \log x^2 + \log|1 - 2v^3| = 2c'$$

$$\Rightarrow \qquad x^2|1 - 2v^3| = e^{2c'}$$

Replacing v by $\frac{y}{x}$, we get

$$|x^3 - 2y^3| = c\,|x| \quad (\text{where } c = e^{2c'})$$

3. Solve the equation $(x^2 + 2xy - y^2)\,dx + (y^2 + 2xy - x^2)\,dy = 0$.

Solution. The given equation can be written as

$$\frac{dy}{dx} = \frac{x^2 + 2xy - y^2}{x^2 - 2xy - y^2} \tag{1}$$

This is a homogeneous equation. Let $y = vx$. Then $\frac{dy}{dx} = v + x\frac{dv}{dx}$. Substituting these values in Eq. (1), we get

$$v + x\frac{dv}{dx} = \frac{1 + 2v - v^2}{1 - 2v - v^2}$$

$$\Rightarrow \qquad x\frac{dv}{dx} = \frac{1 + v + v^2 + v^3}{1 - 2v - v^2}$$

Now, the variables are separable. Hence the solution is

$$\int \frac{1}{x}\,dx = \int \frac{(1 - 2v - v^2)}{1 + v + v^2 + v^3}\,dv + \log|c|$$

$$\Rightarrow \qquad \log|x| = \int \left(\frac{1}{1 + v} - \frac{2v}{1 + v^2}\right)dv + \log|c|$$

$$\Rightarrow \qquad \log|x| = \log|1 + v| - \log|1 + v^2| + \log|c|$$

$$\Rightarrow \qquad |x| = \frac{|c(1+v)|}{|1+v^2|}$$

Replacing v by $\frac{y}{x}$, we get $x^2 + y^2 = |\ c(x+y)\ |$.

EXERCISES

Solve the following differential equations:

1. $y - x\frac{dy}{dx} = x + y\frac{dy}{dx}$
2. $(x^2 + y^2)\,dx - 2xy\,dy = 0$
3. $\frac{dy}{dx} = \frac{x^2 y}{x^3 + y^3}$
4. $\frac{dy}{dx} = \frac{xy + y^2}{2x^2}$
5. $x\,dy - y\,dx = \sqrt{x^2 + y^2}\,dx$
6. $x^2 dy + y\,(x + y)\,dx = 0$
7. $\frac{dy}{dx} = -\frac{x^2 + 3y^2}{3x^2 + y^2}$
8. $(x^2 - y^2)\,dx + 2xy\,dy = 0$
9. $x(x - y)\,dy + y^2 dx = 0$
10. $x(x - y)\,dy = y(x + y)\,dx$
11. $(x^3 - 3xy^2)\,dx = (y^3 - 3x^2 y)\,dy$
12. $x \sin\left(\frac{y}{x}\right)\frac{dy}{dx} = y \sin\left(\frac{y}{x}\right) - x$
13. $2\frac{dy}{dx} = \frac{y}{x} + \frac{y^2}{x^2}$

10.7 Non-homogeneous Equations Reducible to Homogeneous Form

An equation of the form

$$\frac{dy}{dx} = \frac{ax + by + c}{Ax + By + C}$$

is a *non-homogeneous equation* which can be reduced to homogeneous form.

To reduce non-homogeneous equation into homogeneous form, we consider new variables X and Y, related to x and y by the equations

$$x = X + h,\ y = Y + k$$

where h and k are constants which are yet to be chosen. In fact, we choose h and k in such a way that the given equation becomes homogeneous. With the above substitutions, we have

$$\frac{dy}{dx} = \frac{d}{dx}(Y + k) = \frac{dY}{dx} = \frac{dY}{dX}\cdot\frac{dX}{dx} = \frac{dY}{dX}$$

Hence the differential equation reduces to

$$\frac{dY}{dX} = \frac{aX + bY + (ah + bk + c)}{AX + BY + (Ah + Bk + C)}$$

Now, choose h and k such that

$$\left.\begin{aligned} ah + bk + c = 0 \\ Ah + Bk + C = 0 \end{aligned}\right\} \tag{1}$$

Then, the differential equation becomes homogeneous and can be solved by taking the substitution $Y = VX$. Replacing X and Y by $x - h$ and $y - k$, respectively in the solution we get the required solution in the original variables.

Notes

1. The above method is applicable when the determinant $\Delta = \begin{vmatrix} a & b \\ A & B \end{vmatrix} \neq 0$.

2. If the determinant $\Delta = \begin{vmatrix} a & b \\ A & B \end{vmatrix} = 0$, the system (1) has no solutions in the general case and the method described above can not be applied. In this case $\frac{A}{a} = \frac{B}{b} = \lambda$ (say) and hence the given equation has the form

$$\frac{dy}{dx} = \frac{ax + by + c}{\lambda(ax + by) + C}$$

To find the solution of such equations, substitute $ax + by = v$. Then, the given equation reduces to an equation in which variables are separable.

Examples

1. Solve the equation $2x + 3y - 5 + (3x + 2y - 5)\,\frac{dy}{dx} = 0$.

Solution. The given equation can be written as

$$\frac{dy}{dx} = -\frac{2x + 3y - 5}{3x + 2y - 5} \tag{1}$$

Here $\Delta = \begin{vmatrix} 2 & 3 \\ 3 & 2 \end{vmatrix} = -5 \neq 0$. Hence Eq. (1) can be reduced to homogeneous form. Putting $x = X + h$ and $y = Y + k$, then Eq. (1) becomes

$$\frac{dY}{dX} = -\frac{2X + 3Y + (2h + 3k - 5)}{3X + 2Y + (3h + 2k - 5)} \tag{2}$$

Now, choose h and k such that

$$\left.\begin{aligned} 2h + 3k - 5 = 0 \\ 3h + 2k - 5 = 0 \end{aligned}\right\}$$

which give $h = 1$ and $k = 1$. Substituting these values in Eq. (2), we get

$$\frac{dY}{dX} = -\frac{2X + 3Y}{3X + 2Y} \tag{3}$$

Let $Y = VX$. Then $\frac{dY}{dX} = V + X\frac{dV}{dX}$. Substituting these values in Eq. (3), we get

$$V + X\frac{dV}{dX} = -\frac{2 + 3V}{3 + 2V}$$

$$\Rightarrow \quad X\frac{dV}{dX} + \frac{2V^2 + 6V + 2}{2V + 3} = 0$$

Now, the variables are separable. Therefore, on integrating, we get

$$\int \frac{2}{X}\,dX + \int \frac{(2V + 3)}{V^2 + 3V + 1}\,dV = \log|c|$$

$$\Rightarrow \quad 2\log|X| + \log|V^2 + 3V + 1| = \log|c|$$

$$\Rightarrow \quad \log X^2 + \log|V^2 + 3V + 1| = \log|c|$$

$$\Rightarrow \quad X^2|V^2 + 3V + 1| = |c|$$

$$\Rightarrow \quad |Y^2 + 3XY + X^2| = |c|$$

Replacing X by $x - 1$ and Y by $y - 1$, we get

$$|y^2 + 3xy + x^2 - 5x - 5y| = |c|$$

$$\Rightarrow \quad (y^2 + 3xy + x^2 - 5x - 5y)^2 = c^2$$

2. Solve the equation

$$\frac{dy}{dx} = \frac{2x - y + 1}{x + 2y - 3}$$

Solution. Here $\Delta = \begin{vmatrix} 2 & -1 \\ 1 & 2 \end{vmatrix} = 5 \neq 0$. Hence the given equation can be reduced to homogeneous form. Putting $x = X + h$ and $y = Y + k$, the given equation becomes

$$\frac{dY}{dX} = \frac{2X - Y + (2h - k + 1)}{X + 2Y + (h + 2k - 3)} \tag{1}$$

Now, choose h and k such that

$$\left.\begin{aligned} 2h - k + 1 &= 0 \\ h + 2k - 3 &= 0 \end{aligned}\right\}$$

which give $h = -1$ and $k = -1$. Substituting these values in Eq. (1), we get

$$\frac{dY}{dX} = \frac{2X - Y}{X + 2Y} \tag{2}$$

Let $Y = VX$. Then $\frac{dY}{dX} = V + X\frac{dV}{dX}$. Substituting these values in Eq. (2), we get

$$V + X\frac{dV}{dX} = \frac{2 - V}{1 + 2V}$$

$$\Rightarrow \qquad X\frac{dV}{dX} + \frac{2V^2 + 2V - 2}{2V + 1} = 0$$

Now, the variables are separable. Therefore, on integrating, we get

$$\int \frac{2}{X}\,dX + \int \frac{(2V + 1)}{V^2 + V - 1}\,dV = \log|c|$$

$$\Rightarrow \qquad 2\log|X| + \log|V^2 + V - 1| = \log|c|$$

$$\Rightarrow \qquad \log X^2 + \log|V^2 + V - 1| = \log|c|$$

$$\Rightarrow \qquad X^2|V^2 + V - 1| = |c|$$

$$\Rightarrow \qquad |Y^2 + XY - X^2| = |c|$$

Replacing X by $x + 1$ and Y by $y + 1$, we get

$$|y^2 + xy - x^2 - x + 3y| = |c|$$

$$\Rightarrow \qquad (y^2 + xy - x^2 - x + 3y)^2 = c^2$$

3. Solve the equation $(x - 2y - 1)\,dx + (3x - 6y + 2)\,dy = 0$.

Solution. The given equation can be written as

$$\frac{dy}{dx} = -\frac{x - 2y - 1}{3x - 6y + 2} \qquad (1)$$

Here $\Delta = \begin{vmatrix} 1 & -2 \\ 3 & -6 \end{vmatrix} = 0$. Therefore, the given equation can not be reduced to homogeneous form. In order to find the solution we use the substitution $x - 2y = v$. Then $1 - 2\frac{dy}{dx} = \frac{dv}{dx}$. Therefore, Eq. (1) reduces to

$$1 - \frac{dv}{dx} = -\frac{2(v - 1)}{3v + 2}$$

$$\Rightarrow \qquad \frac{dv}{dx} = \frac{5v}{3v + 2}$$

Now, the variables are separable. Therefore, on integrating, we get

$$\int 5\,dx = \int \frac{(3v + 2)}{v}\,dv + c'$$

$$\Rightarrow \quad 5x = 3v + 2 \log |v| + c'$$

$$\Rightarrow \quad 5x - 3(x - 2y) = 2 \log |x - 2y| + c'$$

$$\Rightarrow \quad 2x + 6y = 2 \log |x - 2y| + c'$$

$$\Rightarrow \quad x + 3y = \log |x - 2y| + c \quad \left(\text{where } c = \frac{c'}{2}\right)$$

EXERCISES

Solve the following differential equations:

1. $(x - y)\, dy = (x + y + 1)\, dx$
2. $(x - y - 2)\, dx + (x - 2y - 3)\, dy = 0$
3. $\dfrac{dy}{dx} = \dfrac{y - x + 1}{y + x + 5}$
4. $\dfrac{dy}{dx} = \dfrac{x + y + 1}{2x + 2y + 3}$
5. $\dfrac{dy}{dx} = \dfrac{x + 2y - 3}{2x + y - 3}$
6. $(2x + 3y - 5) \dfrac{dy}{dx} + 3x + 2y - 5 = 0$
7. $(x - y - 2)\, dx - (2x - 2y - 3)\, dy = 0$
8. $(2x + 4y + 3) \dfrac{dy}{dx} = 2y + x + 1$
9. $\dfrac{dy}{dx} = \dfrac{x - y + 3}{2x - 2y + 5}$
10. $\dfrac{dy}{dx} = \dfrac{3y + 2x + 4}{4x + 6y + 5}$
11. $(2x + y + 1)\, dx + (4x + 2y - 1)\, dy = 0$
12. $(6x + 2y - 10) \dfrac{dy}{dx} - 2x - 9y + 20 = 0$
13. $(2x + 2y + 1)\, dy = (x + y + 1)\, dx$
14. $\dfrac{dy}{dx} = \dfrac{x - 2y + 3}{2x - 4y + 5}$
15. $\dfrac{dy}{dx} = \dfrac{7x - 3y - 7}{7y - 3x + 3}$
16. $\dfrac{dy}{dx} = \dfrac{x + 7y + 2}{3x + 5y + 6}$

10.8 Exact Differential Equations

An equation of the form

$$M\, dx + N\, dy = 0$$

where M and N are functions of x and y, is said to be *exact* (or *total*) *differential equation* if the left hand side is a total differential of some function $u(x, y)$, i.e.

$$M\, dx + N\, dy \equiv du \equiv \frac{\partial u}{\partial x} dx + \frac{\partial u}{\partial y} dy$$

10.8.1 Condition for an Equation of First Order be Exact

The necessary and sufficient condition that the differential equation

$$M\, dx + N\, dy = 0 \text{ be exact is } \frac{\partial M}{\partial y} = \frac{\partial N}{\partial x}$$

Assume that the equation $M\, dx + N\, dy = 0$ be exact. Then $M\, dx + N\, dy$ must be a total differential of some function $u(x, y)$, i.e.

$$du = Mdx + Ndy \tag{1}$$

But $$du = \frac{\partial u}{\partial x}dx + \frac{\partial u}{\partial y}dy \tag{2}$$

Therefore, on comparing Eqs. (1) and (2), we get

$$M = \frac{\partial u}{\partial x} \quad \text{and} \quad N = \frac{\partial u}{\partial y}$$

Now $\dfrac{\partial M}{\partial y} = \dfrac{\partial^2 u}{\partial y \partial x}$ and $\dfrac{\partial N}{\partial x} = \dfrac{\partial^2 u}{\partial x \partial y}$. But $\dfrac{\partial^2 u}{\partial y \partial x} = \dfrac{\partial^2 u}{\partial x \partial y}$.

Therefore $$\frac{\partial M}{\partial y} = \frac{\partial N}{\partial x}$$

This proves that the condition is necessary.

Further, let $\dfrac{\partial M}{\partial y} = \dfrac{\partial N}{\partial x}$. We shall prove that the equation $Mdx + Ndy = 0$ is exact.

Assume that $\int M\,dx = U$. Then $\dfrac{\partial U}{\partial x} = M$ and hence by hypothesis

$$\frac{\partial^2 U}{\partial y \partial x} = \frac{\partial M}{\partial y} = \frac{\partial N}{\partial x}$$

which gives $$\frac{\partial N}{\partial x} = \frac{\partial}{\partial x}\left(\frac{\partial U}{\partial y}\right)$$

On integrating, we get $N = \dfrac{\partial U}{\partial y}$ + Terms not containing x

$\Rightarrow$ $$N = \frac{\partial U}{\partial y} + f(y)$$

where $f(y)$ is a function of y alone. Now

$$M\,dx + N\,dy = \frac{\partial U}{\partial x}dx + \left(\frac{\partial U}{\partial y} + f(y)\right)dy$$

$$= d(U + \int f(y)dy) = d(U + F(y))$$

where $F(y)$ is some other function of y alone. Hence, this proves that $M\,dx + N\,dy = 0$ is an exact differential equation.

10.8.2 Rule for Finding the Solution of an Exact Differential Equation of the form $M\,dx + N\,dy = 0$

From above, we have

$$M\,dx + N\,dy = d\,(U + F(y))$$

Therefore, the solution of the equation $M\,dx + N\,dy = 0$ is $U + F(y) = c$.

But $U = \int M dx$ and $F(y) = \int f(y)\,dy$, where $N = \dfrac{\partial U}{\partial y} + f(y)$, i.e.

$$f(y) = N - \frac{\partial U}{\partial y} = N - \frac{\partial}{\partial y}\int M\,dx$$

Hence, the required solution is

$$\int M\,dx + \int\left(N - \frac{\partial}{\partial y}\int M\,dx\right)dy = c$$

Similarly $$\int N\,dy + \int\left(M - \frac{\partial}{\partial x}\int N\,dy\right)dx = c$$

is also a solution.

Working Rule. First integrate M with respect to x by keeping y constant. Then, integrate N with respect to y and retain only those terms which have not already appeared in the integration of M. The sum of the expressions thus obtained equated to an arbitrary constant gives the required solution.

Examples

1. Solve the equation $(3x^2 + 6xy^2)\,dx + (6x^2y + 4y^3)\,dy = 0$.

Solution. Here $M = 3x^2 + 6xy^2$ and $N = 6x^2y + 4y^3$.

Now $$\frac{\partial M}{\partial y} = 12xy \quad \text{and} \quad \frac{\partial N}{\partial x} = 12xy$$

Since $\dfrac{\partial M}{\partial y} = \dfrac{\partial N}{\partial x}$, the given equation is exact. Therefore, integrating M with respect to x by keeping y constant, we get $x^3 + 3x^2y^2$.

Further, integrating N with respect to y by keeping x constant, we get $3x^2y^2 + y^4$.

The term $3x^2y^2$ has already occurred in the integration of M. Therefore, the only terms obtained on integrating N with respect to y is y^4. Hence, the required solution of the given equation is

$$x^3 + 3x^2y^2 + y^4 = c$$

2. Solve the equation

$$\left(\sin y + y \sin x + \frac{1}{x}\right)dx + \left(x \cos y - \cos x + \frac{1}{y}\right)dy = 0$$

Solution. Here $M = \sin y + y \sin x + \dfrac{1}{x}$ and $N = x \cos y - \cos x + \dfrac{1}{y}$. Now

$$\frac{\partial M}{\partial y} = \cos y + \sin x \text{ and } \frac{\partial N}{\partial x} = \cos y + \sin x$$

Since $\frac{\partial M}{\partial y} = \frac{\partial N}{\partial x}$, the given equation is exact. Therefore, integrating M with respect to x by keeping y constant, we get

$$x \sin y - y \cos x + \log |x|$$

Further, integrating N with respect to y by keeping x constant, we get

$$x \sin y - y \cos x + \log |y|$$

The term $x \sin y - y \cos x$ has already occurred in the integration of M. Therefore, the only term obtained on integrating N with respect to y is $\log |y|$. Hence the required solution of the given equation is

$$x \sin y - y \cos x + \log |x| + \log |y| = c$$

$$\Rightarrow \quad x \sin y - y \cos x + \log |xy| = c$$

EXERCISES

Solve the following differential equations:

1. $(x^2 - ay)\, dx = (ax - y^2)\, dy$
2. $\dfrac{dy}{dx} + \dfrac{ax + hy + g}{hx + by + f} = 0$
3. $(e^y + 1) \cos x\, dx + e^y \sin x\, dy = 0$
4. $(x^3 - 3xy^2)\, dx - (3x^2y + y^3)\, dy = 0$
5. $(x^3 + 3xy^2)\, dx + (3x^2y + y^3)\, dy = 0$
6. $(a^2 - 2xy - y^2)\, dx - (x + y)^2\, dy = 0$
7. $\left(1 + e^{\frac{x}{y}}\right) dx + e^{\frac{x}{y}} \left(1 - \frac{x}{y}\right) dy = 0$
8. $\left(y + \frac{y}{x} + \cos y\right) dx + (x + \log x - x \sin y)\, dy = 0, \ (x > 0)$
9. $\cos x (\cos x - \sin \alpha \sin y)\, dx + \cos y (\cos y - \sin \alpha \sin x)\, dy = 0$
10. $(\sin xy + xy \cos xy)\, dx + x^2 \cos xy\, dy = 0$
11. $(x^3 + xy^2)\, dx + (x^2y + y^3)\, dy = 0$
12. $\left(\frac{x}{\sqrt{x^2 + y^2}} + \frac{1}{x} + \frac{1}{y}\right) dx + \left(\frac{y}{\sqrt{x^2 + y^2}} + \frac{1}{y} - \frac{x}{y^2}\right) dy = 0$
13. $\left(2x + \frac{x^2 + y^2}{x^2 y}\right) dx = \frac{(x^2 + y^2)}{xy^2}\, dy$
14. $(3x^2 - 2x - y)\, dx + (2y - x + 3y^2)\, dy = 0$
15. $(3x^2y + y^3)\, dx + (x^3 + 3xy^2)\, dy = 0$
16. $x(3xy - 4y^3 + 6)\, dx + (x^3 - 6x^2y^2 - 1)\, dy = 0$
17. $(\cos 2y - 3x^2y^2)\, dx + (\cos 2y - 2x \sin 2y - 2x^3y)\, dy = 0$
18. $(y^2 - 2xy + 6x)\, dx - (x^2 - 2xy + 2)\, dy = 0$

10.9 Integrating Factors

Given the differential equation $M\,dx + N\,dy = 0$, if $\dfrac{\partial M}{\partial y} = \dfrac{\partial N}{\partial x}$, then the equation is exact and one can find its solution by the method of Sec. 10.8. But if $\dfrac{\partial M}{\partial y} \neq \dfrac{\partial N}{\partial x}$, then the equation is not exact and method of Sec. 10.8 does not apply. What shall we do in such a case? Perhaps we can multiply the non-exact equation by some expression which will transform it into an equivalent exact equation. If so, we can proceed to solve the resulting exact equation by the known method. The expression (or function), which makes the equation exact, is called an *integrating factor*.

Suppose the differential equation

$$M\,dx + N\,dy = 0 \tag{1}$$

is not exact and that μ is an integrating factor of Eq. (1). Then,

$$\mu\,M\,dx + \mu\,N\,dy = 0 \tag{2}$$

is exact equation. Now, using the criterion for exactness, Eq. (2) is exact if and only if

$$\frac{\partial}{\partial y}(\mu M) = \frac{\partial}{\partial x}(\mu N)$$

This equation reduces to

$$N\frac{\partial \mu}{\partial x} - M\frac{\partial \mu}{\partial y} = \left(\frac{\partial M}{\partial y} - \frac{\partial N}{\partial x}\right)\mu \tag{3}$$

Thus μ is an integrating factor of Eq. (1) if and only if it is a solution of Eq. (3) which is a partial differential equation for the general integrating factor μ and we are not in a position to attempt to solve such an equation. Let us instead attempt to determine integrating factors of certain special types.

10.9.1 Special Integrating Factors

Type 1. When $Mx + Ny \neq 0$ and the equation is homogeneous, then $\dfrac{1}{Mx + Ny}$ is an integrating factor of the differential equation $M\,dx + N\,dy = 0$.

Proof. For $x, y > 0$, we have

$$M\,dx + N\,dy = \frac{1}{2}\left[(Mx + Ny)\left(\frac{dx}{x} + \frac{dy}{y}\right) + (Mx - Ny)\left(\frac{dx}{x} - \frac{dy}{y}\right)\right]$$

$$\Rightarrow \quad M dx + N dy = \frac{1}{2}\left[(Mx + Ny)\,d(\log xy) + (Mx - Ny)\,d\left(\log\frac{x}{y}\right)\right]$$

$$\Rightarrow \quad \frac{Mdx + Ndy}{Mx + Ny} = \frac{1}{2} d(\log xy) + \frac{1}{2} \frac{(Mx - Ny)}{(Mx + Ny)} d\left(\log \frac{x}{y}\right)$$

But $Mx + Ny$ is a homogeneous expression. Therefore, $\dfrac{Mx - Ny}{Mx + Ny}$ is also a homogeneous expression and hence a function of $\dfrac{x}{y}$. Since $\dfrac{x}{y} = e^{\log \frac{x}{y}}$, we have

$$\frac{Mdx + Ndy}{Mx + Ny} = \frac{1}{2} d(\log xy) + \frac{1}{2} \phi\left(\log \frac{x}{y}\right) d\left(\log \frac{x}{y}\right)$$

Clearly, this is an exact differential equation.

Note. If $Mx + Ny = 0$, then $\dfrac{M}{y} = -\dfrac{N}{x}$ so that the equation $M\,dx + N\,dy = 0$ reduces to

$$x\,dy - y\,dx = 0$$

The solution of this equation is $x = c\,y$.

Type 2. If the differential equation $M\,dx + N\,dy = 0$ has the form $f(xy)\,y\,dx + F(xy)\,x\,dy = 0$ and $Mx - Ny \neq 0$, then $\dfrac{1}{Mx - Ny}$ is an integrating factor.

Proof. For $x, y > 0$, we have

$$Mdx + Ndy = \frac{1}{2}\left[(Mx + Ny)d(\log xy) + (Mx - Ny)d\left(\log \frac{x}{y}\right)\right]$$

$$\Rightarrow \quad \frac{Mdx + Ny}{Mx - Ny} = \frac{1}{2} \frac{(Mx + Ny)}{(Mx - Ny)} d(\log xy) + \frac{1}{2} d\left(\log \frac{x}{y}\right)$$

But $\quad \dfrac{Mx + Ny}{Mx - Ny} = \dfrac{f(xy)xy + F(xy)xy}{f(xy)xy - F(xy)xy} = \phi(xy) = \psi(\log xy)$

Therefore $\quad \dfrac{Mdx + Ndy}{Mx - Ny} = \dfrac{1}{2} \psi(\log xy) d(\log xy) + \dfrac{1}{2} d\left(\log \dfrac{x}{y}\right)$

Hence $\quad \psi(\log xy)\, d(\log xy) + d\left(\log \dfrac{x}{y}\right) = 0$

is an exact differential equation.

Note. If $Mx - Ny = 0$, then $\dfrac{M}{y} = \dfrac{N}{x}$ and the equation $M\,dx + N\,dy = 0$ reduces to

$$y\,dx + x\,dy = 0$$

which on integration gives the solution $xy = c$.

Type 3. We recall that the linear differential equation

$$\frac{dy}{dx} + Py = Q$$

always possesses the integrating factor $e^{\int P\,dx}$, which depends on x only. Perhaps other equations also have integrating factors which depend only upon x. We, therefore, multiply equation $M\,dx + N\,dy = 0$ by μ when μ depends upon x alone and we get

$$\mu M\,dx + \mu N\,dy = 0$$

This equation is exact if and only if

$$\frac{\partial}{\partial y}(\mu M) = \frac{\partial}{\partial x}(\mu N)$$

Now, M and N are known functions of both x and y, but here the integrating factor μ depends only on x. Thus the above condition reduces to

$$\mu\frac{\partial M}{\partial y} = \mu\frac{\partial N}{\partial x} + N\frac{d\mu}{dx}$$

which gives
$$\frac{d\mu}{\mu} = \frac{1}{N}\left(\frac{\partial M}{\partial y} - \frac{\partial N}{\partial x}\right)dx \tag{4}$$

If $\frac{1}{N}\left(\frac{\partial M}{\partial y} - \frac{\partial N}{\partial x}\right)$ involves the variable y, Eq. (4) involves two dependent variables and we again have difficulties. However, if

$$\frac{1}{N}\left(\frac{\partial M}{\partial y} - \frac{\partial N}{\partial x}\right)$$

depends upon x only, then the variables are separable in Eq. (4). In this case we may integrate to obtain the integrating factor

$$\mu = e^{\int \frac{1}{N}\left(\frac{\partial M}{\partial y} - \frac{\partial N}{\partial x}\right)dx}$$

Similarly, if
$$\frac{1}{M}\left(\frac{\partial N}{\partial x} - \frac{\partial M}{\partial y}\right)$$

depends upon y only, then we may obtain an integrating factor which depends only on y.

We summarize these observations as follows:

(i) If $\frac{1}{N}\left(\frac{\partial M}{\partial y} - \frac{\partial N}{\partial x}\right)$ depends upon x only, then $e^{\int \frac{1}{N}\left(\frac{\partial M}{\partial y} - \frac{\partial N}{\partial x}\right)dx}$ is an integrating factor of Eq. (1).

(ii) If $\frac{1}{M}\left(\frac{\partial N}{\partial x} - \frac{\partial M}{\partial y}\right)$ depends upon y only, then $e^{\int \frac{1}{M}\left(\frac{\partial N}{\partial x} - \frac{\partial M}{\partial y}\right)dy}$ is an integrating factor of Eq. (1).

Remark. For a given differential equation, we have no assurance, in general, that either of the above methods will apply. It may well turn out that $\frac{1}{N}\left(\frac{\partial M}{\partial y} - \frac{\partial N}{\partial x}\right)$ involves y and $\frac{1}{M}\left(\frac{\partial N}{\partial x} - \frac{\partial M}{\partial y}\right)$ involves x for the differential equation under consideration. Then, we must seek other methods. However, since the calculation of the above expressions is generally quite simple, it is often worthwhile to calculate them before trying something more complicated.

Type 4. Let the differential equation be in the form

$$x^a y^b (my\,dx + nx\,dy) + x^r y^s (py\,dx + qx\,dy) = 0$$

where a, b, m, n, r, s, p, q are constants. Then, the integrating factor is $x^h y^k$, where h, k are obtained by applying the condition that after multiplication by $x^h y^k$, the equation must become exact.

Multiplying by $x^h y^k$ the given equation becomes

$$(mx^{a+h}y^{b+k+1} + px^{r+h}y^{s+k+1})\,dx + (nx^{a+h+1}y^{b+k} + qx^{r+h+1}y^{s+k})\,dy = 0$$

This equation will be exact if $\frac{\partial M}{\partial y} = \frac{\partial N}{\partial x}$, i.e.

$$m(b + k + 1)\,x^{a+h}y^{b+k} + p(s + k + 1)\,x^{r+h}\,y^{s+k}$$
$$= n(a + h + 1)\,x^{a+h}y^{b+k} + q(r + h + 1)\,x^{r+h}\,y^{s+k}$$

This will be true when

$$m(b + k + 1) = n(a + h + 1) \quad \text{and} \quad p(s + k + 1) = q(r + h + 1)$$

The required values of h and k are obtained by solving these equations.

Examples

1. Solve the equation $(3y + 4xy^2)\,dx + (2x + 3x^2y)\,dy = 0$.

Solution. Here $M = 3y + 4xy^2$ and $N = 2x + 3x^2y$.

Now $$\frac{\partial M}{\partial y} = 3 + 4xy \quad \text{and} \quad \frac{\partial N}{\partial x} = 2 + 6xy$$

Since $\frac{\partial M}{\partial y} \neq \frac{\partial N}{\partial x}$, the given equation is not exact.

Let $\mu = x^2 y$. Then the corresponding equation of the form $\mu M\, dx + \mu N\, dy = 0$ is

$$(3x^2y^2 + 4x^3y^3)\, dx + (2x^3y + 3x^4y^2)\, dy = 0$$

This equation is exact since

$$\frac{\partial}{\partial y}(\mu M) = 6x^2y + 12x^3y^2 = \frac{\partial}{\partial x}(\mu N)$$

Hence $\mu = x^2y$ is an integrating factor of the given equation. Now, integrating μM with respect to x by keeping y constant, we get

$$x^3y^2 + x^4y^3$$

Further, integrating μN with respect to y by keeping x constant, we get

$$x^3y^2 + x^4y^3$$

Hence the required solution is

$$x^3y^2 + x^4y^3 = c$$

2. Solve the equation $(2x + \tan y)\, dx + (x - x^2 \tan y)\, dy = 0$.

Solution. Here $M = 2x + \tan y$ and $N = x - x^2 \tan y$. Now

$$\frac{\partial M}{\partial y} = \sec^2 y \quad \text{and} \quad \frac{\partial N}{\partial x} = 1 - 2x \tan y$$

Since $\frac{\partial M}{\partial y} \neq \frac{\partial N}{\partial x}$, the equation is not exact. Consider

$$\frac{1}{M}\left(\frac{\partial N}{\partial x} - \frac{\partial M}{\partial y}\right) = \frac{1}{(2x + \tan y)}(1 - 2x \tan y - \sec^2 y) = -\tan y$$

which is a function of y alone. Therefore, integrating factor is

$$e^{\int (-\tan y) dy} = e^{\log \cos y} = \cos y$$

Multiplying the given equation by I.F. $\cos y$, we get

$$(2x \cos y + \sin y)\, dx + (x \cos y - x^2 \sin y)\, dy = 0$$

Now, the differential equation is exact. Henec the solution is

$$x^2 \cos y + x \sin y = c$$

3. Solve the equation $(5xy + 4y^2 + 1)\, dx + (x^2 + 2xy)\, dy = 0$.

Solution. Here $M = 5xy + 4y^2 + 1$ and $N = x^2 + 2xy$.

Now $$\frac{\partial M}{\partial y} = 5x + 8y \quad \text{and} \quad \frac{\partial N}{\partial x} = 2x + 2y$$

Since $\dfrac{\partial M}{\partial y} \neq \dfrac{\partial N}{\partial x}$, the equation is not exact. Consider

$$\frac{1}{N}\left(\frac{\partial M}{\partial y} - \frac{\partial N}{\partial x}\right) = \frac{1}{(x^2 + 2xy)}(5x + 8y - 2x - 2y) = \frac{3}{x}$$

which is a function of x only. Therefore, integrating factor is

$$e^{\int \frac{3}{x} dx} = x^3$$

Multiplying the given equation by I.F. x^3, we get

$$(5x^4y + 4x^3y^2 + x^3)\, dx + (x^5 + 2x^4y)\, dy = 0$$

Now, the differential equation is exact. Hence the solution is

$$4x^5y + 4x^4y^2 + x^4 = c$$

4. Solve the equation $(4xy^2 + 6y)\, dx + (5x^2y + 8x)\, dy = 0$.

Solution. This equation is not exact, but it can be written as

$$xy(4y\, dx + 5x\, dy) + (6y\, dx + 8x\, dy) = 0$$

Therefore, for this equation the integrating factor must be of the form x^hy^k. Hence multiplying the given equation by x^hy^k, we get

$$(4x^{h+1}y^{k+2} + 6x^hy^{k+1})\, dx + (5x^{h+2}y^{k+1} + 8x^{h+1}y^k)\, dy = 0$$

This equation will be exact if $\dfrac{\partial M}{\partial y} = \dfrac{\partial N}{\partial x}$, i.e. if

$$4(k + 2)\, x^{h+1}y^{k+1} + 6(k + 1)\, x^hy^k = 5(h + 2)x^{h+1}y^{k+1} + 8(h + 1)\, x^hy^k$$

Comparing the coefficients of $x^{h+1}y^{k+1}$ and x^hy^k from both sides, we get

$$4(k + 2) = 5(h + 2)$$

and $$6(k + 1) = 8(h + 1)$$

On solving these equations, we get $h = 2$, $k = 3$.

Multiplying the given equation by I.F. x^2y^3, we get

$$(4x^3y^5 + 6x^2y^4)\, dx + (5x^4y^4 + 8x^3y^3)\, dy = 0$$

This equation is exact. Hence the solution is

$$x^4y^5 + 2x^3y^4 = c$$

EXERCISES

Solve the following differential equations:

1. $y(2xy + e^x)\, dx - e^x dy = 0$
2. $x(x - y)\, dy = y(x + y)\, dx$
3. $y(axy + e^x)\, dx - e^x dy = 0$
4. $(xy^2 + 2x^2y^3)\, dx + (x^2y - x^3y^2)\, dy = 0$
5. $ydx - xdy + (1 + x^2)\, dx + x^2 \sin y\, dy = 0$
6. $y(2x^2y + e^x)\, dx - (e^x + y^3)\, dy = 0$
7. $y(1 + xy)\, dx + x(1 - xy)\, dy = 0$
8. $x^2y\, dx - (x^3 + y^3)\, dy = 0$
9. $(x^2y - 2xy^2)\, dx - (x^3 - 3x^2y)\, dy = 0$
10. $(xy \sin xy + \cos xy)\, y\, dx + (xy \sin xy - \cos xy)\, dy = 0$
11. $(x^2y^2 + xy + 1)\, y\, dx + (x^2y^2 - xy + 1)\, x\, dy = 0$
12. $(x^2 + y^2 + 2x)\, dx + 2y\, dy = 0$
13. $\left(y + \frac{y^3}{3} + \frac{x^2}{2}\right) dx + \frac{1}{4}(1 + y^2)\, x\, dy = 0$
14. $(2y\, dx + 3x\, dy) + 2xy\, (3y\, dx + 4x\, dy) = 0$
15. $(xy^3 + y)\, dx + 2\, (x^2y^2 + x + x^4)\, dy = 0$
16. $(xy^2 - x^2)\, dx + (3x^2y^2 + x^2y - 2x^3 + y^2)\, dy = 0$
17. $(2x^2y - 3y^4)\, dx + (3x^3 + 2xy^3)\, dy = 0$
18. $(3x + 2y^2)\, y\, dx + 2x\, (2x + 3y^2)\, dy = 0$
19. $(y^2 + 2x^2y)\, dx + (2x^3 - xy)\, dy = 0$

10.10 Change of Variables

Certain differential equations of order one may not yield at once to the various methods discussed so far. Even then one should not think that the usefulness of those methods has exhausted. It may be possible, in some cases, by some change of variables to transform the equation to one of the forms which we know.

Although there is no special technique of finding a suitable substitution to reduce an equation to a known form, only a closed observation of the differential equation itself may lead us to find a useful transformation. This device is called the change of the dependent or independent or both variables.

Examples

1. Solve the equation $(x + y + 1)\frac{dy}{dx} = 1.$

Solution. Putting $x + y = v$. Then $1 + \frac{dy}{dx} = \frac{dv}{dx}$ and the given equation becomes

$$(v + 1)\left(\frac{dv}{dx} - 1\right) = 1 \quad \Rightarrow \quad \frac{(v + 1)\,dv}{v + 2} = dx$$

Now, the variables are separable. Hence the solution is

$$v - \log |v + 2| = x + c$$

Substituting the value of v, we get

$$y = \log |x + y + 2| + c$$

2. Solve the equation $(x + y + 1)\,dx + (2x + 2y - 1)\,dy = 0$.

Solution. The given equation can be written as

$$\frac{dy}{dx} = -\frac{x + y + 1}{2x + 2y - 1} \tag{1}$$

Let $x + y = v$. Then $1 + \frac{dy}{dx} = \frac{dv}{dx}$ and Eq. (1) reduces to

$$\frac{dv}{dx} = \frac{v - 2}{2v - 1} \quad \Rightarrow \quad \frac{(2v - 1)}{v - 2}\,dv = dx$$

Now, the variables are separable. Therefore, on integrating, we get

$$2v + 3 \log |v - 2| = x + c$$

Substituting the value of v, we get

$$x + 2y + 3 \log |x + y - 2| = c$$

EXERCISES

Solve the following differential equations:

1. $(x + y)^2 \frac{dy}{dx} = a^2$
2. $\frac{dy}{dx} = \sin (x + y) + \cos (x + y)$
3. $(x - y)^2 \frac{dy}{dx} = a^2$
4. $\cos (x + y)\,dy = dx$
5. $\frac{dy}{dx} = (4x + y + 1)^2$
6. $\frac{dy}{dx} = \frac{e^y}{x^2} - \frac{1}{x}$
7. $\frac{dy}{dx} = e^{x-y}(e^x - e^y)$
8. $x\frac{dy}{dx} + y \log y = xy\,e^x$
9. $\frac{(x + y - a)}{(x + y - b)} \frac{dy}{dx} = \frac{x + y + a}{x + y + b}$
10. $\frac{dy}{dx} - \frac{\tan y}{1 + x} = (1 + x)e^x \sec y$
11. $\frac{x\,dx + y\,dy}{x\,dy - y\,dx} = \sqrt{\frac{a^2 - x^2 - y^2}{x^2 + y^2}}$
12. $(x - y^2)\,dx + 2xy\,dy = 0$
13. $\sin y \frac{dy}{dx} = \cos y\,(1 - x \cos y)$
14. $(2x + 3y - 5)\,dy + (2x + 3y - 1)\,dx = 0$

MISCELLANEOUS EXERCISES

Solve the following differential equations:

1. $x^2 \dfrac{dy}{dx} + y = 1$

2. $(x^2 - yx^2)\, dy + (y^2 + xy^2)\, dx = 0$

3. $\sqrt{a + x}\, \dfrac{dy}{dx} + x = 0$

4. $x \dfrac{dy}{dx} + y = x^2 + 3x + 2$

5. $\dfrac{dy}{dx} + \dfrac{2xy}{1 + x^2} = \dfrac{1}{(x^2 + 1)^2}$, given that $y = 0$ when $x = 1$.

6. $\dfrac{dy}{dx} = y \tan x - 2 \sin x$

7. $\sin 2x \dfrac{dy}{dx} = y + \tan x$

8. $x(1 - x^2)\, dy + (2x^2 y - y - ax^3)\, dx = 0$

9. $\dfrac{dy}{dx} = \dfrac{x^2 + xy}{x^2 + y^2}$

10. $(x - y)^2\, dx + 2xy\, dy = 0$

11. $x \cos\left(\dfrac{y}{x}\right)(y\, dx + x\, dy) = y \sin\left(\dfrac{y}{x}\right)(x\, dy - y\, dx)$

12. $(2x - 10y^3) \dfrac{dy}{dx} + y = 0$

13. $L \dfrac{di}{dt} + Ri = E$, where L, R and E are constants; when $t = 0$, $i = 0$.

14. $\dfrac{dy}{dx} + \dfrac{y}{x} = y^2 \sin x$

15. $x^2 y - x^3 \dfrac{dy}{dx} = y^4 \cos x$

16. $2 \dfrac{dy}{dx} - y \sec x = y^3 \tan x$

17. $6y^2 dx - x\,(2x^3 + y)\, dy = 0$

18. $y \sin 2x\, dx - (1 + y^2 + \cos^2 x)\, dy = 0$

19. $(r + \sin\theta - \cos\theta)\, dr + r\,(\sin\theta + \cos\theta)\, d\theta = 0$

20. $(y - xy^2)\, dx - (x + x^2 y)\, dy = 0$

21. $(3x^2 y^4 + 2xy)\, dx + (2x^3 y^3 - x^2)\, dy = 0$

22. $x\,(3y\, dx + 2x\, dy) + 8y^4\,(y\, dx + 3x\, dy) = 0$

23. $(y^3 - 2x^2 y)\, dx + (2xy^2 - x^3)\, dy = 0$

24. $(x^3 e^x - my^2)\, dx + mxy\, dy = 0$

25. $x(x^2 + 1) \dfrac{dy}{dx} = y\,(1 - x^2) + x^3 \log x$

26. $x^2 y \dfrac{dy}{dx} = xy^2 - e^{-\frac{1}{x^3}}$

27. $(x^2 + y^2 + a^2)\, y \dfrac{dy}{dx} + x\,(x^2 + y^2 - a^2) = 0$

28. $x \cos x \dfrac{dy}{dx} + y\,(x \sin x + \cos x) = 1$

29. $y^2 \sin x\, dx + \cos^2 x \log y\, dy = 0$

30. $e^x \sin^3 y + (1 + e^{2x}) \cos y \dfrac{dy}{dx} = 0$

31. $x\sqrt{1 - y^2}\, dx + y\sqrt{1 - x^2}\, dy = 0$, $y = 1$ when $x = 0$.

32. $x \dfrac{dy}{dx} = y(\log y - \log x)$

33. $2x^2 \dfrac{dy}{dx} = x^2 + y^2$

34. $\left(3x^2 \tan y - \dfrac{2y^3}{x^3}\right) dx + \left(x^3 \sec^2 y + 4y^3 + \dfrac{3y^2}{x^2}\right) dy = 0$

35. $\left(\dfrac{y + \sin x \cos^2 xy}{\cos^2 xy}\right) dy + \left(\dfrac{x + \sin y \cos^2 xy}{\cos^2 xy}\right) dy = 0$

36. $(2x^3 - xy^2 - 2y + 3)\, dx - (x^2y + 2x)\, dy = 0$

37. $3y(x^2 - 1)\, dx + (x^3 + 8y - 3x)\, dy = 0$

38. $(xy^2 + x - 2y + 3)\, dx + x^2y\, dy = 2\,(x + y)\, dy$, when $x = 1$, $y = 1$.

39. $(3 + y + 2y^2 \sin^2 x)\, dx + (x + 2xy - y \sin 2x)\, dy = 0$.

40. $(ye^{xy} - 2y^3)\, dx + (xe^{xy} - 6xy^2 - 2y)\, dy = 0$; when $x = 0$, $y = 2$.

41. $y(x^2 + y^2)\, dx + x\,(3x^2 - 5y^2)\, dy = 0$; when $x = 2$, $y = 1$.

42. Show that the curve whose subtangent is m times the abscissa of the point of contact is $cx = y^m$.

43. Show that the curve which is such that the portion of the x-axis cut off by the tangent at its any point is proportional to the ordinate of that point is $y = ce^{\frac{-x}{ky}}$.

11

Equations of the First Order and Higher Degree

11.1 Introduction

In this chapter we shall consider the equations of the first order but not of first degree. To find the solution of such equations we usually denote *dy/dx* by *p* and divide them into three categories as follows:

1. Equations solvable for *p*.
2. Equations solvable for *y*.
3. Equations solvable for *x*.

Apart from these there are certain equations which can be put in a particular form known as Clairaut's form and their solutions can be obtained easily.

11.2 Equations Solvable for *p*

An equation of the first order and degree *n* can be considered as

$$p^n + f_1(x, y)\, p^{n-1} + f_2(x, y)p^{n-2} + \ldots + f_{n-1}(x, y)\, p + f_n(x, y) = 0 \quad (1)$$

We solve this equation for *p*. Let Eq. (1) be factorized as

$$(p - \phi_1(x, y))(p - \phi_2(x, y)) \ldots (p - \phi_n(x, y)) = 0$$

Equating each factor to zero and the resulting equations are of the first order and first degree which can be solved by known methods. Let the solutions of the equations be

$$F_1(x, y, c_1) = 0, \quad F_2(x, y, c_2) = 0, \ldots, F_n(x, y, c_n) = 0 \quad (2)$$

Hence, solution of Eq. (1) can be written as

$$F_1(x, y, c)\, F_2(x, y, c) \ldots F_n(x, y, c) = 0 \quad (3)$$

Remark. There is no loss of generality in replacing the arbitrary constants $c_1, c_2, \ldots, c_n$ by a single arbitrary constant *c*, because every particular solution obtained from Eqs. (2) can also be obtained from Eq. (3) by giving a suitable value to c.

Examples

1. Solve the equation $p^2 - 2px - 8x^2 = 0$.

Solution. Solving the equation for p, we get

$$p = \frac{2x \pm \sqrt{4x^2 + 32x^2}}{2} = \frac{2x \pm 6x}{2}$$

$$\Rightarrow \qquad p = -2x,\ p = 4x$$

Therefore, the corresponding solutions are

$$y = -x^2 + c, \quad y = 2x^2 + c$$

Hence the required solution is

$$(y + x^2 - c)\,(y - 2x^2 - c) = 0$$

2. Solve the equation $xp^2 + (y - 1 - x^2)\,p - x(y - 1) = 0$.

Solution. The given equation can be factorized as

$$(p - x)(xp + y - 1) = 0$$

i.e. $$p - x = 0,\ xp + y - 1 = 0$$

Therefore, the corresponding solutions are

$$2y - x^2 - c = 0, \quad xy - x - c = 0$$

Hence the required solution is

$$(2y - x^2 - c)\,(xy - x - c) = 0$$

3. Solve the equation $p^3 - yp^2 - x^2p + x^2y = 0$.

Solution. The given equation can be factorized as

$$(p - x)\,(p + x)\,(p - y) = 0$$

$$\Rightarrow \qquad p - x = 0,\ p + x = 0,\ p - y = 0$$

Therefore, the corresponding solutions are

$$2y - x^2 - c = 0, \quad 2y + x^2 - c = 0,\ y - ce^x = 0$$

Hence the required solution is

$$(2y - x^2 - c)\,(2y + x^2 - c)\,(y - ce^x) = 0$$

4. Solve the equation $4y^2p^2 + 2pxy(3x + 1) + 3x^3 = 0$.

Solution. The given equation can be factorized as

$$(2yp + x)(2yp + 3x^2) = 0$$

$$\Rightarrow \qquad 2yp + x = 0, \;\; 2yp + 3x^2 = 0$$

Therefore, the corresponding solutions are

$$2y^2 + x^2 - c = 0, \quad y^2 + x^3 - c = 0$$

Hence the required solution is

$$(2y^2 + x^2 - c)(y^2 + x^3 - c) = 0$$

5. Solve the equation $xyp^2 + (x + y)\, p + 1 = 0$.

Solution. The given equation can be factorized as

$$(xp + 1)(yp + 1) = 0$$

$$\Rightarrow \qquad xp + 1 = 0, \quad yp + 1 = 0$$

Therefore, the corresponding soluutions are

$$y + \log|x| - c = 0, \quad y^2 + 2x - c = 0$$

Hence the required solution is

$$(y + \log|x| - c)\,(y^2 + 2x - c) = 0$$

EXERCISES

Solve the following differential equations:

1. $p^2 - 5p + 6 = 0$
2. $p^2 - 9p + 18 = 0$
3. $p^2x^2 - xyp - y^2 = 0$
4. $p^2 + 2py \cot x - y^2 = 0$
5. $x^2p^3 + y(1 + x^2y)p^2 + y^3p = 0$
6. $p^2 + px + py + xy = 0$
7. $p^2 + 2xp - 3x^2 = 0$
8. $x + yp^2 = p(1 + xy)$
9. $x^2p^2 + xyp - 6y^2 = 0$
10. $p^2 - 2p \cosh x + 1 = 0$
11. $xyp^2 - (x^2 - y^2)\, p - xy = 0$
12. $p^3 - p(x^2 + xy + y^2) + xy(x + y) = 0$
13. $x\left(\dfrac{dy}{dx}\right)^2 + (y - x)\dfrac{dy}{dx} - y = 0$
14. $x^2\left(\dfrac{dy}{dx}\right)^2 + 3xy\dfrac{dy}{dx} + 2y^2 = 0$
15. $y\left(\dfrac{dy}{dx}\right)^2 + (x - y)\dfrac{dy}{dx} - x = 0$
16. $x^2\left(\dfrac{dy}{dx}\right)^2 - 2xy\dfrac{dy}{dx} + 2y^2 - x^2 = 0$
17. $p^3 + 2p^2x - p^2y^2 - 2p\, xy^2 = 0$
18. $p^2 + (x - e^x)p - xe^x = 0$
19. $xyp^2 + (x^2 + xy + y^2)p + x^2 + xy = 0$

11.3 Equations Solvable for y

If the differential equation is solvable for y, it can be put in the form

$$y = f(x, p) \tag{1}$$

Differentiating this equation w.r.t. x, we get

$$p = \frac{dy}{dx} = \frac{\partial f}{\partial x} + \frac{\partial f}{\partial p} \cdot \frac{dp}{dx} = \phi\left(x, p, \frac{dp}{dx}\right)$$

This is a differential equation in two variables p and x. Let a possible solution of this equation be

$$F(x, p, c) = 0 \tag{2}$$

The elimination of p between Eqs. (1) and (2) gives the required solution.

Notes

1. In case the elimination of p between Eqs. (1) and (2) is not possible easily, the required solution is obtained by expressing x and y in terms of a parameter p. Thus the solution can be expressed as

$$x = F_1(p, c), \quad y = F_2(p, c)$$

2. The above method is specially useful for differential equations in which x is entirely absent.

Examples

1. Solve the equation $y = 3x + \log p$.

Solution. Differentiating the given equation w.r.t. x, we get

$$p = 3 + \frac{1}{p}\frac{dp}{dx}$$

$$\Rightarrow \qquad dx = \frac{1}{p(p-3)}\, dp$$

Since the variables are separable, on integrating, we get

$$x + c_1 = \frac{1}{3}\log\frac{p-3}{p}$$

$$\Rightarrow \qquad p = \frac{3}{1 - ce^{3x}} \tag{1}$$

Eliminating p from Eq. (1) and the given equation, the required solution is obtained as

$$y = 3x + \log\left|\frac{3}{1 - ce^{3x}}\right|$$

2. Solve the equation $y = (1 + p)x + p^2$.

Solution. Differentiating the given equation w.r.t. x, we get

$$p = 1 + p + x\frac{dp}{dx} + 2p\frac{dp}{dx}$$

$$\Rightarrow \qquad \frac{dx}{dp} + x = -2p \qquad (1)$$

This equation is linear in x. Therefore

$$\text{I.F.} = e^{\int dp} = e^p$$

Multiplying Eq. (1) by the I.F. e^p, we get

$$e^p\frac{dx}{dp} + xe^p = -2pe^p$$

On integrating, we get

$$xe^p = 2(1-p)e^p + c$$

$$\Rightarrow \qquad x = 2(1-p) + c\,e^{-p} \qquad (2)$$

Substituting this value of x in the given equation, we get

$$y = 2 - p^2 + c(1+p)\,e^{-p} \qquad (3)$$

Equations (2) and (3) together give the required solution.

EXERCISES

Solve the following differential equations:

1. $y = 2px - p^2$
2. $y = -px + x^4p^2$
3. $y = \sin p - p\cos p$
4. $y = ap^2 + bp^3$
5. $y = p\sin p + \cos p$
6. $p^3 + p = e^y$
7. $y = p\log p$
8. $y = a + bp + cp^2$
9. $y = xp^2 + p$
10. $y = p(1 + p\cos p)$
11. $e^{p-y} = p^2 - 1$
12. $y = p\tan y + \log\cos p$

11.4 Equations Solvable for x

If the differential equation is solvable for x, it can be put in the form

$$x = f(y, p) \qquad (1)$$

Differentiating this equation w.r.t. y, we get

$$\frac{1}{p} = \frac{\partial f}{\partial y} + \frac{\partial f}{\partial p}\cdot\frac{dp}{dy} = \phi\left(y, p, \frac{dp}{dy}\right)$$

This is a differential equation in two variables p and y. Let a possible solution of this equation be

$$F(y, p, c) = 0 \tag{2}$$

The elimination of p between Eqs. (1) and (2) gives the required solution.

Notes

1. In case the elimination of p between Eqs. (1) and (2) is not possible easily, the required solution is obtained by expressing x and y in terms of a parameter p.
2. This method is specially useful for differential equations in which y is entirely absent.

Examples

1. Solve the equation $x = y + p^2$.

Solution. Differentiating the given equation w.r.t. y, we get

$$\frac{1}{p} = 1 + 2p\frac{dp}{dy}$$

$$\Rightarrow \qquad dy = \frac{2p^2}{1-p}\,dp = -2\left(p + 1 + \frac{1}{p-1}\right)dp$$

Since the variables are separable, on integrating, we get

$$y = -2\left[\frac{p^2}{2} + p + \log(p-1)\right] + c \tag{1}$$

Substituting this value of y in the given equation, we get

$$x = -2\left[\frac{p^2}{2} + p + \log(p-1)\right] + c + p^2 \tag{2}$$

Equations (1) and (2) together give the required solution.

2. Solve the equation $x = yp + ap^2$.

Solution. Differentiating the given equation w.r.t. y, we get

$$\frac{1}{p} = p + y\frac{dp}{dy} + 2ap\frac{dp}{dy}$$

$$\Rightarrow \qquad \frac{dy}{dp} - \frac{py}{1-p^2} = \frac{2ap^2}{1-p^2} \tag{1}$$

This equation is linear in y. Therefore

$$\text{I.F.} = e^{\int \frac{-p}{1-p^2}dp} = \sqrt{1-p^2}$$

Multiplying Eq. (1) by the I.F. $\sqrt{1-p^2}$, we get

$$\sqrt{1-p^2}\,\frac{dy}{dp} - \frac{py}{\sqrt{1-p^2}} = \frac{2ap^2}{\sqrt{1-p^2}}$$

On integrating, we get

$$y\sqrt{1-p^2} = a\sin^{-1}p - ap\sqrt{1-p^2} + c$$

$$\Rightarrow \qquad y = -ap + \frac{a\sin^{-1}p + c}{\sqrt{1-p^2}} \qquad (2)$$

Substituting this value of y in the given equation, we get

$$x = \frac{p}{\sqrt{1-p^2}}(a\sin^{-1}p + c) \qquad (3)$$

Equations (2) and (3) together give the required solution.

3. Solve the equation $y = 3px + y^2p^2$.

Solution. Solving the given equation for x, we get

$$3x = \frac{y}{p} - y^2 p$$

Differentiating this equation w.r.t. y, we get

$$\frac{3}{p} = \frac{1}{p} - \frac{y}{p^2}\frac{dp}{dy} - 2yp - y^2\frac{dp}{dy}$$

$$\Rightarrow \qquad \left(2 + \frac{y}{p}\frac{dp}{dy}\right)\left(\frac{1}{p} + yp\right) = 0$$

Omitting the second factor $\frac{1}{p} + yp = 0$, we get

$$2 + \frac{y}{p}\frac{dp}{dy} = 0$$

$$\Rightarrow \qquad \frac{2}{y}dy + \frac{1}{p}dp = 0$$

On integrating, we get

$$y^2p = c$$

$$\Rightarrow \qquad p = \frac{c}{y^2}$$

Substituting this value of p in the given equation the required solution is obtained as

$$y^3 = 3cx + c^2$$

4. Solve the equation $y^2 \log y = xyp + p^2$, $(y > 0)$.

Solution. Solving the given equation for x, we get

$$x = \frac{y \log y}{p} - \frac{p}{y}$$

Differentiating this equation w.r.t. y, we get

$$\frac{1}{p} = \frac{1}{p}(1 + \log y) - \frac{(y \log y)}{p^2}\frac{dp}{dy} - \frac{1}{y}\frac{dp}{dy} + \frac{p}{y^2}$$

$$\Rightarrow \qquad \left(p - y\frac{dp}{dy}\right)(y^2 \log y + p^2) = 0$$

Omitting the second factor $y^2 \log y + p^2 = 0$*, we get

$$p - y\frac{dp}{dy} = 0$$

On integrating, we get

$$p = cy$$

Substituting this value of p in the given equation the required solution is obtained as

$$\log y = cx + c^2, \quad (y > 0)$$

EXERCISES

Solve the following differential equations:

1. $x = ap + bp^2$
2. $p^2 - 2xp + 1 = 0$
3. $x = \log p + \sin p$
4. $x = p + \sin p$
5. $xp^3 = a + bp$
6. $x(1 + p^2)^{3/2} = a$
7. $p^3 - p(y + 3) + x = 0$
8. $x = y + a \log p$

11.5 The Clairaut Equation

An equation of the form

$$y = xp + f(p) \tag{1}$$

is known as *Clairaut equation.*

---

*The factor $y^2 \log y + p^2 = 0$ gives the singular solution of the equation.

To solve this equation, differentiate it w.r.t. x to get

$$p = p + (x + f'(p)) \frac{dp}{dx}$$

$$\Rightarrow \qquad (x + f'(p)) \frac{dp}{dx} = 0$$

$$\Rightarrow \qquad x + f'(p) = 0 \text{ or } \frac{dp}{dx} = 0$$

The second equation, on integrating, gives

$$p = c \tag{2}$$

Eliminating p between Eqs. (1) and (2), the required solution is obtained as

$$y = cx + f(c)$$

Remark. If we eliminate p between $x + f'(p) = 0$ and the original Eq. (1), we obtain a solution which does not contain any arbitrary constant, and is not a particular case of the solution $y = cx + f(c)$. Such a solution is called a *singular solution*.

Note. Sometimes an equation can be reduced to Clairaut equation by a suitable substitution.

11.6 The Lagrange Equation

An equation of the form

$$y = x\,\phi(p) + f(p)$$

is known as *Lagrange equation*.

To solve this equation, differentiate it w.r.t. x to get

$$p = \phi(p) + x\phi'(p) \frac{dp}{dx} + f'(p) \frac{dp}{dx}$$

$$\Rightarrow \qquad \frac{dx}{dp} + \frac{x\phi'(p)}{\phi(p) - p} = -\frac{f'(p)}{\phi(p) - p}$$

This equation is linear in x. Solving this equation, we get

$$x = \psi(p, c)$$

Hence, we obtain the general solution of the original equation in the parametric form

$$x = \psi(p, c),\ y = \psi(p, c)\,\phi(p) + f(p)$$

p being a parameter.

Note. The Lagrange and Clairaut equations are particular cases of the equation considered in Sec. 11.3.

Examples

1. Solve the equation $y = xp + a\sqrt{1 + p^2}$.

Solution. This equation is Clairaut equation. Hence, the solution is

$$y = cx + a\sqrt{1 + c^2}$$

where c is an arbitrary constant.

2. Solve the equation $y = xp^2 - \dfrac{1}{p}$.

Solution. This equation is Lagrange equation. Differentiating it wr.t. x, we get

$$p = p^2 + \left(2xp + \frac{1}{p^2}\right)\frac{dp}{dx}$$

$$\Rightarrow \qquad \frac{dx}{dp} + \frac{2x}{p-1} = -\frac{1}{p^3(p-1)} \tag{1}$$

This equation is linear in x. Therefore

$$\text{I.F.} = e^{\int \frac{2}{p-1}dp} = e^{2\log(p-1)} = e^{\log(p-1)^2} = (p-1)^2$$

Multiplying Eq. (1) by the integrating factor $(p - 1)^2$, we get

$$(p-1)^2\frac{dx}{dp} + 2(p-1)x = -\frac{p-1}{p^3}$$

On integrating, we get

$$x(p-1)^2 = c_1 + \frac{1}{p} - \frac{1}{2p^2}$$

$$\Rightarrow \qquad x(p-1)^2 = \frac{2c_1p^2 + 2p - 1}{2p^2}$$

$$\Rightarrow \qquad x = \frac{cp^2 + 2p - 1}{2p^2(p-1)^2} \tag{2}$$

Substituting this value of x in the given equation, we get

$$y = \frac{cp^2 + 2p - 1}{2(p-1)^2} - \frac{1}{p} \tag{3}$$

Equations (2) and (3) together give the required solution.

3. Solve the equation $(px - y)(py + x) = h^2p$.

Solution. Let $x^2 = u$ and $y^2 = v$. Then

$$2x\,dx = du \text{ and } 2y\,dy = dv$$

Therefore $\dfrac{2ydy}{2xdx} = \dfrac{dv}{du} \Rightarrow \dfrac{y}{x}p = \dfrac{dv}{du} \quad \Rightarrow \quad p = \dfrac{x}{y}\dfrac{dv}{du} = \dfrac{u^{1/2}}{v^{1/2}}\dfrac{dv}{du}$

Substituting these values in the given equation, it reduces to

$$\left(\frac{u^{1/2}}{v^{1/2}}\frac{dv}{du}\cdot u^{1/2} - v^{1/2}\right)\left(\frac{u^{1/2}}{v^{1/2}}\frac{dv}{du}\cdot v^{1/2} + u^{1/2}\right) = h^2\frac{u^{1/2}}{v^{1/2}}\frac{dv}{du}$$

$$\Rightarrow \qquad \left(u\frac{dv}{du} - v\right)\left(\frac{dv}{du} + 1\right) = h^2\frac{dv}{du} \quad \Rightarrow \quad v = u\frac{dv}{du} - \frac{h^2\dfrac{dv}{du}}{\dfrac{dv}{du} + 1}$$

This equation is Clairaut equation. Hence the solution is

$$v = cu - \frac{ch^2}{c+1}$$

$$\Rightarrow \qquad y^2 = cx^2 - \frac{ch^2}{c+1}$$

4. Solve the equation $y = 2px + p^n$.

Solution. This equation is Lagrange equation. Differentiating it w.r.t x, we get

$$p = 2p + (2x + np^{n-1})\frac{dp}{dx}$$

$$\Rightarrow \qquad \frac{dx}{dp} + \frac{2}{p}x = -np^{n-2} \tag{1}$$

This equation is linear in x. Therefore

$$\text{I.F.} = e^{\int 2/p\,dp} = e^{2\log p} = e^{\log p^2} = p^2$$

Multiplying Eq. (1) by the I.F. p^2, we get

$$p^2\frac{dx}{dp} + 2px = -np^n$$

On integrating, we get

$$xp^2 = -\frac{n}{n+1}p^{n+1} + c$$

$$\Rightarrow \qquad x = \frac{c}{p^2} - \frac{n}{n+1}p^{n-1} \tag{2}$$

Substituting this value of x in the given equation, we get

$$y = \frac{2c}{p} - \frac{(n-1)}{(n+1)}p^n \tag{3}$$

Equations (2) and (3) together give the required solution.

EXERCISES

Solve the following differential equations:

1. $y = 2xp + \log p$
2. $y = xp + \dfrac{a}{2p}$
3. $y = xp + ap(1 - p)$
4. $xp^2 - yp + a = 0$
5. $y = xp + (1 + p^2)^{3/2}$
6. $y = xp + \dfrac{a}{p^2}$
7. $y = x(1 + p) + p^2$
8. $y = xp + p^2$
9. $(y - px)(p - 1) = p$
10. $p = \log(px - y)$
11. $y = xp + \sin^{-1} p$
12. $xp^2 - yp - p + 1 = 0$
13. $(px - y)(px + y) = 2p$
14. $p^2 \cos^2 y + p \sin x \cos x \cos y - \sin y \cos^2 x = 0$

MISCELLANEOUS EXERCISES

Solve the following differential equations:

1. $xp^2 = e^{1/p}$
2. $y = 2px + y^2p^3$
3. $xp^2 + (y - x)p - y = 0$
4. $y = xp + p^3$
5. $x^2(y - px) = yp^2$
6. $x = p^2 - 2p + 2$
7. $y = 2xp + \sin p$
8. $y = xp^2 + p$
9. $y = 2xp + yp^2$
10. $y = \dfrac{3}{2}xp + e^p$
11. $y = \sin^{-1} p + \log(1 + p^2)$
12. $p^4 - (x + 2y + 1)p^3 + (x + 2y + 2xy)p^2 - 2xyp = 0$
13. $3x^4p^2 - xp - y = 0$
14. $16y^3p^2 - 4xp + y = 0$
15. $4(xp^2 + yp) = y^4$
16. $9(y + xp \log p) = (2 + 3 \log p)p^3$
17. $e^{3x}(p - 1) + p^3e^{2y} = 0$
18. $y = x(p + \sqrt{1 + p^2})$
19. $x + \dfrac{p}{\sqrt{1 + p^2}} = a$
20. $yp^2 - 2xp + y = 0$
21. $y - 2px = f(xp^2)$
22. $y = 2xp + 4yp^2$
23. $(y - px)^2 = a^2(1 + p^2)$
24. $p = \tan(x - p/(1 + p^2))$
25. $xyp^2 + p(3x^2 - 2y^2) - 6xy = 0$
26. $axyp^2 + (x^2 - ay^2 - b)p - xy = 0$
27. $p^2x^2 + py(2x + y) + y^2 = 0$
28. $y = p \tan p + \log \cos p$
29. $x(x - 2)p^2 + (2y - 2xy - x + 2)p + y^2 + y = 0$
30. $(x^2 + x)p^2 + (x^2 + x - 2xy - y)p + y^2 - xy = 0$

12

Trajectories

12.1 Introduction

A curve which cuts every member of a given family of curves in accordance with some given law is called a *trajectory* of the given family of curves.

A trajectory is said to be an *orthogonal trajectory* if it cuts every member of the family at right angles. Thus, a line through the common centre of a system of concentric circles cuts every member of the system at right angles. The line is, therefore, an orthogonal trajectory of the system of concentric circles. There being an infinite number of such lines, the orthogonal trajectories of the given family of concentric circles themselves constitute a family. This is true in general, that the orthogonal trajectories of a family of curves themselves form a family. Now, we shall discuss to find their differential equation.

12.2 Differential Equation of the Orthogonal Trajectories of a Single Parameter Family of Curves

12.2.1 Cartesian Coordinates

Let the equation of the family of curves be

$$f(x, y, c) = 0 \tag{1}$$

where c is the parameter and let the differential equation of the family be

$$F\left(x, y, \frac{dy}{dx}\right) = 0 \tag{2}$$

Let P be a point of intersection of a curve C of the family (1) and an orthogonal trajectory T of the family. We shall denote P by (x, y) or (X, Y) accordingly as it is taken as a point on the curve C or on the trajectory T. At P, the slope of C is $\dfrac{dy}{dx}$ and that of T is $\dfrac{dX}{dY}$. But the two curves C and T intersect at right angles. Therefore

$$\frac{dy}{dx} = -\frac{dY}{dX}$$

Also $x = X$, $y = Y$. Substituting for x, y and $\frac{dy}{dx}$ in Eq. (2), we get

$$F\left(X, Y, -\frac{dY}{dX}\right) = 0$$

Changing (X, Y) into (x, y) the differential equation to the family of orthogonal trajectories is

$$F\left(x, y, -\frac{dy}{dx}\right) = 0 \tag{3}$$

Working Rule. Find the differential equation of the given family of curves and then change $\frac{dy}{dx}$ into $-\frac{dx}{dy}$. The resulting equation is the differential equation of the orthogonal trajectories. Solve Eq. (3) to obtain the family of orthogonal trajectories.

12.2.2 Polar Coordinates

Let the equation of the family of curves be

$$f(r, \theta, c) = 0 \tag{1}$$

where c is the parameter and let the differential equation of the family be

$$F\left(r, \theta, \frac{dr}{d\theta}\right) = 0 \tag{2}$$

Let P be a point of intersection of a curve C of the family (1) and an orthogonal trajectory T of the family. We shall denote P by (r, θ) or (ρ, α) accordingly as it is considered as a point on curve C or on the trajectory T. Let ϕ and ϕ' be the angles which the tangents at P to the curve C and the trajectory T, respectively, make with the radius vector OP. Then

$$\tan \phi = \frac{rd\theta}{dr} \quad \text{and} \quad \tan \phi' = \rho\frac{d\alpha}{d\rho}$$

But the two curves C and T intersect at right angles. Therefore

$$|\phi - \phi'| = \frac{\pi}{2}$$

$$\Rightarrow \qquad \phi = \phi' + \frac{\pi}{2}$$

$$\Rightarrow \qquad \tan\phi = \tan\left(\phi' + \frac{\pi}{2}\right)$$

$$\Rightarrow \qquad \tan\phi\tan\phi' = -1$$

$$\Rightarrow \qquad r\frac{d\theta}{dr}\cdot\rho\frac{d\alpha}{d\rho} = -1$$

$$\Rightarrow \qquad \frac{dr}{d\theta} = -r\rho\frac{d\alpha}{d\rho} = -\rho^2\frac{d\alpha}{d\rho}$$

since $r = \rho$, $\theta = \alpha$ at P.

Substituting for r, θ and $\frac{dr}{d\theta}$ in Eq. (2), we get

$$F\left(\rho, \alpha, -\rho^2\frac{d\alpha}{d\rho}\right) = 0$$

Changing (ρ, α) into (r, θ) the differential equation to the family of orthogonal trajectories is

$$F\left(r, \theta, -r^2\frac{d\theta}{dr}\right) = 0 \tag{3}$$

Working Rule. Find the differential equation of the given family of curves and change $\frac{dr}{d\theta}$ into $-r^2\frac{d\theta}{dr}$. The resulting equation is the differential equation of the orthogonal trajectories. Solve Eq. (3) to obtain the family of orthogonal trajectories.

Examples

1. Find the orthogonal trajectories of a system of concentric circles.

Solution. Let the common centre of the concentric circles be at the origin. Then, the equation of the family of circles is

$$x^2 + y^2 = r^2 \tag{1}$$

where r is the parameter.

Differentiating Eq. (1) w.r.t. x, we get

$$x + y\frac{dy}{dx} = 0 \tag{2}$$

This equation is free from parameter and hence is the differential equation of the family. Changing $\frac{dy}{dx}$ into $-\frac{dx}{dy}$, we get the differential equation of the orthogonal trajectories as

$$x - y\frac{dx}{dy} = 0 \Rightarrow \frac{dy}{y} = \frac{dx}{x} \tag{3}$$

Integrating Eq. (3), the required equation of the family of orthogonal trajectories is obtained as

$$y = cx$$

which represents a system of lines passing through the origin.

2. Find the orthogonal trajectory of the curves

$$\frac{x^2}{a^2 + \lambda} + \frac{y^2}{b^2 + \lambda} = 1 \tag{1}$$

λ being the parameter of the family.

Solution. Differentiating Eq. (1) w.r.t. x, we get

$$\frac{x}{a^2 + \lambda} + \frac{y}{b^2 + \lambda}\frac{dy}{dx} = 0 \tag{2}$$

To find the differential equation of the given family of curves, we eliminate λ from Eqs. (1) and (2). From Eq. (2)

$$\lambda = -\frac{b^2 x + a^2 y\dfrac{dy}{dx}}{x + y\dfrac{dy}{dx}}$$

Therefore $$a^2 + \lambda = \frac{(a^2 - b^2)x}{x + y\dfrac{dy}{dx}}$$

and $$b^2 + \lambda = -\frac{(a^2 - b^2)y\dfrac{dy}{dx}}{x + y\dfrac{dy}{dx}}$$

Substituting these values in Eq. (1), the differential equation of the given family of curves is

$$\left(x - y\frac{dx}{dy}\right)\left(x + y\frac{dy}{dx}\right) = a^2 - b^2 \tag{3}$$

Changing $\dfrac{dy}{dx}$ into $-\dfrac{dx}{dy}$, we get the differential equation of the orthogonal trajectories as

$$\left(x + y\frac{dy}{dx}\right)\left(x - y\frac{dx}{dy}\right) = a^2 - b^2 \tag{4}$$

Eq. (4) is same as Eq. (3). Hence the family of curves represented by Eq. (1) is self orthogonal, i.e. every member of the given family intersects its own members orthogonally.

3. Find the orthogonal trajectories of the family of cardioides

$$r = a(1 - \cos\ \theta) \tag{1}$$

a being the parameter.

Solution. Differentiating Eq. (1) w.r.t. θ, we get

$$\frac{dr}{d\theta} = a \sin \theta \tag{2}$$

To find the differential equation of the given family of curves, we eliminate a from Eqs. (1) and (2) and get

$$r = \frac{1}{\sin \theta}\frac{dr}{d\theta}(1 - \cos \theta)$$

$$\Rightarrow \qquad r\frac{d\theta}{dr} = \frac{1 - \cos \theta}{\sin \theta} = \tan \frac{\theta}{2} \tag{3}$$

Changing $r\dfrac{d\theta}{dr}$ into $-\dfrac{1}{r}\dfrac{dr}{d\theta}$, we get the differential equation of the orthogonal trajectories as

$$-\frac{1}{r}\frac{dr}{d\theta} = \tan \frac{\theta}{2}$$

$$\Rightarrow \qquad \frac{1}{r}dr = -\tan \frac{\theta}{2}d\theta$$

On integration, we obtain the required equation of the orthogonal trajectories as

$$\log r = 2 \log \cos \frac{\theta}{2} + \log c$$

$$\Rightarrow \qquad \log r = \log \cos^2 \frac{\theta}{2} + \log c$$

$$\Rightarrow \qquad r = b(1 + \cos\ \theta)$$

This is another family of coaxal cardioides.

EXERCISES

Find the orthogonal trajectories of the following family of curves:

1. $xy = c^2$, c being the parameter
2. $x^2 = cy$, c being the parameter
3. $ay^2 = x^3$, a being the parameter
4. Find the orthogonal trajectories of the family of coaxal circles

$$x^2 + y^2 + 2gx + c = 0$$

where g is the parameter.

5. Show that the system of parabolas

$$y^2 = 4a(x + a)$$

is self-orthogonal.

6. Show that the orthogonal trajectories of the family of conics

$$y^2 - x^2 + 4xy - 2cx = 0$$

consist of a family of cubics with the common asymptote $x + y = 0$.

7. Prove that the families of curves given by the equations

$$y^2 + 3x^2 = 2ax$$

$$y^3 = b(y^2 - x^2)$$

where a and b are arbitrary parameters, intersect at right angles.

Find the orthogonal trajectories of the following families of curves:

8. $r = a\theta$
9. $r\theta = a$
10. $r = a(1 + \cos\theta)$
11. $r = e^{a\theta}$
12. $r^n = a^n \cos n\theta$
13. $r^n = a^n \sin n\theta$
14. $r^n \sin n\theta = a^n$
15. $r = c\sin^2\theta$
16. $r = a + \sin 5\theta$
17. $r = \sin\theta\tan\theta$
18. $r = a(\sec\theta + \tan\theta)$
19. $r = 2a(\sin\theta + \cos\theta)$
20. $r = \dfrac{a}{1 + 2\cos\theta}$
21. Find the equation of the system of orthogonal trajectories of a system of confocal and coaxal parabolas

$$r = \frac{2a}{1 + \cos\theta}$$

13

Linear Differential Equations with Constant Coefficients

13.1 Introduction

A linear differential equation of nth order has the form

$$\frac{d^n y}{dx^n} + P_1 \frac{d^{n-1} y}{dx^{n-1}} + P_2 \frac{d^{n-2} y}{dx^{n-2}} + \ldots + P_{n-1} \frac{dy}{dx} + P_n y = Q \qquad (1)$$

where $P_1, P_2, \ldots, P_n$ are functions of x or constants and do not contain y or derivatives of y.

If $P_1, P_2, \ldots, P_n$ are constants and Q is any function x, the equations of the form given by (1) are called linear differential equations with constant coefficients. This chapter considers such equations.

13.2 Linear Differential Equations with Constant Coefficients

Let the equation be

$$\frac{d^n y}{dx^n} + a_1 \frac{d^{n-1} y}{dx^{n-1}} + a_2 \frac{d^{n-2} y}{dx^{n-2}} + \ldots + a_n y = Q$$

where $a_1, a_2, \ldots, a_n$ are constants and Q is a function of x or a constant.

The above equation may also be written as

$$(D^n + a_1 D^{n-1} + a_2 D^{n-2} + \ldots + a_n) y = Q$$

where D stands for d/dx. In short form it can also be written as

$$f(D)\, y = Q \qquad (1)$$

where $f(D) \equiv D^n + a_1 D^{n-1} + a_2 D^{n-2} + \ldots + a_n$. Since Eq. (1) is of the nth order, its general solution will contain n arbitrary constants. A solution free from arbitrary constants is called a *particular integral* (P.I.).

13.2.1 Theorem

If $y = v$ is a particular integral of

$$f(D)y = Q \qquad (1)$$

and $y = u$ is the general solution of

$$f(D)\,y = 0 \qquad (2)$$

where u contains n independent arbitrary constants, then $y = u + v$ is the general solution of Eq. (1).

Since $y = v$ is a solution of Eq. (1) and $y = u$ is a solution of Eq. (2), we have

$$f(D)v = Q \text{ and } f(D)\,u = 0$$

Therefore $\quad f(D)\,(u + v) = f(D)u + f(D)v = 0 + Q = Q$

Hence $y = u + v$ is a solution of Eq. (1). But this solution contains n independent arbitrary constants. Therefore, it is the general solution of Eq. (1).

The general solution u of Eq. (2) is called the *complementary function* (C.F.) of the original Eq. (1). Thus, for any linear equation, the *general solution* is given by

$$y = \text{C.F.} + \text{P.I.}$$

13.2.2 Theorem

If $y = y_1$ is a solution of $f(D)y = 0$, then $y = c_1y_1$ is also a solution, where c_1 is an arbitrary constant.

We have

$$f(D)\,(c_1y_1) = c_1\,f(D)y_1 = 0 \qquad (\because f(D)y_1 = 0)$$

Hence $y = c_1y_1$ is a solution of $f(D)y = 0$.

13.2.3 Theorem

If $y = y_1, y = y_2, \ldots, y = y_n$ are n independent solutions of $f(D)y = 0$, then

$$y = c_1y_1 + c_2y_2 + \ldots + c_n\,y_n$$

where $c_1, c_2, \ldots, c_n$ are n arbitrary constants, is also a solution.

Since $y = y_1, y = y_2, \ldots, y = y_n$ are solutions of $f(D)y = 0$, we have

$$f(D)y_1 = 0, f(D)y_2 = 0, \ldots, f(D)y_n = 0$$

Now

$$f(D)(c_1y_1 + c_2y_2 + \ldots + c_n\,y_n) = f(D)(c_1y_1)$$
$$+ f(D)\,(c_2y_2) + \ldots + f(D)(c_ny_n)$$
$$= c_1f(D)y_1 + c_2f(D)y_2 + \ldots + c_nf(D)y_n = 0$$

Hence $y = c_1y_1 + c_2y_2 + \ldots + c_ny_n$ is also a solution of $f(D)y = 0$.

13.3 The Complementary Function

Consider the equation

$$f(D)y = 0 \tag{1}$$

Let $m_1, m_2, \ldots, m_n$ be the roots of the equation

$$f(m) = 0 \tag{2}$$

called the *auxiliary equation*. Then, Eq. (1) can be written as

$$(D - m_1)(D - m_2) \ldots (D - m_n)y = 0 \tag{3}$$

This equation has n linear equations given by

$$(D - m_1)\,y = 0, \quad (D - m_2)\,y = 0, \ldots, \quad (D - m_n)y = 0 \tag{4}$$

whose solutions are the solutions of Eq. (3) and hence the solutions of Eq. (1).

Now, consider the equation

$$(D - m_r)y = 0$$

$$\Rightarrow \qquad \frac{dy}{dx} - m_r y = 0$$

$$\Rightarrow \qquad \frac{1}{y}\,dy = m_r dx$$

On integrating, we get $y = e^{m_r x}$. Hence

$$y_1 = e^{m_1 x}, y_2 = e^{m_2 x}, \ldots, y_n = e^{m_n x}$$

are the particular integrals of Eqs. (4) and therefore of Eq. (1). There arise three cases:

***Case I.* Roots of the auxiliary equation are all distinct**

Let the roots $m_1, m_2, \ldots, m_n$ be all distinct. Then, the functions $e^{m_1 x}, e^{m_2 x}, \ldots, e^{m_n x}$ are linearly independent. Hence the general solution of Eq. (1) is

$$y = \text{C.F.} = c_1 e^{m_1 x} + c_2 e^{m_2 x} + \ldots + c_n e^{m_n x} \tag{5}$$

***Case II.* Roots of the auxiliary equation are complex roots**

If the auxiliary Eq. (2) has complex roots, the above form (5) may be avoided in the case of these complex indices and the corresponding terms may be expressed in terms of trigonometric functions by using the relations

$$e^{ix} = \cos x + i \sin x \text{ and } e^{-ix} = \cos x - i \sin x$$

Since the complex roots occur in conjugate pairs, we consider $m_1 = \alpha + i\beta$ and $m_2 = \alpha - i\beta$, the corresponding terms of the C.F. are

$$c_1 e^{(\alpha+i\beta)x} + c_2 e^{(\alpha-i\beta)x} = e^{\alpha x}(c_1 e^{i\beta x} + c_2 e^{-i\beta x})$$

$$= e^{\alpha x}\,[c_1(\cos \beta x + i \sin \beta x) + c_2(\cos \beta x - i \sin \beta x)]$$

$$= e^{\alpha x}\,[(c_1 + c_2) \cos \beta x + i\,(c_1 - c_2) \sin \beta x]$$

$$= e^{\alpha x}(A \cos \beta x + B \sin \beta x)$$

where A and B are new arbitrary constants.

***Case III.* Roots of the auxiliary equation are repeated**

Let $m_1 = m_2 = \ldots m_r$, the other roots being all distinct. Then, the first r terms of the form (5) can be combined into a single term $(c_1 + c_2 + \ldots + c_r)\, e^{m_1 x}$, i.e. $ce^{m_1 x}$. The solution will, therefore, contain $n - r + 1$ arbitrary constants and will not be general.

Since the repeated root of the auxiliary equation corresponds to a factor $(m - m_1)^r$, the Eq. (1) may be written as

$$(D - m_1)^r\,(D - m_{r+1}) \ldots (D - m_n)y = 0$$

Therefore, we find the general solution of

$$(D - m_1)^r\, y = 0 \tag{6}$$

the solution corresponding to other operators being found as before.

For the solution of Eq. (6), let $y = ve^{m_1 x}$. Then

$$(D - m_1)y = (D - m_1)\, ve^{m_1 x} = D(ve^{m_1 x}) - m_1 ve^{m_1 x}$$

$$= e^{m_1 x} Dv + ve^{m_1 x} \cdot m_1 - m_1 ve^{m_1 x} = e^{m_1 x} Dv$$

Operating on both sides by $(D - m_1)$, we get

$$(D - m_1)^2 y = (D - m_1)\,(e^{m_1 x} Dv) = e^{m_1 x} D(Dv) = e^{m_1 x} D^2 v$$

Therefore, repeating the operation r times, we get

$$(D - m_1)^r\; y = e^{m_1 x} D^r v$$

Hence, Eq. (6) transforms to

$$e^{m_1 x} D^r v = 0 \Rightarrow D^r v = 0$$

Particular solutions of this equation are

$$v = 1, \quad v = x, \quad v = x^2, \ldots, \quad v = x^{r-1}$$

and these are linearly independent. Hence the general solution of the equation $D^r v = 0$ is

$$v = c_1 + c_2 x + \ldots + c_r x^{r-1}$$

and therefore of Eq. (6) is

$$y = (c_1 + c_2 x + c_3 x^2 + \ldots + c_r x^{r-1})\, e^{m_1 x}$$

Thus, when r roots of the auxiliary equation are equal to m_1 each and the remaining roots are all distinct, then

$$\text{C.F.} = (c_1 + c_2 x + c_3 x^2 + \ldots + c_r x^{r-1})e^{m_1 x} + c_{r+1} e^{m_{r+1} x} + c_n e^{m_n x}$$

Examples

1.Solve the equation $(D^2 + 2D - 15)y = 0$.

Solution. The auxiliary equation is

$$m^2 + 2m - 15 = 0$$

$$\Rightarrow \qquad (m - 3)(m + 5) = 0$$

$$\Rightarrow \qquad m = 3, -5$$

Hence, the general solution of the equation is

$$y = c_1 e^{3x} + c_2 e^{-5x}$$

2. Solve the equation $(D^3 - 6D^2 + 11D - 6)y = 0$.

Solution. The auxiliary equation is

$$m^3 - 6m^2 + 11m - 6 = 0$$

$$\Rightarrow \qquad (m - 1)(m - 2)(m - 3) = 0$$

$$\Rightarrow \qquad m = 1, 2, 3$$

Hence, the general solution of the equation is

$$y = c_1 e^x + c_2 e^{2x} + c_3 e^{3x}$$

3. Solve the equation

$$\frac{d^2 x}{dt^2} - \frac{dx}{dt} - 12x = 0$$

given that when $t = 0$, $x = 3$ and $\dfrac{dx}{dt} = 5$.

Solution. The auxiliary equation is

$$m^2 - m - 12 = 0$$

$$\Rightarrow \quad (m + 3)(m - 4) = 0$$

$$\Rightarrow \quad m = -3, 4$$

Hence, the general solution of the equation is

$$x = c_1 e^{-3t} + c_2 e^{4t} \tag{1}$$

When $t = 0$, $x = 3$, we have

$$3 = c_1 + c_2 \tag{2}$$

Differentiating Eq. (1), we get

$$\frac{dx}{dt} = -3c_1 e^{-3t} + 4c_2 e^{4t}$$

When $t = 0$, $\frac{dx}{dt} = 5$, we get

$$5 = -3c_1 + 4c_2 \tag{3}$$

Solving Eqs. (2) and (3), we get

$$c_1 = 1 \text{ and } c_2 = 2$$

Hence, the required solution is

$$x = e^{-3t} + 2e^{4t}$$

4. Solve the equation $(D^2 - 2D + 5)y = 0$.

Solution. The auxiliary equation is

$$m^2 - 2m + 5 = 0$$

$$\Rightarrow \quad m = \frac{2 \pm \sqrt{4 - 20}}{2} = 1 \pm 2i$$

Hence, the general solution of the equation is

$$y = e^x(c_1 \cos 2x + c_2 \sin 2x)$$

5. Solve the equation $(D^3 - 2D^2 + 4D - 8)y = 0$.

Solution. The auxiliary equation is

$$m^3 - 2m^2 + 4m - 8 = 0$$

$$\Rightarrow \quad (m - 2)(m^2 + 4) = 0$$

$$\Rightarrow \quad m = 2, \pm 2i$$

Hence the general solution of the equation is

$$y = c_1e^{2x} + c_2 \cos 2x + c_3 \sin 2x$$

6. Solve the equation $(D^3 - 5D^2 + 7D - 3)y = 0$.

Solution. The auxiliary equation is

$$m^3 - 5m^2 + 7m - 3 = 0$$

$$\Rightarrow \qquad (m - 1)^2 (m - 3) = 0$$

$$\Rightarrow \qquad m = 1, 1, 3$$

Hence, the general solution of the equation is

$$y = (c_1 + c_2x)e^x + c_3e^{3x}$$

7. Solve the equation $(D^3 + 3D^2 + 3D + 1)y = 0$.

Solution. The auxiliary equation is

$$m^3 + 3m^2 + 3m + 1 = 0$$

$$\Rightarrow \qquad (m + 1)^3 = 0 \Rightarrow m = -1, -1, -1$$

Hence, the general solution of the equation is

$$y = (c_1 + c_2x + c\ x^2)e^{-x}$$

8. Solve the equation $(D^4 + 8D^2 + 16)y = 0$.

Solution. The auxiliary equation is

$$m^4 + 8m^2 + 16 = 0$$

$$\Rightarrow \qquad (m^2 + 4)^2 = 0$$

$$\Rightarrow \qquad m = \pm\, 2i, \pm\, 2i$$

Hence, the general solution of the equation is

$$y = (c_1 + c_2x) \cos 2x + (c_3 + c_4x) \sin 2x$$

9. Solve the equation $(D^3 - 3D - 2)y = 0$

when $\qquad x = 0, \quad y = 0, \quad \dfrac{dy}{dx} = 9, \quad \dfrac{d^2y}{dx^2} = 0$

Solution. The auxiliary equation is

$$m^3 - 3m - 2 = 0$$

$$\Rightarrow \qquad (m + 1)^2(m - 2) = 0$$

$$\Rightarrow \qquad m = -1, -1, 2$$

Hence, the general solution of the equation is

$$y = (c_1 + c_2x)e^{-x} + c_3e^{2x}$$

When $x = 0$, $y = 0$, we have

$$0 = c_1 + c_3 \tag{1}$$

Now $$\frac{dy}{dx} = -(c_1 + c_2x)e^{-x} + c_2e^{-x} + 2c_3e^{2x}$$

When $x = 0$, $\frac{dy}{dx} = 9$. Therefore

$$9 = -c_1 + c_2 + 2c_3 \tag{2}$$

Further $$\frac{d^2y}{dx^2} = (c_1 + c_2x)e^{-x} - c_2e^{-x} - c_2e^{-x} + 4c_3e^{2x}$$

When $x = 0$, $\frac{d^2y}{dx^2} = 0$. Therefore

$$0 = c_1 - 2c_2 + 4c_3 \tag{3}$$

On solving Eqs. (1), (2) and (3), we get

$$c_1 = -2;\ c_2 = 3,\ c_3 = 2$$

Hence, the required solution is

$$y = (3x - 2)e^{-x} + 2e^{2x}$$

10. Solve the equation

$$\frac{d^3y}{dx^3} + \frac{d^2y}{dx^2} + 4\frac{dy}{dx} + 4y = 0$$

when $$x = 0,\quad y = 0,\quad \frac{dy}{dx} = -1,\quad \frac{d^2y}{dx^2} = 5$$

Solution. The auxiliary equation is

$$m^3 + m^2 + 4m + 4 = 0$$

$$\Rightarrow \quad (m + 1)(m^2 + 4) = 0$$

$$\Rightarrow \quad m = -1, \pm 2i$$

Hence, the general solution of the equation is

$$y = c_1e^{-x} + c_2 \cos 2x + c_3 \sin 2x$$

When $x = 0$, $y = 0$, we have

$$0 = c_1 + c_2 \tag{1}$$

Now $$\frac{dy}{dx} = -c_1 e^{-x} - 2c_2 \sin 2x + 2c_3 \cos 2x$$

When $x = 0$, $\frac{dy}{dx} = -1$. Therefore

$$-1 = -c_1 + 2c_3 \tag{2}$$

Further $$\frac{d^2y}{dx^2} = c_1 e^{-x} - 4c_2 \cos 2x - 4c_3 \sin 2x$$

When $x = 0$, $\frac{d^2y}{dx^2} = 5$. Therefore

$$5 = c_1 - 4c_2 \tag{3}$$

On solving Eqs. (1), (2) and (3), we get

$$c_1 = 1,\ c_2 = -1,\ c_3 = 0$$

Hence, the required solution is

$$y = e^{-x} - \cos 2x$$

EXERCISES

Solve the following differential equations:

1. $\frac{d^2y}{dx^2} + 3\frac{dy}{dx} + 2y = 0$
2. $(3D^3 + 5D^2 - 2D)y = 0$
3. $(D^3 - 4D^2 + D + 6)y = 0$
4. $(D^3 + D^2 - 2D)y = 0$
5. $(D^3 + 2D^2 - 5D - 6)y = 0$
6. $(D^3 - 9D^2 + 23D - 15)y = 0$
7. $(D^3 - 3D^2 - D + 3)y = 0$
8. $\frac{d^2x}{dt^2} - 3\frac{dx}{dt} + 2x = 0$, given that when $t = 0$, $x = 0$ and $\frac{dx}{dt} = 0$.
9. $\frac{d^3x}{dt^3} - 2\frac{d^2x}{dt^2} - 3\frac{dx}{dt} = 0$
10. $(4D^4 - 15D^2 + 5D + 6)y = 0$
11. $(D^2 + 8D + 25)y = 0$
12. $(D^2 - 4D + 13)y = 0$
13. $(D^3 - D^2 + D - 1)y = 0$
14. $(D^3 - 3D^2 + 9D + 13)y = 0$
15. $(D^4 + 2D^3 + 10D^2)y = 0$
16. $(D^4 + 18D^2 + 81)y = 0$
17. $(D^4 + 8D^2 + 16)y = 0$
18. $(D^4 + 5D^2 + 4)y = 0$
19. $(D^4 - 4D^3 + 14D^2 - 20D + 25)y = 0$
20. $(D^3 + 7D^2 + 19D + 13)y = 0$; when $x = 0$, $y = 0$, $\frac{dy}{dx} = 2$, $\frac{d^2y}{dx^2} = -12$.
21. $(D^2 - 6D + 9)y = 0$
22. $(D^3 - 4D^2 + 4D)y = 0$
23. $(D^2 - 8D + 16)y = 0$
24. $(D^3 - 6D^2 + 12D - 8)y = 0$
25. $(D^3 - 3D^2 + 4)y = 0$
26. $(D^4 + 2D^3 + D^2)y = 0$
27. $2\frac{d^4y}{dx^4} - 3\frac{d^3y}{dx^3} - 2\frac{d^2y}{dx^2} = 0$
28. $\frac{d^4y}{dx^4} + 3\frac{d^3y}{dx^3} - 6\frac{d^2y}{dx^2} - 28\frac{dy}{dx} - 24y = 0$
29. $(D^5 - 2D^4 + D^3)y = 0$
30. $(D^5 - 2D^3 - 2D^2 - 3D - 2)y = 0$

13.4 The Particular Integral

Let $\frac{1}{f(D)}Q$ denote some function of x which when operated upon by $f(D)$ gives Q. Then, clearly this function is a particular solution of the differential equation

$$f(D)y = Q$$

since this equation is satisfied if we take

$$y = \frac{1}{f(D)}Q$$

Thus, we have $$f(D)\left(\frac{1}{f(D)}Q\right) = Q$$

Hence, the operator $\frac{1}{f(D)}$ is the inverse of the operator $f(D)$.

In particular, we have

$$D\left(\frac{1}{D}Q\right) = Q, \text{ i.e. } \frac{d}{dx}\left(\frac{1}{D}Q\right) = Q$$

Hence $$\frac{1}{D}Q = \int Q\,dx$$

Note. In finding $\int Q\,dx$, there is no need to add the constant of integration, since we want only a particular integral.

13.4.1 Finding the Particular Integral $\frac{1}{D-m}Q$

Let $\frac{1}{D-m}Q = v$. Then, we have

$$(D-m)v = (D-m)\left(\frac{1}{D-m}Q\right) = Q$$

$$\Rightarrow \quad \frac{dv}{dx} - mv = Q$$

This is a linear differential equation of first order and its solution is

$$ve^{-mx} = \int Qe^{-mx}\,dx \quad \text{(omitting the constant of integration)}$$

$$\Rightarrow \quad v = e^{mx}\int Qe^{-mx}\,dx$$

$$\Rightarrow \frac{1}{D-m}Q = e^{mx}\int Qe^{-mx}\,dx$$

13.4.2 Finding the Particular Integral $\frac{1}{f(D)}Q$

First Method. Let

$$f(D) = (D - m_1)(D - m_2)\ldots(D - m_n)$$

Then $$\frac{1}{f(D)}Q = \frac{1}{(D - m_1)(D - m_2)\ldots(D - m_n)}Q$$

$$= \frac{1}{(D - m_1)(D - m_2)\ldots(D - m_{n-1})}\left[e^{m_n x}\int Qe^{-m_n x}\,dx\right]$$

Proceeding in this manner, we will obtain the required particular integral.

Second Method. Resolving into partial fractions, we have

$$\frac{1}{f(D)}Q = \frac{1}{(D - m_1)(D - m_2)\ldots(D - m_n)}Q$$

$$= \left[\frac{A_1}{D - m_1} + \frac{A_2}{D - m_2} + \ldots + \frac{A_n}{D - m_n}\right]Q$$

$$= A_1 e^{m_1 x}\int Qe^{-m_1 x}\,dx + A_2 e^{m_2 x}\int Qe^{-m_2 x}\,dx + \ldots + A_n e^{m_n x}\int Qe^{-m_n x}\,dx$$

Examples

1. Solve the equation $(D^2 - 3D + 2)y = e^x$.

Solution. The auxiliary equation is

$$m^2 - 3m + 2 = 0$$

$$\Rightarrow \quad (m - 1)(m - 2) = 0$$

$$\Rightarrow \quad m = 1, 2$$

Hence, the complementary function is

$$c_1 e^x + c_2 e^{2x}$$

Now, particular integral is

$$\frac{1}{(D^2 - 3D + 2)}e^x = \frac{1}{(D - 2)(D - 1)}e^x$$

$$= \frac{1}{(D - 2)}\left[e^x \int e^x \cdot e^{-x}\,dx\right] = \frac{1}{D - 2}xe^x$$

$$= e^{2x}\int xe^x \cdot e^{-2x}\,dx = e^{2x}\int xe^{-x}\,dx$$

$$= e^{2x}[-(x + 1)e^{-x}] = -(x + 1)e^x$$

Thus, the general solution of the given equation is

$$y = c_1 e^x + c_2 e^{2x} - (x + 1)e^x$$

$$\Rightarrow \quad y = (c_1 - 1)e^x + c_2 e^{2x} - xe^x$$

$$\Rightarrow \quad y = c_1' e^x + c_2 e^{2x} - xe^x$$

2. Solve the equation $(D^2 + a^2)y = \sec ax$.

Solution. The auxiliary equation is

$$m^2 + a^2 = 0 \Rightarrow m = \pm ai$$

Therefore $\qquad \text{C.F.} = c_1 \cos ax + c_2 \sin ax$

Further

$$\text{P.I.} = \frac{1}{D^2 + a^2} \sec ax = \frac{1}{2ai}\left[\frac{1}{D - ai} - \frac{1}{D + ai}\right]\sec ax$$

$$= \frac{1}{2ai}\left[\frac{1}{D - ai} \sec ax - \frac{1}{D + ai} \sec ax\right]$$

$$= \frac{1}{2ai}\left[e^{iax}\int e^{-iax} \sec ax\, dx - e^{-iax}\int e^{iax} \sec ax dx\right]$$

$$= \frac{1}{2ai}\left[e^{iax}\int \frac{(\cos ax - i \sin ax)}{\cos ax} dx - e^{-iax}\int \frac{(\cos ax + i \sin ax)}{\cos ax} dx\right]$$

$$= \frac{1}{2ai}\left[e^{iax}\int (1 - i \tan ax)dx - e^{-iax}\int (1 + i \tan ax)dx\right]$$

$$= \frac{1}{2ai}\left[e^{iax}\left(x + \frac{i}{a} \log \cos ax\right) - e^{-iax}\left(x - \frac{i}{a} \log \cos ax\right)\right]$$

$$= \frac{x}{2ai}[e^{iax} - e^{-iax}] + \frac{1}{2ai} \cdot \frac{i}{a}(e^{iax} + e^{-iax}) \log \cos ax$$

$$= \frac{x}{a} \sin ax + \frac{1}{a^2} \cos ax \log \cos ax$$

Hence, the general solution of the given equation is

$$y = c_1 \cos ax + c_2 \sin ax + \frac{x}{a} \sin ax + \frac{1}{a^2} \cos ax \log \cos ax$$

EXERCISES

Solve the following differential equations:

1. $(D^2 - 5D + 6)y = e^{4x}$
2. $(D^2 - 3D + 2)y = e^{5x}$
3. $(D^2 + 2D + 1)y = 2e^{2x}$
4. $(D^2 - 1)y = e^x$
5. $(D^2 - 2D + 1)y = e^x$
6. $(D^2 + 1)y = \cos 2x$
7. $(D^2 + 4)y = \tan 2x$
8. $(D^2 + 9)y = \sec 3x$
9. $(D^2 + D - 2)y = 2x$
10. $(D^2 + 1)y = \sec^2 x$

13.5 Particular Integral when Q is of the Form e^{ax}

Case I. $f(a) \neq 0$. By successive differentiation, we have

$$De^{ax} = ae^{ax}$$

$$D^2e^{ax} = a^2e^{ax}$$

$$\cdots\cdots\cdots\cdots$$

$$\cdots\cdots\cdots\cdots$$

$$D^ne^{ax} = a^ne^{ax}$$

In general, $$f(D)e^{ax} = f(a)e^{ax}$$

Operating upon both sides by $\dfrac{1}{f(D)}$, we get

$$\frac{1}{f(D)}[f(D)e^{ax}] = \frac{1}{f(D)}[f(a)e^{ax}]$$

$$\Rightarrow \qquad e^{ax} = f(a)\frac{1}{f(D)}e^{ax}$$

$$\Rightarrow \qquad \frac{1}{f(D)}e^{ax} = \frac{1}{f(a)}e^{ax}$$

Case II. $f(a) = 0$. Then $(D - a)$ is a factor of $f(D)$. Let

$$f(D) = (D - a)^r\, \phi(D)$$

where $\phi(a) \neq 0$. Then

$$\frac{1}{f(D)}e^{ax} = \frac{1}{(D-a)^r\, \phi(D)}e^{ax}$$

$$= \frac{1}{(D-a)^r}\cdot\frac{1}{\phi(a)}e^{ax} = \frac{1}{\phi(a)}\cdot\frac{x^r}{\lfloor r}e^{ax}$$

Examples

1. Solve the equation $(D^2 - 3D - 4)y = e^x$.

Solution. The auxiliary equation is

$$m^2 - 3m - 4 = 0$$

$$\Rightarrow \quad (m + 1)(m - 4) = 0$$

$$\Rightarrow \quad m = -1, 4$$

Therefore $\quad \text{C.F.} = c_1e^{-x} + c_2e^{4x}$

Further $\quad \text{P.I.} = \dfrac{1}{D^2 - 3D - 4}e^x = e^x \dfrac{1}{1^2 - 3 \times 1 - 4} = -\dfrac{1}{6}e^x$

Hence, the general solution of the given equation is

$$y = c_1e^{-x} + c_2e^{4x} - \frac{1}{6}e^x$$

2. Solve the equation $(D^2 - 4D + 4)y = e^{2x}$.

Solution. The auxiliary equation is

$$m^2 - 4m + 4 = 0$$

$$\Rightarrow \quad (m - 2)^2 = 0$$

$$\Rightarrow \quad m = 2, 2$$

Therefore $\quad \text{C.F.} = (c_1 + c_2x)e^{2x}$

Further $\text{P.I.} = \dfrac{1}{(D-2)^2}e^{2x} = \dfrac{1}{(D-2)} \cdot \dfrac{1}{(D-2)}e^{2x} \qquad (\because f(2) = 0)$

$$= \frac{1}{(D-2)}e^{2x}\int e^{2x} \cdot e^{-2x}\,dx$$

$$= \frac{1}{(D-2)}xe^{2x} = e^{2x}\int xe^{2x} \cdot e^{-2x}\,dx = e^{2x} \cdot \frac{x^2}{2}$$

Hence, the general solution of the given equation is

$$y = \left(c_1 + c_2x + \frac{x^2}{2}\right)e^{2x}$$

EXERCISES

Solve the following differential equations:

1. $(D^2 - 4D + 4)y = e^x$
2. $(D^2 + 3D - 4)y = e^{2x}$
3. $(D^2 + 3D - 4)y = e^x$
4. $(D^2 - 2kD + k^2)y = e^x$
5. $(D^2 + D + 1)y = e^{-x}$
6. $(D^2 - 4D + 3)y = 2e^{3x}$
7. $(3D^2 + D - 14)y = 13e^{2x}$
8. $(D^4 - 18D^2 + 81)y = 36e^{3x}$
9. $D^2(D + 1)(D - 1)^3y = e^x$
10. $\dfrac{d^2y}{dx^2} - 2\dfrac{dy}{dx} - 3y = 2e^x$ when $x = 0$, $y = 2$ and $\dfrac{dy}{dx} = 4$.

13.6 Particular Integral when Q is of the Form sin ax and cos ax

By successive differentiation, we have

$$D \sin ax = a \cos ax$$
$$D^2 \sin ax = - a^2 \sin ax$$
$$D^3 \sin ax = - a^3 \cos ax$$
$$D^4 \sin ax = a^4 \sin ax = (-a^2)^2 \sin ax$$
$$\dots\dots\dots\dots\dots\dots\dots\dots$$
$$\dots\dots\dots\dots\dots\dots\dots\dots$$
$$(D^2)^n \sin ax = (- a^2)^n \sin ax$$

Hence if $f(D)$ contains only even powers of D, we denote it by $\phi(D^2)$. Then

$$\phi(D^2) \sin ax = \phi(-a^2) \sin ax$$

Operating upon both sides by $\frac{1}{\phi(D^2)}$, we get

$$\frac{1}{\phi(D^2)}[\phi(D^2) \sin ax] = \frac{1}{\phi(D^2)}[\phi(-a^2) \sin ax]$$

$$\Rightarrow \qquad \sin ax = \phi(-a^2)\frac{1}{\phi(D^2)} \sin ax$$

$$\Rightarrow \qquad \frac{1}{\phi(D^2)} \sin ax = \frac{1}{\phi(-a^2)} \sin ax$$

Similarly, it can be shown that

$$\frac{1}{\phi(D^2)} \cos ax = \frac{1}{\phi(-a^2)} \cos ax$$

Note. The above method fails if $\phi(-a^2) = 0$.

Remark. When $f(D)$ contains odd powers of D also, the method to be followed is as shown in the following examples.

Examples

1. Solve the equation $(D^2 - 4)y = e^x + \sin 2x$.

Solution. The auxiliary equation is

$$m^2 - 4 = 0 \Rightarrow m = 2, -2$$

Therefore $\qquad$ $\text{C.F.} = c_1e^{2x} + c_2e^{-2x}$

Further $\quad \text{P.I.} = \dfrac{1}{(D^2 - 4)} e^x + \dfrac{1}{(D^2 - 4)} \sin 2x$

$$= \frac{1}{(1^2 - 4)} e^x + \frac{1}{(-2^2 - 4)} \sin 2x = -\frac{1}{3} e^x - \frac{1}{8} \sin 2x$$

Hence, the general solution of the given equation is

$$y = c_1 e^{2x} + c_2 e^{-2x} - \frac{1}{3} e^x - \frac{1}{8} \sin 2x$$

2. Solve the equation $(D^4 - 1)y = \sin 2x$.

Solution. The auxiliary equation is

$$m^4 - 1 = 0$$

$$\Rightarrow \qquad (m^2 - 1)(m^2 + 1) = 0$$

$$\Rightarrow \qquad m = 1, -1, \pm i$$

Therefore $\quad \text{C.F.} = c_1 e^x + c_2 e^{-x} + c_3 \cos x + c_4 \sin x$

Further $\quad \text{P.I.} = \dfrac{1}{(D^2 - 1)(D^2 + 1)} \sin 2x$

$$= \frac{1}{(-2^2 - 1)(-2^2 + 1)} \sin 2x = \frac{1}{15} \sin 2x$$

Hence, the general solution of the given equation is

$$y = c_1 e^x + c_2 e^{-x} + c_3 \cos x + c_4 \sin x + \frac{1}{15} \sin 2x$$

3. Solve the equation $(D^3 + D^2 - D - 1)y = \cos 2x$.

Solution. The auxiliary equation is

$$m^3 + m^2 - m - 1 = 0$$

$$\Rightarrow \qquad (m + 1)^2(m - 1) = 0$$

$$\Rightarrow \qquad m = -1, -1, 1$$

Therefore $\quad \text{C.F.} = (c_1 + c_2 x)e^{-x} + c_3 e^x$

Further $\quad \text{P.I.} = \dfrac{1}{(D^2 - 1)(D + 1)} \cos 2x = \dfrac{1}{(-2^2 - 1)(D + 1)} \cos 2x$

$$= -\frac{1}{5(D + 1)} \cos 2x = -\frac{(D - 1)}{5(D + 1)(D - 1)} \cos 2x$$

$$= -\frac{(D-1)}{5(D^2-1)}\cos 2x = -\frac{(D-1)}{5(-2^2-1)}\cos 2x$$

$$= \frac{1}{25}(D\cos 2x - \cos 2x) = \frac{1}{25}(-2\sin 2x - \cos 2x)$$

Hence, the general solution of the given equation is

$$y = (c_1 + c_2 x)e^{-x} + c_3 e^x - \frac{2}{25}\sin 2x - \frac{1}{25}\cos 2x$$

4. Solve the equation $(D^2 - 4D + 4)y = \sin 2x$.

Solution. The auxiliary equation is

$$m^2 - 4m + 4 = 0 \Rightarrow m = 2, 2$$

Therefore $\qquad \text{C.F.} = (c_1 + c_2 x)e^{2x}$

Further $\text{P.I.} = \dfrac{1}{(D^2 - 4D + 4)}\sin 2x$

$$= \frac{1}{(-2^2 - 4D + 4)}\sin 2x = -\frac{1}{4D}\sin 2x = -\frac{1}{4}\cdot\frac{D}{D^2}\sin 2x$$

$$= -\frac{D}{4(-2^2)}\sin 2x = \frac{1}{16}(2\cos 2x) = \frac{1}{8}\cos 2x$$

Hence, the general solution of the given equation is

$$y = (c_1 + c_2 x)e^{2x} + \frac{1}{8}\cos 2x$$

5. Solve the equation $(D^2 - 3D + 2)y = \cos 3x$.

Solution. The auxiliary equation is

$$m^2 - 3m + 2 = 0 \Rightarrow m = 1, 2$$

Therefore $\qquad \text{C.F.} = c_1 e^x + c_2 e^{2x}$

Further $\text{P.I.} = \dfrac{1}{(D^2 - 3D + 2)}\cos 3x = \dfrac{1}{(-3^2 - 3D + 2)}\cos 3x$

$$= -\frac{1}{(3D+7)}\cos 3x = -\frac{(3D-7)}{(3D+7)(3D-7)}\cos 3x$$

$$= -\frac{(3D-7)}{(9D^2-49)}\cos 3x = -\frac{(3D\cos 3x - 7\cos 3x)}{9(-3^2) - 49}$$

$$= -\frac{(-9\sin 3x - 7\cos 3x)}{-130} = -\frac{9}{130}\sin 3x - \frac{7}{130}\cos 3x$$

Hence, the general solution of the given equation is

$$y = c_1e^x + c_2e^{2x} - \frac{9}{130}\sin 3x - \frac{7}{130}\cos 3x$$

EXERCISES

Solve the following differential equations:

1. $(D^2 + 4)y = \cos 3x$
2. $(D^2 + 1)y = \sin x$
3. $(D^2 - 2D - 3)y = 2e^{2x} + 3\sin x$
4. $(D^2 - 2D + 1)y = 2e^{-x} + \sin 2x + 2\cos 2x$
5. $(D^2 + 4)y = 2e^{-x} + \sin x + 2\cos 2x$
6. $(D^2 - 5D + 6)y = \sin 3x$
7. $(D^2 - 8D + 9)y = 40\sin 5x$
8. $(D^2 - 3D + 2)y = 6e^{3x} + 40\sin 2x$
9. $(D^2 - 2D + 5)y = 26\sin 3x$
10. $(D^2 + 16)y = 24\sin 4x$
11. $\dfrac{d^2y}{dx^2} - 2\dfrac{dy}{dx} - 3y = 2e^x - 10\sin x$ when $x = 0, y = 2, \dfrac{dy}{dx} = 4$.
12. $\dfrac{d^2y}{dx^2} + 4\dfrac{dy}{dx} + 13y = 5\sin 2x$ when $x = 0, y = 1, \dfrac{dy}{dx} = -2$.

13.7 Particular Integral when Q is of the Form x^m, m being a Positive Integer

To evaluate $\dfrac{1}{f(D)}x^m$, we expand $\dfrac{1}{f(D)}$ in ascending powers of D as far as the term D^m and then operate on x^m by the various powers of D. It is not necessary to carry the expansion beyond D^m because $D^{m+1}x^m = 0$, $D^{m+2}x^m = 0$, and so on. Sometimes it may be convenient to express $\dfrac{1}{f(D)}$ into its partial fractions and then expand each term into ascending powers of D.

Examples

1. Solve the equation $(D^2 + D - 6)y = x$.

Solution. The auxiliary equation is

$$m^2 + m - 6 = 0$$

$$\Rightarrow \quad (m - 2)(m + 3) = 0$$

$$\Rightarrow \quad m = 2, -3$$

Therefore $\qquad \text{C.F.} = c_1e^{2x} + c_2e^{-3x}$

Further $\qquad \text{P.I.} = \dfrac{1}{(D^2 + D - 6)}x = \dfrac{1}{-6\left(1 - \dfrac{D + D^2}{6}\right)}x$

$$= -\frac{1}{6}\left[1 - \frac{1}{6}(D + D^2)\right]^{-1}(x)$$

$$= -\frac{1}{6}\left[1 + \frac{1}{6}(D + D^2) + \ldots\right](x)$$

$$= -\frac{1}{6}\left(x + \frac{1}{6}\right) = -\frac{1}{6}x - \frac{1}{36}$$

Hence, the general solution of the given equation is

$$y = c_1e^{2x} + c_2e^{-3x} - \frac{1}{6}x - \frac{1}{36}$$

2. Solve the equation $(D^3 + 8)y = x^4 + 2x + 1$.

Solution. The auxiliary equation is

$$m^3 + 8 = 0$$

$\Rightarrow \qquad (m + 2)(m^2 - 2m + 4) = 0$

$\Rightarrow \qquad m = -2, m = \dfrac{2 \pm \sqrt{4 - 16}}{2} = 1 \pm i\sqrt{3}$

Therefore $\quad \text{C.F.} = c_1e^{-2x} + e^x(c_2 \cos \sqrt{3}x + c_3 \sin \sqrt{3}x)$

Further

$$\text{P.I.} = \frac{1}{(8 + D^3)}(x^4 + 2x + 1) = \frac{1}{8\left(1 + \dfrac{D^3}{8}\right)}(x^4 + 2x + 1)$$

$$= \frac{1}{8}\left(1 + \frac{D^3}{8}\right)^{-1}(x^4 + 2x + 1)$$

$$= \frac{1}{8}\left(1 - \frac{D^3}{8} + \frac{D^6}{64} - \ldots\right)(x^4 + 2x + 1)$$

$$= \frac{1}{8}\left(x^4 + 2x + 1 - \frac{1}{8}\cdot 24x\right) = \frac{1}{8}(x^4 - x + 1)$$

Hence, the general solution of the given equation is

$$y = c_1 e^{-2x} + e^{x}(c_2 \cos\sqrt{3}x + c_3 \sin\sqrt{3}x) + \frac{1}{8}(x^4 - x + 1)$$

3. Solve the equation $(D^3 + 3D^2 + 2D)y = x^2$.

Solution. The auxiliary equation is

$$m^3 + 3m^2 + 2m = 0$$

$$\Rightarrow \qquad m(m+1)(m+2) = 0$$

$$\Rightarrow \qquad m = 0, -1, -2$$

Therefore $\qquad$ C.F. $= c_1 + c_2 e^{-x} + c_3 e^{-2x}$

Further P.I. $= \dfrac{1}{(2D + 3D^2 + D^3)} x^2 = \dfrac{1}{2D\left[1 + \frac{1}{2}(3D + D^2)\right]} x^2$

$$= \frac{1}{2D}\left[1 + \frac{1}{2}(3D + D^2)\right]^{-1}(x^2)$$

$$= \frac{1}{2D}\left[1 - \frac{1}{2}(3D + D^2) + \frac{1}{4}(3D + D^2)^2 - \ldots\right](x^2)$$

$$= \frac{1}{2D}\left[1 - \frac{3}{2}D + \frac{7}{4}D^2 + \ldots\right](x^2)$$

$$= \frac{1}{2D}\left[x^2 - \frac{3}{2}\cdot 2x + \frac{7}{4}\cdot 2\right] = \frac{1}{2}\cdot\frac{1}{D}\left(x^2 - 3x + \frac{7}{2}\right)$$

$$= \frac{1}{2}\int\left(x^2 - 3x + \frac{7}{2}\right)dx = \frac{1}{2}\left(\frac{x^3}{3} - \frac{3x^2}{2} + \frac{7x}{2}\right)$$

Hence, the general solution of the given equation is

$$y = c_1 + c_2 e^{-x} + c_3 e^{-2x} + \frac{1}{2}\left(\frac{x^3}{3} - \frac{3x^2}{2} + \frac{7x}{2}\right)$$

EXERCISES

Solve the following differential equations:

1. $(D^2 - 5D + 6)y = x$
2. $(D^2 - 2D + 2)y = x$
3. $(D^2 + D)y = x^3 + 2x^2$
4. $(D^2 + D + 1)y = x^3 + 2x^2$
5. $(D^2 - 4)y = x^2$
6. $(D^2 - 2D + 1)y = x - 1$
7. $(D^2 + D - 2)y = x + \sin x$
8. $(D^3 - D^2 - 6D)y = 1 + x^2$

9. $(D^3 + 2D^2 + D)y = e^{2x} + x^2 + x$ 10. $(D^2 - 3D + 2)y = 2x^2 + 1$

11. $(D^2 + 4D + 5)y = 50x + 13e^{3x}$

12. $\dfrac{d^2y}{dx^2} + 3\dfrac{dy}{dx} = -18y$ when $x = 0$, $y = 0$, $\dfrac{dy}{dx} = 5$.

13. $\dfrac{d^2x}{dt^2} + 4\dfrac{dx}{dt} + 5x = 10$ when $t = 0$, $x = 0$, $\dfrac{dx}{dt} = 0$.

13.8 Particular Integral when Q is of the Form $e^{ax}V$, where V is a Function of x

Let V_1 be any function of x. Then, by successive differentiation, we have

$$D(e^{ax}V_1) = e^{ax}DV_1 + ae^{ax}V_1 = e^{ax}(D + a)V_1$$

$$D^2(e^{ax}V_1) = D(e^{ax}(D + a)V_1) = e^{ax}D(D + a)V_1 + ae^{ax}(D + a)V_1$$

$$= e^{ax}(D + a)^2V_1$$

....

....

In general $$D^n(e^{ax}V_1) = e^{ax}(D + a)^nV_1$$

Therefore $$f(D)(e^{ax}V_1) = e^{ax}f(D + a)V_1 \quad (1)$$

Now, let $f(D + a)V_1 = V$. Then, V will also be a function of x since V_1 is a function of x, and

$$V_1 = \frac{1}{f(D + a)}V$$

Substituting this value of V_1 in Eq. (1), we get

$$f(D)\left[e^{ax}\frac{1}{f(D + a)}V\right] = e^{ax}V$$

Therefore $$\frac{1}{f(D)}\left[f(D)e^{ax}\frac{1}{f(D + a)}V\right] = \frac{1}{f(D)}e^{ax}V$$

$$\Rightarrow \quad \frac{1}{f(D)}e^{ax}V = e^{ax}\frac{1}{f(D + a)}V$$

Examples

1. Solve the equation $(D^2 + 3D + 2)y = e^{2x}\sin x$.

Solution. The auxiliary equation is

$$m^2 + 3m + 2 = 0$$

$$\Rightarrow \qquad (m+1)(m+2) = 0$$

$$\Rightarrow \qquad m = -1, -2$$

Therefore $\quad$ C.F. $= c_1e^{-x} + c_2e^{-2x}$

Further $\quad$ P.I. $= \dfrac{1}{(D^2 + 3D + 2)} e^{2x} \sin x$

$$= e^{2x}\frac{1}{[(D+2)^2 + 3(D+2) + 2]} \sin x = e^{2x}\frac{1}{(D^2 + 7D + 12)} \sin x$$

$$= e^{2x}\frac{1}{(-1^2 + 7D + 12)} \sin x = e^{2x}\frac{1}{(7D + 11)} \sin x$$

$$= e^{2x}\frac{(7D - 11)}{(7D + 11)(7D - 11)} \sin x = e^{2x}\frac{(7D - 11)}{(49D^2 - 121)} \sin x$$

$$= e^{2x}\frac{(7D - 11)}{[49(-1^2) - 121]} \sin x = e^{2x}\frac{(7\cos x - 11\sin x)}{-170}$$

$$= -\frac{1}{170} e^{2x}(7\cos x - 11\sin x)$$

Hence, the general solution of the given equation is

$$y = c_1e^{-x} + c_2e^{-2x} - \frac{1}{170} e^{2x}(7\cos x - 11\sin x)$$

2. Solve the equation $(D^2 - 3D + 2)y = e^x$.

Solution. The auxiliary equation is

$$m^2 - 3m + 2 = 0$$

$$\Rightarrow \qquad (m-1)(m-2) = 0$$

$$\Rightarrow \qquad m = 1, 2$$

Therefore $\quad$ C.F. $= c_1e^x + c_2e^{2x}$

Further $\quad$ P.I. $= \dfrac{1}{(D^2 - 3D + 2)} e^x = \dfrac{1}{(D^2 - 3D + 2)} e^x \cdot 1$

$$= e^x\frac{1}{[(D+1)^2 - 3(D+1) + 2]}(1) = e^x\frac{1}{(D^2 - D)}(1)$$

$$= e^x\frac{1}{-D}(1-D)^{-1}(1) = -e^x\frac{1}{D}(1 + D + D^2 + \ldots)(1)$$

$$= -e^x\frac{1}{D}(1) = -xe^x$$

Hence, the general solution of the given equation is

$$y = c_1e^x + c_2e^{2x} - xe^x$$

3. Solve the equation

$$\frac{d^3y}{dx^3} + 3\frac{d^2y}{dx^2} + 3\frac{dy}{dx} + y = e^{-x}$$

Solution. The auxiliary equation is

$$m^3 + 3m^2 + 3m + 1 = 0$$

$$\Rightarrow \qquad (m+1)^3 = 0 \Rightarrow m = -1, -1, -1$$

Therefore $\qquad \text{C.F.} = (c_1 + c_2x + c_3x^2)e^{-x}$

Further $\qquad \text{P.I.} = \frac{1}{(D+1)^3}e^{-x} = e^{-x}\frac{1}{[(D-1)+1]^3} \quad (1)$

$$= e^{-x}\frac{1}{D^3}(1) = e^{-x}\cdot\frac{x^3}{6}$$

Hence, the general solution of the given equation is

$$y = (c_1 + c_2x + c_3x^2)e^{-x} + \frac{x^3}{6}e^{-x}$$

4. Solve the equation $(D^2 - 4D + 4)y = x^3e^{2x}$.

Solution. The auxiliary equation is

$$m^2 - 4m + 4 = 0$$

$$\Rightarrow (m-2)^2 = 0 \Rightarrow m = 2, 2$$

Therefore $\qquad \text{C.F.} = (c_1 + c_2x)e^{2x}$

Further $\qquad \text{P.I.} = \frac{1}{(D-2)^2}x^3e^{2x} = e^{2x}\frac{1}{(D+2-2)^2}x^3$

$$= e^{2x}\frac{1}{D^2}x^3 = \frac{x^5}{20}e^{2x}$$

Hence, the general solution of the given equation is

$$y = (c_1 + c_2x)\,e^{2x} + \frac{x^5}{20}e^{2x}$$

5. Solve the equation $(D^2 - 4D + 1)y = e^{2x}\sin 2x$.

Solution. The auxiliary equation is

$$m^2 - 4m + 1 = 0$$

$$\Rightarrow \qquad m = \frac{4 \pm \sqrt{16-4}}{2} = 2 \pm \sqrt{3}$$

Therefore C.F. $= c_1 e^{(2+\sqrt{3})x} + c_2 e^{(2-\sqrt{3})x}$

Further P.I. $= \dfrac{1}{(D^2 - 4D + 1)} e^{2x} \sin 2x$

$$= e^{2x} \frac{1}{[(D+2)^2 - 4(D+2) + 1]} \sin 2x$$

$$= e^{2x} \frac{1}{(D^2 - 3)} \sin 2x = e^{2x} \frac{1}{(-2^2 - 3)} \sin 2x = -\frac{1}{7} e^{2x} \sin 2x$$

Hence, the general solution of the given equation is

$$y = c_1 e^{(2+\sqrt{3})x} + c_2 e^{(2-\sqrt{3})x} - \frac{1}{7} e^{2x} \sin 2x$$

6. Solve the equation $(D^2 + 1)y = x^2 \sin 2x$.

Solution. The auxiliary equation is

$$m^2 + 1 = 0 \Rightarrow m = \pm i$$

Therefore C.F. $= c_1 \cos x + c_2 \sin x$

Further P.I. $= \dfrac{1}{(D^2 + 1)} x^2 \sin 2x$

$$= \text{Imaginary part of } \frac{1}{(D^2 + 1)} x^2 e^{2ix}$$

Now

$$\frac{1}{(D^2 + 1)} x^2 e^{2ix} = e^{2ix} \frac{1}{[(D + 2i)^2 + 1]} x^2 = e^{2ix} \frac{1}{(D^2 + 4Di - 3)} x^2$$

$$= -\frac{1}{3} e^{2ix} \left(1 - \frac{4Di + D^2}{3}\right)^{-1} (x^2)$$

$$= -\frac{1}{3} e^{2ix} \left[1 + \frac{4Di + D^2}{3} + \frac{(4Di + D^2)^2}{9} + \ldots\right] (x^2)$$

$$= -\frac{1}{3} e^{2ix} \left[1 + \frac{4i}{3} D - \frac{13}{9} D^2 + \ldots\right] (x^2)$$

$$= -\frac{1}{3} e^{2ix} \left(x^2 + \frac{8ix}{3} - \frac{26}{9}\right)$$

$$= -\frac{1}{3} (\cos 2x + i \sin 2x) \left(x^2 - \frac{26}{9} + \frac{8xi}{3}\right)$$

Hence P.I. $= -\dfrac{1}{3}\left(x^2 - \dfrac{26}{9}\right) \sin 2x - \dfrac{8x}{9} \cos 2x$

Thus, the general solution of the given equation is

$$y = c_1 \cos x + c_2 \sin x - \frac{1}{3}\left(x^2 - \frac{26}{9}\right) \sin 2x - \frac{8}{9}x \cos 2x$$

EXERCISES

Solve the following differential equations:

1. $(D^2 + D - 2)y = e^x$
2. $(D^2 - 2D + 1)y = x^2e^{3x}$
3. $(D^2 - 2D + 5)y = e^{2x} \sin x$
4. $(D^2 - 1)y = 8xe^x$
5. $(D^3 + 3D^2 - 4)y = xe^{-2x}$
6. $(D^2 - 2D + 4)y = e^x \cos x$
7. $(D^3 - 7D - 6)y = e^{2x}(1 + x)$
8. $(D^2 + 4D - 12)y = e^{2x}(x - 1)$
9. $(D^2 - 4D + 4)y = x^3e^{2x}$
10. $(D^2 - 2D + 4)y = e^x \sin x$
11. $(D^2 - 1)y = e^x \sin 3x$
12. $(D^4 - 1)y = e^x \cos x$
13. $(D^3 + 3D^2 + 3D + 1)y = x^2e^{-x}$
14. $(D^2 + 2D + 4)y = e^x \sin 2x$
15. $(D^2 - 4D + 3)y = e^x \cos 2x + \cos 3x$
16. $(D^2 + 2D + 1)y = x \sin x$
17. $(D^2 + 4)y = x \sin x$
18. $(D^4 - 1)y = x^2 \sin x$
19. $(D^2 + 3D + 2)y = x \cos 2x$
20. $(D^2 - 2D + 1)y = xe^x \sin x$

MISCELLANEOUS EXERCISES

Solve the following differential equations:

1. $(D^2 - 3D - 4)y = 0$
2. $(2D^2 + 5D + 2)y = 0$
3. $(D^3 - D^2 - 12D)y = 0$
4. $(D^3 - 3D^2 - D + 3)y = 0$
5. $(D^3 + 2D^2 - 8D)y = 0$
6. $\frac{d^3x}{dt^3} - 7\frac{dx}{dt} + 6x = 0$
7. $(10D^3 + D^2 - 7D + 2)y = 0$
8. $(D^4 - 2D^3 - 13D^2 + 38D - 24)y = 0$
9. $(D^2 + D - 2)y = 2x - 40 \cos 2x$
10. $(D^3 + D^2 + 3D - 5)y = 5 \sin 2x + 10x^2 - 3x + 7$
11. $(D^2 + D - 6)y = 10e^{2x} - 18e^{3x} - 6x - 11$
12. $(D^2 - D - 2)y = 6x + 6e^{-x}$
13. $\frac{d^2y}{dx^2} - 4\frac{dy}{dx} + 3y = 9x^2 + 4$ when $x = 0, y = 6, \frac{dy}{dx} = 8.$
14. $\frac{d^2y}{dx^2} + y = 3x^2 - 4 \sin x$ when $x = 0, y = 0, \frac{dy}{dx} = 1.$
15. $(D^4 + D^2)y = 3x^2 + 4 \sin x - 2 \cos x$
16. $\frac{d^3y}{dx^3} - 4\frac{d^2y}{dx^2} + \frac{dy}{dx} + 6y = 2e^x$ when $x = 0, y = 0, \frac{dy}{dx} = 0$ and $\frac{d^2y}{dx^2} = 0.$
17. $\frac{d^2y}{dx^2} - 4\frac{dy}{dx} + 29y = 0$ when $x = 0, y = 0, \frac{dy}{dx} = 5.$
18. $\frac{d^2y}{dx^2} - 6\frac{dy}{dx} + 25y = 0$ when $x = 0, y = -3, \frac{dy}{dx} = -1.$
19. $4\frac{d^4y}{dx^4} + 4\frac{d^3y}{dx^3} - 3\frac{d^2y}{dx^2} - 2\frac{dy}{dx} + y = 0$

20. $(D^4 - 7D^3 + 18D^2 - 20D + 8)y = 0$
21. $(D^4 - D^3 - 3D^2 + D + 2)y = 0$
22. $(D^4 + 3D^3 + 2D^2)y = 0$ when $x = 0$, $y = 0$, $\frac{dy}{dx} = 3$, $\frac{d^2y}{dx^2} = -5$, $\frac{d^3y}{dx^3} = 9$.
23. $(D + 2)(D - 1)^3 y = e^x$
24. $\frac{d^2y}{dx^2} + a^2 y = \cos ax$
25. $(D^3 - 6D^2 + 12D - 8)y = xe^{2x} + x^2e^{3x}$
26. $(D^2 - 2D + 1)y = x \sin x$
27. $(D^4 + 1)y = x^2 \cos x$
28. $(D^2 - 4D + 4)y = 8x^2e^{2x} \sin 2x$

14

Homogeneous Linear Differential Equations with Variable Coefficients

14.1 Homogeneous Linear Equations

An equation of the form

$$x^n \frac{d^n y}{dx^n} + a_1 x^{n-1} \frac{d^{n-1} y}{dx^{n-1}} + \ldots + a_{n-1} x \frac{dy}{dx} + a_n y = Q \qquad (1)$$

where $a_1, a_2, \ldots, a_n$ are constants and Q is a function of x called the *homogeneous linear differential equation.*

To solve such type of equations we generally change the independent variable x to z by substituting $x = e^z$. We have

$$\frac{dx}{dz} = e^z = x$$

Now $\dfrac{dy}{dz} = \dfrac{dy}{dx}\dfrac{dx}{dz} = x\dfrac{dy}{dx}$

$$\frac{d^2 y}{dz^2} = \frac{d}{dz}\left(\frac{dy}{dz}\right) = \frac{d}{dx}\left(x\frac{dy}{dx}\right)\frac{dx}{dz} = \left(x\frac{d^2 y}{dx^2} + \frac{dy}{dx}\right)x = x^2\frac{d^2 y}{dx^2} + x\frac{dy}{dx}$$

$$\frac{d^3 y}{dz^3} = \frac{d}{dx}\left(x^2\frac{d^2 y}{dx^2} + x\frac{dy}{dx}\right)\frac{dx}{dz} = x^3\frac{d^3 y}{dx^3} + 3x^2\frac{d^2 y}{dx^2} + x\frac{dy}{dx}$$

Writing D for $\dfrac{d}{dz}$, we get

$$x\frac{dy}{dx} = Dy$$

$$x^2\frac{d^2 y}{dx^2} = D^2 y - Dy = D(D-1)y$$

$$x^3 \frac{d^3y}{dx^3} = D^3y - 3(D^2 - D)y - Dy = D(D-1)(D-2)y$$

. . . .

. . . .

$$x^n \frac{d^ny}{dx^n} = D(D-1)(D-2)\ldots(D-n+1)y$$

Substituting these values of $x\frac{dy}{dx}, x^2\frac{d^2y}{dx^2}, \ldots$ in Eq. (1), it reduces to a linear differential equation with constant coefficients, which can be solved by the methods of the previous chapter.

Examples

1. Solve the equation

$$x^3\frac{d^3y}{dx^3} + 2x^2\frac{d^2y}{dx^2} + 3x\frac{dy}{dx} - 3y = x^2 + x$$

Solution. Substituting $x = e^z$ and $D \equiv \frac{d}{dz}$, the given equation transforms to

$$[D(D-1)(D-2) + 2D(D-1) + 3D - 3]y = e^{2z} + e^z$$

$$\Rightarrow \qquad (D^3 - D^2 + 3D - 3)y = e^{2z} + e^z$$

The auxiliary equation is

$$m^3 - m^2 + 3m - 3 = 0$$

$$\Rightarrow \qquad (m-1)(m^2+3) = 0$$

$$\Rightarrow \qquad m = 1, \pm i\sqrt{3}$$

Therefore $\qquad$ $\text{C.F.} = c_1e^z + c_2\cos\sqrt{3}z + c_3\sin\sqrt{3}z$

Further

$$\text{P.I.} = \frac{1}{(D^3 - D^2 + 3D - 3)}e^{2z} + \frac{1}{(D^3 - D^2 + 3D - 3)}e^z$$

$$= \frac{1}{(D^2+3)(D-1)}e^{2z} + \frac{1}{(D^2+3)(D-1)}e^z$$

$$= \frac{1}{(4+3)\cdot 1}e^{2z} + \frac{1}{(1+3)(D-1)}e^z$$

$$= \frac{1}{7}e^{2z} + \frac{1}{4}e^z\frac{1}{(D+1-1)}(1) = \frac{1}{7}e^{2z} + \frac{1}{4}ze^z$$

Hence, the solution of the transformed equation is

$$y = c_1e^z + c_2 \cos \sqrt{3}\,z + c_3 \sin \sqrt{3}\,z + \frac{1}{7}e^{2z} + \frac{1}{4}ze^z$$

Replacing z by $\log x$, the general solution of the given equation is

$$y = c_1x + c_2 \cos (\sqrt{3} \log x) + c_3 \sin (\sqrt{3} \log x) + \frac{1}{7}x^2 + \frac{1}{4}x \log x$$

2. Solve the equation

$$x^2 \frac{d^2y}{dx^2} - x\frac{dy}{dx} - 3y = x^2 \log x$$

Solution. Substituting $x = e^z$ and $D \equiv \dfrac{d}{dz}$, the given equation transforms to

$$[D(D-1) - D - 3]\, y = e^{2z} \cdot z$$

$\Rightarrow$
$$(D^2 - 2D - 3)\, y = ze^{2z}$$

The auxiliary equation is

$$m^2 - 2m - 3 = 0$$

$\Rightarrow$
$$(m+1)(m-3) = 0$$

$\Rightarrow$
$$m = -1, 3$$

Therefore
$$\text{C.F.} = c_1e^{-z} + c_2e^{3z}$$

Further

$$\text{P.I.} = \frac{1}{(D+1)(D-3)}ze^{2z} = e^{2z}\frac{1}{(D+2+1)(D+2-3)}z$$

$$= e^{2z}\frac{1}{(D+3)(D-1)}z = e^{2z}\frac{1}{(D^2+2D-3)}z$$

$$= -\frac{1}{3}e^{2z}\left(1 - \frac{2D+D^2}{3}\right)^{-1}(z)$$

$$= -\frac{1}{3}e^{2z}\left[1 + \frac{2D+D^2}{3} + \ldots\right](z) = -\frac{1}{3}e^{2z}\left(z + \frac{2}{3}\right)$$

Hence, the solution of the transformed equation is

$$y = c_1e^{-z} + c_2e^{3z} - \frac{1}{3}e^{2z}\left(z + \frac{2}{3}\right)$$

Replacing z by $\log x$, the general solution of the given equation is

$$y = c_1 x^{-1} + c_2 x^3 - \frac{1}{9} x^2 (3 \log x + 2)$$

3. Solve the equation

$$x^3 \frac{d^3 y}{dx^3} - 3x^2 \frac{d^2 y}{dx^2} + 6x \frac{dy}{dx} - 6y = (\log x)^2$$

Solution. Substituting $x = e^z$ and $D \equiv \frac{d}{dz}$, the given equation transforms to

$$[D(D-1)(D-2) - 3D(D-1) + 6D - 6]\, y = z^2$$

$$\Rightarrow \qquad (D^3 - 6D^2 + 11D - 6)\, y = z^2$$

The auxiliary equation is

$$m^3 - 6m^2 + 11m - 6 = 0$$

$$\Rightarrow \qquad (m-1)(m-2)(m-3) = 0$$

$$\Rightarrow \qquad m = 1, 2, 3$$

Therefore $\qquad \text{C.F.} = c_1 e^z + c_2 e^{2z} + c_3 e^{3z}$

Further

$$\text{P.I.} = \frac{1}{(-6 + 11D - 6D^2 + D^3)} z^2 = -\frac{1}{6}\left(1 - \frac{11D - 6D^2 + D^3}{6}\right)^{-1} (z^2)$$

$$= -\frac{1}{6}\left[1 + \frac{11D - 6D^2 + D^3}{6} + \frac{(11D - 6D^2 + D^3)^2}{36} + \ldots\right](z^2)$$

$$= -\frac{1}{6}\left(1 + \frac{11D}{6} + \frac{85D^2}{36} \ldots\right)(z^2) = -\frac{1}{6}\left(z^2 + \frac{11}{3} z + \frac{85}{18}\right)$$

Hence, the solution of the transformed equation is

$$y = c_1 e^z + c_2 e^{2z} + c_3 e^{3z} - \frac{1}{6} z^2 - \frac{11}{18} z - \frac{85}{108}$$

Replacing z by $\log x$, the general solution of the given equation is

$$y = c_1 x + c_2 x^2 + c_3 x^3 - \frac{1}{6} (\log x)^2 - \frac{11}{18} \log x - \frac{85}{108}$$

EXERCISES

Solve the following differential equations:

1. $x^2 \frac{d^2 y}{dx^2} - 2y = x^2 + \frac{1}{x}$

2. $x^2 \frac{d^2 y}{dx^2} - x \frac{dy}{dx} + y = 2 \log x$

3. $x^3\frac{d^3y}{dx^3} - x^2\frac{d^2y}{dx^2} + 2x\frac{dy}{dx} - 2y = x^3 + 3x$

4. $x^3\frac{d^3y}{dx^3} + 3x^2\frac{d^2y}{dx^2} + x\frac{dy}{dx} = 24x^2$
5. $x^2\frac{d^2y}{dx^2} - 3x\frac{dy}{dx} + 5y = x^2 \sin(\log x)$

6. $x^4\frac{d^3y}{dx^3} + 2x^3\frac{d^2y}{dx^2} - x^2\frac{dy}{dx} + xy = 1$ [*Hint:* Divide throughout by x]

7. $x^2\frac{d^2y}{dx^2} + x\frac{dy}{dx} - 4y = x^3$
8. $x^2\frac{d^2y}{dx^2} - 2x\frac{dy}{dx} + 2y = x + x^2\log x + x^3$

9. $x^3\frac{d^3y}{dx^3} + 2x^2\frac{d^2y}{dx^2} + 2y = 10\left(x + \frac{1}{x}\right)$

10. $\frac{d^3y}{dx^3} - \frac{4}{x}\frac{d^2y}{dx^2} + \frac{5}{x^2}\frac{dy}{dx} - \frac{2y}{x^3} = 1$
11. $x^2\frac{d^2y}{dx^2} - x\frac{dy}{dx} + 2y = x\log x$

12. $x^2\frac{d^2y}{dx^2} + 7x\frac{dy}{dx} + 13y = \log x$
13. $x^2\frac{d^2y}{dx^2} + 4x\frac{dy}{dx} + 2y = x + \log x$

14.2 Alternate Method

Consider the homogeneous linear equation

$$x^n\frac{d^ny}{dx^n} + a_1x^{n-1}\frac{d^{n-1}y}{dx^{n-1}} + \ldots + a_{n-1}x\frac{dy}{dx} + a_n y = Q \qquad (1)$$

where $a_1, a_2, \ldots, a_n$ are constants and Q is any function of x.

Here we denote the operator $x\frac{d}{dx}$ by θ, therefore $\theta y = x\frac{dy}{dx}$ and

$$\theta^2 y = x\frac{d}{dx}\left(x\frac{dy}{dx}\right) = x\left(\frac{dy}{dx} + x\frac{d^2y}{dx^2}\right) = \theta y + x^2\frac{d^2y}{dx^2}$$

$$\Rightarrow \qquad x^2\frac{d^2y}{dx^2} = \theta(\theta - 1)y$$

Hence, the given Eq. (1) transforms to

$$f(\theta)y = Q \qquad (2)$$

Assume that $y = cx^m$ is a solution of the equation $f(\theta)y = 0$, then substitution shows that we must have

$$cx^m f(m) = 0$$

Hence, $y = cx^m$ will be a solution of $f(\theta)y = 0$, if m is a root of the auxiliary equation

$$f(m) = 0 \qquad (3)$$

Case I. Let Eq. (3) have n distinct roots, say, $m_1, m_2, \ldots, m_n$. Then

$$\text{C.F.} = c_1 x^{m_1} + c_2 x^{m_2} + \ldots + c_n x^{m_n}$$

Case II. Let Eq. (3) have r repeated roots. Then, the corresponding solution is

$$y = [c_1 + c_2 \log x + c_3 (\log x)^2 + \ldots + c_r (\log x)^{r-1}]\ x^{m_1}$$

Case III. Let roots of Eq. (3) be $\alpha \pm i\beta$. Then, the corresponding solution is

$$y = x^{\alpha}\ [c_1 \cos(\beta \log x) + c_2 \sin(\beta \log x)]$$

Particular Integral

The function $\dfrac{1}{f(\theta)} Q$ is a particular integral of Eq. (2), because the equation is satisfied when this is substituted for y in Eq. (2).

Now, consider $\dfrac{1}{(\theta - m)} Q$ and let $\dfrac{1}{(\theta - m)} Q = v$. Then $(\theta - m)\, v = Q$

$$\Rightarrow \qquad x\frac{dv}{dx} - mv = Q$$

$$\Rightarrow \qquad \frac{dv}{dx} - \frac{m}{x} v = \frac{Q}{x}$$

This is a linear differential equation of first order and its solution is

$$vx^{-m} = \int \frac{Q}{x} \cdot x^{-m}\, dx$$

Hence

$$\frac{1}{(\theta - m)} Q = x^m \int Q x^{-m-1}\, dx \tag{4}$$

Further, let

$$f(\theta) = (\theta - m_1)(\theta - m_2) \ldots (\theta - m_n)$$

then breaking $\dfrac{1}{f(\theta)}$ into partial fractions, the particular integral

$$\frac{1}{f(\theta)} Q = \left[\frac{A_1}{\theta - m_1} + \frac{A_2}{\theta - m_2} + \ldots + \frac{A_n}{\theta - m_n}\right] Q$$

Note. The value of the right hand side is obtained by using the relation (4).

Particular Case. If $Q = x^m$, then it can be easily shown that

$$\frac{1}{f(\theta)} x^m = \frac{1}{f(m)} x^m$$

since

$$f(\theta)\, x^m = f(m)\, x^m.$$

Note. If $f(\theta) = 0$ has a root $\theta = m$, then $f(m) = 0$, hence this method fails.

Examples

1. Solve the equation

$$x^2 \frac{d^2y}{dx^2} - 4x\frac{dy}{dx} + 6y = x^4$$

Solution. Since $x\dfrac{dy}{dx} = \theta y$ and $x^2\dfrac{d^2y}{dx^2} = \theta(\theta - 1)y$, the given equation reduces to

$$[\theta(\theta - 1) - 4\theta + 6]\, y = x^4$$

$\Rightarrow$
$$(\theta^2 - 5\theta + 6)\, y = x^4$$

The auxiliary equation is

$$m^2 - 5m + 6 = 0$$

$\Rightarrow$
$$m = 2,\ 3$$

Therefore
$$\text{C.F.} = c_1x^2 + c_2x^3$$

Further
$$\text{P.I.} = \frac{1}{(\theta^2 - 5\theta + 6)}x^4 = \frac{1}{4^2 - 5 \times 4 + 6}x^4 = \frac{1}{2}x^4$$

Hence, the general solution of the given equation is

$$y = c_1x^2 + c_2x^3 + \frac{1}{2}x^4$$

2. Solve the equation

$$x^2\frac{d^2y}{dx^2} + x\frac{dy}{dx} - y = x^3$$

Solution. Since $x\dfrac{dy}{dx} = \theta y$ and $x^2\dfrac{d^2y}{dx^2} = \theta(\theta - 1)\, y$, the given equation reduces to

$$[\theta(\theta - 1) + \theta - 1]\, y = x^3$$

$\Rightarrow$
$$(\theta^2 - 1)\, y = x^3$$

The auxiliary equation is

$$m^2 - 1 = 0$$

$\Rightarrow$
$$m = 1,\ -1$$

Therefore
$$\text{C.F.} = c_1x + c_2x^{-1}$$

Further $\quad$ P.I. $= \dfrac{1}{(\theta^2 - 1)} x^3 = \dfrac{1}{3^2 - 1} x^3 = \dfrac{1}{8} x^3$

Hence, the general solution of the given equation is

$$y = c_1 x + c_2 x^{-1} + \frac{1}{8} x^3$$

3. Solve the equation

$$x^2 \frac{d^2 y}{dx^2} + 2x \frac{dy}{dx} - 20\, y = (x + 1)^2$$

Solution. Since $x\dfrac{dy}{dx} = \theta y$ and $x^2 \dfrac{d^2 y}{dx^2} = \theta(\theta - 1)\, y$, the given equation reduces to

$$[\theta(\theta - 1) + 2\theta - 20]\; y = (x + 1)^2$$

$$\Rightarrow \qquad (\theta^2 + \theta - 20)\; y = (x + 1)^2$$

The auxiliary equation is

$$m^2 + m - 20 = 0$$

$$\Rightarrow \qquad m = 4, -5$$

Therefore $\quad$ C.F. $= c_1 x^4 + c_2 x^{-5}$

Further

$$\text{P.I.} = \frac{1}{(\theta^2 + \theta - 20)} (x + 1)^2 = \frac{1}{(\theta^2 + \theta - 20)} (x^2 + 2x + 1)$$

$$= \frac{x^2}{2^2 + 2 - 20} + \frac{2x}{1^2 + 1 - 20} + \frac{1}{0^2 + 0 - 20} = \frac{x^2}{-14} + \frac{2x}{-18} + \frac{1}{-20}$$

Hence, the general solution of the given equation is

$$y = c_1 x^4 + c_2 x^{-5} - \left(\frac{x^2}{14} + \frac{x}{9} + \frac{1}{20} \right)$$

4. Solve the equation

$$x^2 \frac{d^2 y}{dx^2} + 3x \frac{dy}{dx} + y = \frac{1}{(1 - x)^2}$$

Solution. Since $x\dfrac{dy}{dx} = \theta y$ and $x^2 \dfrac{d^2 y}{dx^2} = \theta(\theta - 1)\, y$, the given equation reduces to

$$[\theta(\theta - 1) + 3\theta + 1]\, y = \frac{1}{(1 - x)^2}$$

$$\Rightarrow \qquad (\theta^2 + 2\theta + 1)\, y = \frac{1}{(1 - x)^2}$$

The auxiliary equation is

$$m^2 + 2m + 1 = 0 \quad \Rightarrow \quad m = -1, -1$$

Therefore $\qquad \text{C.F.} = (c_1 + c_2 \log x)\, x^{-1}$

Further

$$\text{P.I.} = \frac{1}{(\theta + 1)^2}(1 - x)^{-2} = \frac{1}{(\theta + 1)} x^{-1} \int x^{1-1}(1 - x)^{-2}\, dx$$

$$= \frac{1}{(\theta + 1)} x^{-1} \int (1 - x)^{-2}\, dx = \frac{1}{(\theta + 1)} x^{-1}[(1 - x)^{-1}]$$

$$= x^{-1} \int x^{1-1} x^{-1}(1 - x)^{-1}\, dx = x^{-1} \int \left(\frac{1}{x} + \frac{1}{1 - x}\right) dx$$

$$= x^{-1} (\log |x| - \log |1 - x|) = x^{-1} \log \left|\frac{x}{1 - x}\right|$$

Hence, the general solution of the given equation is

$$y = (c_1 + c_2 \log x)\, x^{-1} + x^{-1} \log \left|\frac{x}{1 - x}\right|$$

EXERCISES

Solve the following differential equations:

1. $x^2 \dfrac{d^2y}{dx^2} - 4x \dfrac{dy}{dx} + 6y = x$
2. $x^2 \dfrac{d^2y}{dx^2} + x \dfrac{dy}{dx} - 4y = x^2$
3. $x^2 \dfrac{d^2y}{dx^2} - 3x \dfrac{dy}{dx} + 4y = 2x^2$
4. $x^2 \dfrac{d^2y}{dx^2} - 2x \dfrac{dy}{dx} - 4y = x^4$
5. $x^2 \dfrac{d^2y}{dx^2} + 7x \dfrac{dy}{dx} + 5y = x^5$
6. $x^2 \dfrac{d^2y}{dx^2} + 5x \dfrac{dy}{dx} + 4y = x^4$
7. $x^2 \dfrac{d^2y}{dx^2} + 7x \dfrac{dy}{dx} + 5y = 2x^6$
8. $x^2 \dfrac{d^2y}{dx^2} + 3x \dfrac{dy}{dx} + y = (1 - x)^2$
9. $x^2 \dfrac{d^2y}{dx^2} - 2x \dfrac{dy}{dx} + 2y = x^{-1}$
10. $x^2 \dfrac{d^2y}{dx^2} - 3x \dfrac{dy}{dx} + 4y = x^m$

14.3 Equations Reducible to Homogeneous Linear Form

An equation of the form

$$(a+bx)^n \frac{d^n y}{dx^n} + a_1 (a+bx)^{n-1} \frac{d^{n-1} y}{dx^{n-1}} + \ldots + a_{n-1}(a+bx)\frac{dy}{dx} + a_n y = Q \tag{1}$$

where $a_1, a_2, \ldots, a_n$ are constants and Q is a function of x, can be transformed into homogeneous linear equation with constant coefficients by changing the independent variable from x to z, by the substitution $a + bx = z$. Then, we have

$$\frac{dy}{dx} = \frac{dy}{dz}\frac{dz}{dx} = b\frac{dy}{dz}$$

$$\frac{d^2 y}{dx^2} = \frac{d}{dx}\left(\frac{dy}{dx}\right) = \frac{d}{dz}\left(b\frac{dy}{dz}\right)\frac{dz}{dx} = b^2\frac{d^2 y}{dz^2}$$

$$\cdots\cdots\cdots\cdots\cdots\cdots\cdots\cdots$$

$$\cdots\cdots\cdots\cdots\cdots\cdots\cdots\cdots$$

$$\frac{d^n y}{dx^n} = b^n \frac{d^n y}{dz^n}$$

Substituting these values in Eq. (1) and dividing throughout by b^n, the given equation reduces to

$$z^n \frac{d^n y}{dz^n} + \frac{a_1}{b} z^{n-1} \frac{d^{n-1} y}{dz^{n-1}} + \frac{a_2}{b^2} z^{n-2} \frac{d^{n-2} y}{dz^{n-2}} + \ldots + \frac{a_{n-1}}{b^{n-1}} z\frac{dy}{dz} + \frac{a_n}{b^n} y = F(z)$$

This equation can now be solved by the methods of Sec. 14.1 and 14.2.

Examples

1. Solve the equation

$$(1+x)^2 \frac{d^2 y}{dx^2} + (1+x)\frac{dy}{dx} + y = 4\cos(\log(1+x))$$

Solution. Let $1 + x = z$. Then

$$\frac{dy}{dx} = \frac{dy}{dz}\frac{dz}{dx} = \frac{dy}{dz}$$

$$\frac{d^2 y}{dx^2} = \frac{d}{dx}\left(\frac{dy}{dx}\right) = \frac{d}{dz}\left(\frac{dy}{dz}\right)\frac{dz}{dx} = \frac{d^2 y}{dz^2}$$

Substituting these values in the given equation, we get

$$z^2\frac{d^2 y}{dz^2} + z\frac{dy}{dz} + y = 4\cos\log z$$

Putting $z = e^u$ and $D \equiv \dfrac{d}{du}$, the above equation transforms to

$$[D(D-1) + D + 1]\, y = 4 \cos u$$

$$\Rightarrow \qquad (D^2 + 1)y = 4 \cos u$$

The auxiliary equation is

$$m^2 + 1 = 0 \Rightarrow m = \pm i$$

Therefore

$$\text{C.F.} = c_1 \cos u + c_2 \sin u = c_1 \cos(\log z) + c_2 \sin(\log z)$$
$$= c_1 \cos(\log(1 + x)) + c_2 \sin(\log(1 + x))$$

Further $\quad$ $\text{P.I.} = \dfrac{1}{D^2+1} 4 \cos u = 4\left[\text{Real part of } \dfrac{1}{D^2+1} e^{iu}\right]$

Now

$$\frac{1}{D^2+1} e^{iu} = e^{iu} \frac{1}{(D+i)^2+1}(1) = e^{iu} \frac{1}{D^2 + 2Di - 1 + 1}(1)$$

$$= e^{iu} \frac{1}{2Di\left(1 + \dfrac{D}{2i}\right)}(1) = e^{iu} \frac{1}{2Di}\left(1 + \frac{D}{2i}\right)^{-1}(1)$$

$$= e^{iu} \frac{1}{2Di}\left(1 - \frac{D}{2i} + \ldots\right)(1) = e^{iu} \frac{1}{2i} \cdot \frac{1}{D}(1)$$

$$= \frac{e^{iu}}{2i} u = \frac{u(\cos u + i \sin u)}{2i}$$

Hence

$$\text{P.I.} = 4 \cdot \frac{1}{2} u \sin u = 2 \log z \sin \log z = 2 \log(1 + x) \sin(\log(1 + x))$$

Thus, the general solution of the given equation is

$$y = c_1 \cos(\log(1+x)) + c_2 \sin(\log(1+x)) + 2 \log(1+x) \sin(\log(1 + x))$$

2. Solve the equation

$$(3x + 2)^2 \frac{d^2y}{dx^2} + 3(3x + 2) \frac{dy}{dx} - 36y = 3x^2 + 4x + 1$$

Solution. Let $3x + 2 = z$. Then

$$\frac{dy}{dx} = \frac{dy}{dz} \frac{dz}{dx} = 3 \frac{dy}{dz}$$

$$\frac{d^2y}{dx^2} = \frac{d}{dx}\left(\frac{dy}{dx}\right) = \frac{d}{dz}\left(3\frac{dy}{dz}\right)\frac{dz}{dx} = 9\frac{d^2y}{dz^2}$$

Substituting these values in the given equation, we get

$$3^2 z^2 \frac{d^2y}{dz^2} + 3.3z\frac{dy}{dz} - 36y = 3\left(\frac{z-2}{3}\right)^2 + 4\left(\frac{z-2}{3}\right) + 1$$

$$\Rightarrow \qquad z^2\frac{d^2y}{dz^2} + z\frac{dy}{dz} - 4y = \frac{1}{27}(z^2 - 1)$$

Let $z\frac{dy}{dz} = \theta y$. Then $z^2\frac{d^2y}{dz^2} = \theta(\theta - 1)y$ and hence the above equation reduces to

$$[\theta(\theta - 1) + \theta - 4]\, y = \frac{1}{27}(z^2 - 1)$$

$$\Rightarrow \qquad (\theta^2 - 4)y = \frac{1}{27}z^2 - \frac{1}{27}$$

The auxiliary equation is

$$m^2 - 4 = 0 \;\Rightarrow\; m = 2, -2$$

Therefore $\quad$ C.F. $= c_1z^2 + c_2z^{-2} = c_1(3x + 2)^2 + c_2(3x + 2)^{-2}$

Further

$$\text{P.I.} = \frac{1}{(\theta^2 - 4)}\left(\frac{z^2}{27}\right) - \frac{1}{(\theta^2 - 4)}\left(\frac{1}{27}\right)$$

$$= \frac{1}{27}\cdot\frac{1}{(\theta^2 - 4)}z^2 - \frac{1}{27}\cdot\frac{1}{(0^2 - 4)} = \frac{1}{27}\cdot\frac{1}{(\theta - 2)(\theta + 2)}(z^2) + \frac{1}{108}$$

$$= \frac{1}{27}\cdot\frac{1}{(\theta - 2)}z^{-2}\int z^2 \cdot z^{2-1}\,dz + \frac{1}{108} = \frac{1}{27}\cdot\frac{1}{(\theta - 2)}z^{-2}\cdot\frac{z^4}{4} + \frac{1}{108}$$

$$= \frac{1}{108}\cdot\frac{1}{(\theta - 2)}z^2 + \frac{1}{108} = \frac{1}{108}z^2\int z^2\cdot z^{-2-1}\,dz + \frac{1}{108}$$

$$= \frac{1}{108}z^2\log|z| + \frac{1}{108} = \frac{1}{108}[(3x + 2)^2\log|3x + 2| + 1]$$

Hence, the general solution of the given equation is

$$y = c_1(3x + 2)^2 + c_2(3x + 2)^{-2} + \frac{1}{108}[(3x + 2)^2\log|3x + 2| + 1]$$

EXERCISES

Solve the following differential equations:

1. $(2x-1)^3 \frac{d^3y}{dx^3} + (2x-1)\frac{dy}{dx} - 2y = 0$

2. $(x+a)^2 \frac{d^2y}{dx^2} - 4(x+a)\frac{dy}{dx} + 6y = x$

3. $(5+2x)^2 \frac{d^2y}{dx^2} - 6(5+2x)\frac{dy}{dx} + 8y = 0$

4. $(x+1)^2 \frac{d^2y}{dx^2} - 3(x+1)\frac{dy}{dx} + 4y = x^2$

5. $(1+2x)^2 \frac{d^2y}{dx^2} - 6(1+2x)\frac{dy}{dx} + 16y = 8(1+2x)^2$

MISCELLANEOUS EXERCISES

Solve the following differential equations:

1. $x^3 \frac{d^3y}{dx^3} + 6x^2 \frac{d^2y}{dx^2} + 4x\frac{dy}{dx} - 4y = 0$

2. $x^3 \frac{d^3y}{dx^3} - 3x^2 \frac{d^2y}{dx^2} + 7x\frac{dy}{dx} - 8y = 0$

3. $x^2 \frac{d^2y}{dx^2} + x\frac{dy}{dx} + y = x$

4. $x^2 \frac{d^2y}{dx^2} + 4x\frac{dy}{dx} + 2y = e^x$

5. $x^2 \frac{d^2y}{dx^2} + x\frac{dy}{dx} - y = x^m$

6. $x^2 \frac{d^2y}{dx^2} + 2x\frac{dy}{dx} = \log x$

7. $x^2 \frac{d^2y}{dx^2} + 5x\frac{dy}{dx} + 4y = x \log x$

8. $x^3 \frac{d^3y}{dx^3} + 3x^2 \frac{d^2y}{dx^2} + x\frac{dy}{dx} + y = x + \log x$

9. $x^2 \frac{d^2y}{dx^2} - x\frac{dy}{dx} + 4y = \cos(\log x) + x \sin(\log x)$

10. $x^4 \frac{d^4y}{dx^4} + 2x^3 \frac{d^3y}{dx^3} + x^2 \frac{d^2y}{dx^2} - x\frac{dy}{dx} + y = \log x$

11. $x^2 \frac{d^2y}{dx^2} - 6y = x^3 \log x$

12. $(x+3)^2 \dfrac{d^2y}{dx^2} - 4(x+3)\dfrac{dy}{dx} + 6y = x$

13. $(x+1)^2 \dfrac{d^2y}{dx^2} + (x+1)\dfrac{dy}{dx} = (2x+3)(2x+4)$

14. $x^2 \dfrac{d^2y}{dx^2} - 3x\dfrac{dy}{dx} + y = \dfrac{\log x \cdot \sin(\log x) + 1}{x}$

15. $x^3 \dfrac{d^3y}{dx^3} + 6x^2 \dfrac{d^2y}{dx^2} + 8x\dfrac{dy}{dx} + 2y = x^2 + 3x - 4$

16. $x^4 \dfrac{d^4y}{dx^4} + 6x^3 \dfrac{d^3y}{dx^3} + 9x^2 \dfrac{d^2y}{dx^2} + 3x\dfrac{dy}{dx} + y = (1 + \log x)^2$

15

Linear Differential Equations of Second Order

15.1 Introduction

In this chapter we shall consider linear differential equations of the form

$$\frac{d^2y}{dx^2} + P\frac{dy}{dx} + Qy = R \qquad (1)$$

where P, Q and R are functions of x.

Since the general method of solving such equations can not be given, we shall consider some particular cases in which the integral can be obtained.

15.2 Complete Solution in Terms of Known Integral Belonging to the Complementary Function

If an integral included in the complementary function of a second order linear differential equation be known, its order can be depressed and the complete solution can be obtained.

Let $y = u$ be an integral in the complementary function of Eq. (1). Putting $y = uv$. Then

$$y_1 = u_1v + uv_1$$

and

$$y_2 = u_2v + 2u_1v_1 + uv_2$$

Substituting these values of y, y_1 and y_2 in Eq. (1), we get

$$uv_2 + (2u_1 + Pu)\, v_1 + (u_2 + Pu_1 + Qu)\, v = R \qquad (2)$$

But $y = u$ is a solution of Eq. (1). Therefore, the coefficient of v will vanish. Hence, Eq. (2) becomes

$$v_2 + \left(\frac{2u_1}{u} + P\right) v_1 = \frac{R}{u}$$

This is a linear equation in v_1. Hence v_1 can be obtained and then

$$y = u \int v_1\, dx + c_1 u$$

is the integral of Eq. (1). The other constant of integration will occur in expression for v_1. Thus we get the complete solution.

Note. The following rules will help in finding the complementary function.

(i) If $P + Qx = 0$, then $y = x$ is a part of C.F.
(ii) If $1 + P + Q = 0$, then $y = e^x$ is a part of C.F.
(iii) If $1 + \frac{P}{a} + \frac{Q}{a^2} = 0$, then $y = e^{ax}$ is a part of C.F.
(iv) If $1 - P + Q = 0$, then $y = e^{-x}$ is a part of C.F.
(v) If $2 + 2Px + Qx^2 = 0$, then $y = x^2$ is a part of C.F.
(vi) If $m\,(m - 1) + P\,mx + Qx^2 = 0$, then $y = x^m$ is a part of C.F.

Examples

1. Solve the equation

$$(1 - x^2)\frac{d^2y}{dx^2} + x\frac{dy}{dx} - y = x(1 - x^2)^{\frac{3}{2}} \tag{1}$$

Solution. Here $P + Qx = 0$, therefore $y = x$ is a solution of the equation

$$(1 - x^2)\frac{d^2y}{dx^2} + x\frac{dy}{dx} - y = 0$$

Let $y = vx$. Then

$$\frac{dy}{dx} = v + x\frac{dv}{dx}$$

and

$$\frac{d^2y}{dx^2} = x\frac{d^2v}{dx^2} + 2\frac{dv}{dx}$$

Substituting these values of y, $\frac{dy}{dx}$ and $\frac{d^2y}{dx^2}$ in Eq. (1), we get

$$\frac{d^2v}{dx^2} + \left(\frac{2}{x} + \frac{x}{1 - x^2}\right)\frac{dv}{dx} = (1 - x^2)^{\frac{1}{2}}$$

Putting $\frac{dv}{dx} = p$, this equation reduces to

$$\frac{dp}{dx} + \left(\frac{2}{x} + \frac{x}{1 - x^2}\right)p = (1 - x^2)^{\frac{1}{2}} \tag{2}$$

This is a linear equation. Therefore

$$\text{I.F.} = e^{\int\left(\frac{2}{x}+\frac{x}{1-x^2}\right)dx} = \frac{x^2}{(1-x^2)^{\frac{1}{2}}}$$

Multiplying Eq. (2) by the I.F. $\dfrac{x^2}{(1-x^2)^{\frac{1}{2}}}$, we get

$$\frac{x^2}{(1-x^2)^{\frac{1}{2}}}\frac{dp}{dx} + \left(\frac{2x}{(1-x^2)^{\frac{1}{2}}} + \frac{x^3}{(1-x^2)^{3/2}}\right)p = x^2$$

On integrating, we get

$$p \cdot \frac{x^2}{\sqrt{1-x^2}} = \frac{1}{3}x^3 + c_1$$

$$\Rightarrow \qquad p = \frac{dv}{dx} = \frac{1}{3}x\sqrt{1-x^2} + c_1\frac{\sqrt{1-x^2}}{x^2}$$

Further, on integrating, we get

$$v = -\frac{1}{9}(1-x^2)^{\frac{3}{2}} + c_1\left(\sin^{-1}x + \frac{\sqrt{1-x^2}}{x}\right) + c_2$$

Hence, the complete solution is

$$y = vx = -\frac{1}{9}x(1-x^2)^{\frac{3}{2}} + c_1(x\sin^{-1}x + \sqrt{1-x^2}) + c_2 x$$

2. Solve the equation

$$x^2\frac{d^2y}{dx^2} - 2x(1+x)\frac{dy}{dx} + 2(1+x)y = x^3$$

Solution. The given equation can be written as

$$\frac{d^2y}{dx^2} - \frac{2(1+x)}{x}\frac{dy}{dx} + \frac{2(1+x)}{x^2}y = x \tag{1}$$

Here $P = -\dfrac{2(1+x)}{x}$ and $Q = \dfrac{2(1+x)}{x^2}$. Since $P + Qx = 0$, $y = x$ is a solution of the equation

$$\frac{d^2y}{dx^2} - \frac{2(1+x)}{x}\frac{dy}{dx} + \frac{2(1+x)}{x^2}y = 0$$

Let $y = vx$. Then

$$\frac{dy}{dx} = v + x\frac{dv}{dx}$$

and
$$\frac{d^2y}{dx^2} = x\frac{d^2v}{dx^2} + 2\frac{dv}{dx}$$

Substituting these values of y, $\frac{dy}{dx}$ and $\frac{d^2y}{dx^2}$ in Eq. (1), we get

$$x\frac{d^2v}{dx^2} + 2\frac{dv}{dx} - \frac{2(1+x)}{x}\left(x\frac{dv}{dx} + v\right) + \frac{2(1+x)}{x^2}vx = x$$

$$\Rightarrow \qquad x\frac{d^2v}{dx^2} - 2x\frac{dv}{dx} = x$$

$$\Rightarrow \qquad \frac{d^2v}{dx^2} - 2\frac{dv}{dx} = 1$$

Putting $\frac{dv}{dx} = p$, this equation reduces to

$$\frac{dp}{dx} - 2p = 1 \tag{2}$$

This is a linear equation. Therefore

$$\text{I.F.} = e^{\int -2dx} = e^{-2x}$$

Multiplying Eq. (2) by the I.F. e^{-2x}, we get

$$e^{-2x}\frac{dp}{dx} - 2pe^{-2x} = e^{-2x}$$

On integrating, we get

$$pe^{-2x} = -\frac{1}{2}e^{-2x} + c_1$$

$$\Rightarrow \qquad p = \frac{dv}{dx} = -\frac{1}{2} + c_1e^{2x}$$

Further, on integrating, we get

$$v = -\frac{1}{2}x + \frac{c_1}{2}e^{2x} + c_2$$

Hence, the complete solution is

$$y = vx = -\frac{1}{2}x^2 + \frac{c_1}{2}xe^{2x} + c_2x$$

3. Solve the equation

$$\frac{d^2y}{dx^2} - \cot x \frac{dy}{dx} - (1 - \cot x)\, y = e^x \sin x$$

Solution. Here $P = -\cot x$ and $Q = -(1 - \cot x)$. Since $1 + P + Q = 0$, $y = e^x$ is a part of C.F. Let $y = ve^x$. Then

$$\frac{dy}{dx} = e^x\left(\frac{dv}{dx} + v\right) \quad \text{and} \quad \frac{d^2y}{dx^2} = e^x\left(\frac{d^2v}{dx^2} + \frac{2\,dv}{dx} + v\right)$$

Substituting these values of y, $\frac{dy}{dx}$ and $\frac{d^2y}{dx^2}$ in the given equation, it reduces to

$$e^x\left[\left(\frac{d^2v}{dx^2} + 2\frac{dv}{dx} + v\right) - \cot x\left(\frac{dv}{dx} + v\right) - (1 - \cot x)\, v\right] = e^x \sin x$$

$$\Rightarrow \qquad \frac{d^2v}{dx^2} + (2 - \cot x)\frac{dv}{dx} = \sin x$$

Putting $\frac{dv}{dx} = p$, this equation reduces to

$$\frac{dp}{dx} + (2 - \cot x)\, p = \sin x \tag{1}$$

This is a linear equation. Therefore

$$\text{I.F.} = e^{\int (2-\cot x)dx} = e^{2x - \log \sin x} = \frac{e^{2x}}{\sin x}$$

Multiplying Eq. (1) by the I.F. $\frac{e^{2x}}{\sin x}$, we get

$$\frac{e^{2x}}{\sin x}\frac{dp}{dx} + (2 - \cot x)\; p\,\frac{e^{2x}}{\sin x} = e^{2x}$$

On integrating, we get

$$p\,\frac{e^{2x}}{\sin x} = \frac{1}{2}e^{2x} + c_1$$

$$\Rightarrow \qquad p = \frac{dv}{dx} = \frac{1}{2}\sin x + c_1 e^{-2x}\sin x$$

Further, on integrating, we get

$$v = -\frac{1}{2}\cos x - \frac{1}{5}c_1 e^{-2x}(\cos x + 2\sin x) + c_2$$

Hence, the complete solution is

$$y = ve^x = -\frac{1}{2}e^x \cos x - \frac{1}{5}c_1e^{-x}(\cos x + 2\sin x) + c_2e^x$$

EXERCISES

Solve the following differential equations:

1. $(x+2)\frac{d^2y}{dx^2} - (2x+5)\frac{dy}{dx} + 2y = (x+1)e^x$

2. $x\frac{d^2y}{dx^2} - (2x-1)\frac{dy}{dx} + (x-1)y = 0$

3. $\frac{d^2y}{dx^2} - x^2\frac{dy}{dx} + xy = x$

4. $\frac{d^2y}{dx^2} - \frac{3}{x}\frac{dy}{dx} + \frac{3}{x^2}y = 2x - 1$

5. $x^2\frac{d^2y}{dx^2} - (x^2+2x)\frac{dy}{dx} + (x+2)y = x^3e^x$

6. $(x-1)\frac{d^2y}{dx^2} - x\frac{dy}{dx} + y = (x-1)^2$

7. $(x+1)\frac{d^2y}{dx^2} - 2(x+3)\frac{dy}{dx} + (x+5)y = e^x$

8. $(3-x)\frac{d^2y}{dx^2} - (9-4x)\frac{dy}{dx} + (6-3x)y = 0$

9. $x\frac{d^2y}{dx^2} + (1-x)\frac{dy}{dx} - y = e^x$

10. $(x\sin x + \cos x)\frac{d^2y}{dx^2} - x\cos x\frac{dy}{dx} + y\cos x = \sin x(x\sin x + \cos x)^2$

15.3 Removal of the First Derivative: Normal Form

If an integral included in the complementary function can not be determined easily, it is sometimes useful to reduce Eq. (1) of Sec. 15.1 to the normal form in which the term containing the first derivative is missing. To achieve the goal we put $y = uv$ and proceeding on the same lines as in Sec. 15.2, we get

$$v_2 + \left(\frac{2u_1}{u} + P\right)v_1 + \left(\frac{u_2}{u} + \frac{Pu_1}{u} + Q\right)v = \frac{R}{u}$$

Now, we choose u such that the first derivative is removed, i.e.

$$\frac{2u_1}{u} + P = 0$$

This gives $u = e^{-\frac{1}{2}\int P\,dx}$. Hence the above equation becomes

$$\frac{d^2v}{dx^2} + Q_1 v = R_1$$

where $Q_1 = Q - \frac{1}{2}\frac{dP}{dx} - \frac{1}{4}P^2$ and $R_1 = Re^{\frac{1}{2}\int P\,dx}$. If the value of Q_1 is a constant or a constant divided by x^2, the above equation becomes integrable.

Note. The equation $\frac{d^2v}{dx^2} + Q_1 v = R_1$ is called the *normal form* of Eq. (1) of Sec. 15.1.

Examples

1. Solve the equation

$$\frac{d^2y}{dx^2} - 2\tan x\frac{dy}{dx} + 5y = e^x \sec x$$

Solution. Here $P = -2\tan x$, $Q = 5$ and $R = e^x \sec x$, and hence

$$u = e^{-\frac{1}{2}\int P\,dx} = e^{\int \tan x} = e^{\log \sec x} = \sec x$$

Now, putting $y = uv$, the given equation becomes

$$\frac{d^2v}{dx^2} + Q_1 v = R_1$$

where $$Q_1 = Q - \frac{1}{2}\frac{dP}{dx} - \frac{1}{4}P^2 = 5 + \sec^2 x - \tan^2 x = 6$$

and $$R_1 = Re^{\frac{1}{2}\int P\,dx} = e^x \sec x \cdot e^{\frac{1}{2}\int (-2\tan x)\,dx} = e^x \sec x \cos x = e^x$$

Hence after removing the first derivative term, the given equation becomes

$$\frac{d^2v}{dx^2} + 6v = e^x$$

The auxiliary equation is

$$m^2 + 6 = 0 \Rightarrow m = \pm\sqrt{6}\,i$$

Therefore $$\text{C.F.} = c_1 \cos\sqrt{6}\,x + c_2 \sin\sqrt{6}\,x$$

Further $$\text{P.I.} = \frac{1}{D^2 + 6}e^x = \frac{e^x}{7}$$

Thus the complete solution is

$$y = uv = \sec x\,(c_1 \cos \sqrt{6}\, x + c_2 \sin \sqrt{6}\, x) + \frac{1}{7} e^x \sec x$$

2. Solve the equation

$$x^2 \frac{d^2y}{dx^2} - 2(x^2 + x)\frac{dy}{dx} + (x^2 + 2x + 2)y = 0$$

Solution. Dividing by x^2, the given equation is written in the standard form as

$$\frac{d^2y}{dx^2} - 2\left(1 + \frac{1}{x}\right)\frac{dy}{dx} + \left(1 + \frac{2}{x} + \frac{2}{x^2}\right)y = 0$$

Here $P = -2\left(1 + \frac{1}{x}\right)$, $Q = 1 + \frac{2}{x} + \frac{2}{x^2}$ and $R = 0$, and hence

$$u = e^{-\frac{1}{2}\int P\,dx} = e^{\int \left(1+\frac{1}{x}\right)dx} = e^{x+\log x} = xe^x$$

Now, putting $y = uv$, the given equation becomes

$$\frac{d^2v}{dx^2} + Q_1 v = 0$$

where $Q_1 = Q - \frac{1}{2}\frac{dP}{dx} - \frac{1}{4}P^2 = 1 + \frac{2}{x} + \frac{2}{x^2} - \frac{1}{x^2} - \left(1 + \frac{1}{x}\right)^2 = 0.$

Hence, the transformed equation is

$$\frac{d^2v}{dx^2} = 0$$

$$\Rightarrow \qquad \frac{dv}{dx} = c_1$$

$$\Rightarrow \qquad v = c_1 x + c_2$$

Thus the complete solution of the given equation is

$$y = uv = xe^x\,(c_1 x + c_2) = e^x(c_1 x^2 + c_2 x)$$

3. Solve the equation

$$\frac{d^2y}{dx^2} - 4x\frac{dy}{dx} + (4x^2 - 1)y = -3e^{x^2} \sin 2x$$

Solution. Here $P = -4x$, $Q = 4x^2 - 1$ and $R = -3e^{x^2} \sin 2x$, and hence to remove the first derivative term, we choose

$$u = e^{-\frac{1}{2}\int P\,dx} = e^{\int 2x\,dx} = e^{x^2}$$

Now, putting $y = uv$, the given equation becomes

$$\frac{d^2v}{dx^2} + Q_1 v = R_1$$

where $Q_1 = Q - \frac{1}{2}\frac{dP}{dx} - \frac{1}{4}P^2 = 4x^2 - 1 - \frac{1}{2}(-4) - \frac{1}{4}\cdot 16x^2 = 1$

and $R_1 = Re^{\frac{1}{2}\int P dx} = -3e^{x^2}\sin 2x \cdot e^{\int(-2x)dx}$

$$= -3e^{x^2}\sin 2x \cdot e^{-x^2} = -3\sin 2x$$

Hence, the transformed equation is

$$\frac{d^2v}{dx^2} + v = -3\sin 2x$$

The auxiliary equation is

$$m^2 + 1 = 0 \quad \Rightarrow \quad m = \pm i$$

Therefore $\qquad$ C.F. $= c_1 \cos x + c_2 \sin x$

Further $\quad$ P.I. $= \dfrac{1}{(D^2+1)}(-3\sin 2x) = -\dfrac{3\sin 2x}{(-2^2+1)} = \sin 2x$

Hence $\qquad v = c_1 \cos x + c_2 \sin x + \sin 2x$

Thus the complete solution of the given equation is

$$y = uv = e^{x^2}(c_1 \cos x + c_2 \sin x + \sin 2x)$$

EXERCISES

Solve the following differential equations:

1. $\dfrac{d^2y}{dx^2} - 4x\dfrac{dy}{dx} + (4x^2 - 3)y = e^{x^2}$

2. $\dfrac{d^2y}{dx^2} - 2x\dfrac{dy}{dx} + (x^2 + 2)y = e^{\frac{1}{2}(x^2+2x)}$

3. $\dfrac{d^2y}{dx^2} + 4x\dfrac{dy}{dx} + 4x^2 y = 0$

4. $\dfrac{d^2y}{dx^2} - 2bx\dfrac{dy}{dx} + b^2x^2 y = 0$

5. $\dfrac{d^2y}{dx^2} + \dfrac{2}{x}\dfrac{dy}{dx} + n^2 y = 0$

6. $\dfrac{d^2y}{dx^2} + 2x\dfrac{dy}{dx} + (x^2 + 1)y = x(x^2 + 3)$

7. $\dfrac{d^2y}{dx^2} - \dfrac{2}{x}\dfrac{dy}{dx} + \left(1 + \dfrac{2}{x^2}\right) y = xe^x$

8. $4x^2 \dfrac{d^2y}{dx^2} + 4x^5 \dfrac{dy}{dx} + (x^8 + 6x^4 + 4)\, y = 0$

9. $x^2 \dfrac{d^2y}{dx^2} - 2x(3x-2)\dfrac{dy}{dx} + 3x(3x-4)\, y = e^{3x}$

10. $\dfrac{d^2y}{dx^2} + 2x\dfrac{dy}{dx} + (x^2+5)\, y = xe^{-\frac{1}{2}x^2}$

11. $\dfrac{d^2y}{dx^2} - 4x\dfrac{dy}{dx} + 4x^2 y = e^{x^2}$

12. $4x\dfrac{d^2y}{dx^2} - 4(x+2)\dfrac{dy}{dx} + (4+x)\, y = 0$

13. $\dfrac{d^2y}{dx^2} - 2\tan x\dfrac{dy}{dx} + 5y = 0$

15.4 Change of Independent Variable

Sometimes by changing the independent variable, the equation reduces to a form which is easily integrable.

Let the equation of the second order be

$$\frac{d^2y}{dx^2} + P\frac{dy}{dx} + Qy = R \tag{1}$$

where P, Q and R are functions of x. Let the independent variable x be changed to a variable z, z being a given function of x. Then

$$\frac{dy}{dx} = \frac{dy}{dz}\frac{dz}{dx}$$

and $$\frac{d^2y}{dx^2} = \frac{d}{dx}\left(\frac{dy}{dx}\right) = \frac{d}{dx}\left(\frac{dy}{dz}\frac{dz}{dx}\right) = \frac{d}{dx}\left(\frac{dy}{dz}\right)\frac{dz}{dx} + \frac{dy}{dz}\cdot\frac{d^2z}{dx^2}$$

$$= \frac{d}{dz}\left(\frac{dy}{dz}\right)\left(\frac{dz}{dx}\right)^2 + \frac{dy}{dz}\cdot\frac{d^2z}{dx^2} = \frac{d^2y}{dz^2}\left(\frac{dz}{dx}\right)^2 + \frac{dy}{dz}\cdot\frac{d^2z}{dx^2}$$

Substituting these values of $\dfrac{dy}{dx}$ and $\dfrac{d^2y}{dx^2}$ in Eq. (1), we get

$$\frac{d^2y}{dz^2}\left(\frac{dz}{dx}\right)^2 + \frac{dy}{dz}\cdot\frac{d^2z}{dx^2} + P\left(\frac{dy}{dz}\frac{dz}{dx}\right) + Qy = R$$

$$\Rightarrow \quad \frac{d^2y}{dz^2}\left(\frac{dz}{dx}\right)^2 + \left(\frac{d^2z}{dx^2} + P\frac{dz}{dx}\right)\frac{dy}{dz} + Qy = R$$

$$\Rightarrow \qquad \frac{d^2y}{dz^2} + \frac{\left(\dfrac{d^2z}{dx^2} + P\dfrac{dz}{dx}\right)}{\left(\dfrac{dz}{dx}\right)^2}\frac{dy}{dz} + \frac{Qy}{\left(\dfrac{dz}{dx}\right)^2} = \frac{R}{\left(\dfrac{dz}{dx}\right)^2}$$

This equation can be written as

$$\frac{d^2y}{dz^2} + P_1\frac{dy}{dz} + Q_1y = R_1 \qquad (2)$$

where $P_1 = \left(\dfrac{d^2z}{dx^2} + P\dfrac{dz}{dx}\right)\Big/\left(\dfrac{dz}{dx}\right)^2$, $Q_1 = Q\Big/\left(\dfrac{dz}{dx}\right)^2$ and $R_1 = R\Big/\left(\dfrac{dz}{dx}\right)^2$.

Clearly, P_1, Q_1 and R_1 are functions of x as shown above and can be easily expressed as functions of z by the given relation between z and x. After getting Eq. (2) we choose z such that Eq. (2) can be easily integrated. For this purpose, we consider the following cases.

Case I. When $P_1 = 0$. If we choose z such that $P_1 = 0$, i.e.

$$\frac{d^2z}{dx^2} + P\frac{dz}{dx} = 0$$

then, we have

$$\frac{d}{dx}\left(\frac{dz}{dx}\right) + P\frac{dz}{dx} = 0$$

$$\Rightarrow \qquad \frac{dz}{dx} = e^{-\int P\,dx} \quad \Rightarrow \quad z = \int (e^{-\int P\,dx})\,dx$$

In this case Eq. (2) becomes

$$\frac{d^2y}{dz^2} + Q_1y = R_1$$

This equation can be solved if Q_1 is a constant or of the form $\dfrac{\lambda}{z^2}$ (λ is constant).

Case II. When $Q_1 = a^2$. If we choose z such that $Q_1 = a^2$, then

$$\frac{Q}{\left(\dfrac{dz}{dx}\right)^2} = a^2$$

$$\Rightarrow \qquad a\frac{dz}{dx} = \sqrt{Q}$$

$$\Rightarrow \qquad az = \int \sqrt{Q}\,dx$$

In this case Eq. (2) becomes

$$\frac{d^2y}{dz^2} + P_1 \frac{dy}{dz} + a^2 y = R_1$$

If P_1 comes out to be constant, then the above equation can be easily integrated.

Examples

1. Solve the equation

$$\frac{d^2y}{dx^2} + \cot x \frac{dy}{dx} + 4y \operatorname{cosec}^2 x = 0$$

Solution. Here $P = \cot x$, $Q = 4 \operatorname{cosec}^2 x$, $R = 0$. Changing the independent variable from x to z, the given equation becomes

$$\frac{d^2y}{dz^2} + P_1 \frac{dy}{dz} + Q_1 y = 0$$

where $P_1 = \left(\dfrac{d^2z}{dx^2} + P\dfrac{dz}{dx}\right) \Big/ \left(\dfrac{dz}{dx}\right)^2$ and $Q_1 = Q \Big/ \left(\dfrac{dz}{dx}\right)^2$.

Choose z such that $Q_1 = a^2 = 1$ (say), then

$$\left(\frac{dz}{dx}\right)^2 = Q = 4 \operatorname{cosec}^2 x$$

$$\Rightarrow \qquad \frac{dz}{dx} = 2 \operatorname{cosec} x \Rightarrow z = 2 \log \tan \frac{x}{2}$$

Therefore $$P_1 = \frac{-2 \operatorname{cosec} x \cot x + 2 \cot x \operatorname{cosec} x}{4 \operatorname{cosec}^2 x} = 0$$

Hence, the transformed equation is

$$\frac{d^2y}{dz^2} + y = 0$$

The auxiliary equation is

$$m^2 + 1 = 0 \quad \Rightarrow \quad m = \pm i$$

Therefore $$y = c_1 \cos z + c_2 \sin z$$

Hence, the solution of the given equation is

$$y = c_1 \cos\left(2 \log \tan \frac{x}{2}\right) + c_2 \sin\left(2 \log \tan \frac{x}{2}\right)$$

2. Solve the equation $x\dfrac{d^2y}{dx^2} - \dfrac{dy}{dx} - 4x^3y = 8x^3 \sin x^2$.

Solution. Dividing by x, the given equation can be written in the standard form as

$$\frac{d^2y}{dx^2} - \frac{1}{x}\frac{dy}{dx} - 4x^2y = 8x^2 \sin x^2$$

Here $P = -\dfrac{1}{x}$, $Q = -4x^2$ and $R = 8x^2 \sin x^2$. Changing the independent variable from x to z, the above equation becomes

$$\frac{d^2y}{dz^2} + P_1\frac{dy}{dz} + Q_1y = R_1$$

where $P_1 = \left(\dfrac{d^2z}{dx^2} + P\dfrac{dz}{dx}\right)\Big/\left(\dfrac{dz}{dx}\right)^2$, $Q_1 = Q\Big/\left(\dfrac{dz}{dx}\right)^2$ and $R_1 = R\Big/\left(\dfrac{dz}{dx}\right)^2$.

Choose z such that $Q_1 = -1$ (say), then

$$\left(\frac{dz}{dx}\right)^2 = -Q = 4x^2$$

$$\Rightarrow \qquad \frac{dz}{dx} = 2x \quad \Rightarrow \quad z = x^2$$

Therefore

$$P_1 = \frac{2 - \frac{1}{x}\cdot 2x}{4x^2} = 0 \quad \text{and} \quad R_1 = \frac{8x^2 \sin x^2}{4x^2} = 2\sin z$$

Hence, the transformed equation is

$$\frac{d^2y}{dz^2} - y = 2\sin z$$

The auxiliary equation is

$$m^2 - 1 = 0 \quad \Rightarrow \quad m = 1, -1$$

Therefore $\qquad$ $\text{C.F.} = c_1e^z + c_2e^{-z}$

Further $\qquad$ $\text{P.I.} = \dfrac{1}{(D^2 - 1)}2\sin z = \dfrac{2\sin z}{-1^2 - 1} = -\sin z$

Hence, the solution of the given equation is

$$y = c_1e^z + c_2e^{-z} - \sin z$$

i.e. $\qquad$ $y = c_1e^{x^2} + c_2e^{-x^2} - \sin x^2$

3. Solve the equation

$$x^6 \frac{d^2y}{dx^2} + 3x^5 \frac{dy}{dx} + a^2 y = \frac{1}{x^2}$$

Solution. Dividing by x^6, the given equation can be written in the standard form as

$$\frac{d^2y}{dx^2} + \frac{3}{x}\frac{dy}{dx} + \frac{a^2}{x^6} y = \frac{1}{x^8}$$

Here $P = \frac{3}{x}$, $Q = \frac{a^2}{x^6}$ and $R = \frac{1}{x^8}$. Changing the independent variable from x to z, the above equation becomes

$$\frac{d^2y}{dz^2} + P_1 \frac{dy}{dz} + Q_1 y = R_1$$

where $P_1 = \left(\frac{d^2z}{dx^2} + P\frac{dz}{dx}\right) \Big/ \left(\frac{dz}{dx}\right)^2$, $Q_1 = Q \Big/ \left(\frac{dz}{dx}\right)^2$ and $R_1 = R \Big/ \left(\frac{dz}{dx}\right)^2$.

Choose z such that $Q = a^2$ (say), then

$$a^2\left(\frac{dz}{dx}\right)^2 = Q = \frac{a^2}{x^6}$$

$$\Rightarrow \qquad \frac{dz}{dx} = \frac{1}{x^3} \quad \Rightarrow \quad z = -\frac{1}{2x^2}$$

$$\text{Therefore } P_1 = \frac{-\frac{3}{x^4} + \frac{3}{x}\cdot\frac{1}{x^3}}{\frac{1}{x^6}} = 0 \quad \text{and} \quad R_1 = \frac{\frac{1}{x^8}}{\frac{1}{x^6}} = \frac{1}{x^2} = -2z$$

Hence, the transformed equation is

$$\frac{d^2y}{dz^2} + a^2 y = -2z$$

The auxiliary equation is

$$m^2 + a^2 = 0 \quad \Rightarrow \quad m = \pm ai$$

Therefore $\qquad$ C.F. $= c_1 \cos az + c_2 \sin az$

Further $\qquad$ $\text{P.I.} = \frac{1}{(D^2 + a^2)}(-2z) = -\frac{2}{a^2}\left(1 + \frac{D^2}{a^2}\right)^{-1}(z)$

$$= -\frac{2}{a^2}\left(1 - \frac{D^2}{a^2} + \ldots\right)(z) = -\frac{2z}{a^2}$$

Hence, the solution of the given equation is

$$y = c_1 \cos az + c_2 \sin az - \frac{2z}{a^2}$$

i.e.
$$y = c_1 \cos\left(-\frac{a}{2x^2}\right) + c_2 \sin\left(-\frac{a}{2x^2}\right) + \frac{1}{a^2x^2}$$

$\Rightarrow$
$$y = c_1 \cos\frac{a}{2x^2} - c_2 \sin\frac{a}{2x^2} + \frac{1}{a^2x^2}$$

4. Solve the equation

$$\frac{d^2y}{dx^2} - (8e^{2x} + 2)\frac{dy}{dx} + 4e^{4x}y = e^{6x}$$

Solution. Here $P = -8e^{2x} - 2$, $Q = 4e^{4x}$ and $R = e^{6x}$. Changing the independent variable from x to z, the above equation becomes

$$\frac{d^2y}{dz^2} + P_1\frac{dy}{dz} + Q_1 y = R_1$$

where $P_1 = \left(\dfrac{d^2z}{dx^2} + P\dfrac{dz}{dx}\right)\Big/\left(\dfrac{dz}{dx}\right)^2$, $Q_1 = Q\Big/\left(\dfrac{dz}{dx}\right)^2$ and $R_1 = R\Big/\left(\dfrac{dz}{dx}\right)^2$.

Choose z such that $Q = 1$ (say), then

$$\left(\frac{dz}{dx}\right)^2 = Q = 4e^{4x}$$

$\Rightarrow$
$$\frac{dz}{dx} = 2e^{2x} \quad\Rightarrow\quad z = e^{2x}$$

Therefore
$$P_1 = \frac{4e^{2x} - (8e^{2x} + 2)\cdot 2e^{2x}}{4e^{4x}} = -4$$

and
$$R_1 = \frac{e^{6x}}{4e^{4x}} = \frac{1}{4}e^{2x} = \frac{1}{4}z$$

Hence, the transformed equation is

$$\frac{d^2y}{dz^2} - 4\frac{dy}{dz} + y = \frac{1}{4}z$$

The auxiliary equation is

$$m^2 - 4m + 1 = 0 \quad\Rightarrow\quad m = \frac{4 \pm \sqrt{16 - 4}}{2} = 2 \pm \sqrt{3}$$

Therefore $$\text{C.F.} = c_1 e^{(2+\sqrt{3})z} + c_2 e^{(2-\sqrt{3})z}$$

Further $$\text{P.I.} = \frac{1}{(D^2 - 4D + 1)}\left(\frac{1}{4}z\right) = \frac{1}{4}(1 - 4D + D^2)^{-1}(z)$$

$$= \frac{1}{4}(1 + 4D - D^2 \ldots)(z) = \frac{1}{4}(z + 4) = \frac{1}{4}z + 1$$

Hence, the solution of the given equation is

$$y = c_1 e^{(2+\sqrt{3})z} + c_2 e^{(2-\sqrt{3})z} + \frac{1}{4}z + 1$$

i.e. $$y = c_1 e^{(2+\sqrt{3})e^{2x}} + c_2 e^{(2-\sqrt{3})e^{2x}} + \frac{1}{4}e^{2x} + 1$$

EXERCISES

Solve the following differential equations:

1. $\dfrac{d^2y}{dx^2} + \dfrac{2}{x}\dfrac{dy}{dx} + \dfrac{a^2}{x^4}y = 0$

2. $(1 + x^2)^2 \dfrac{d^2y}{dx^2} + 2x(1 + x^2)\dfrac{dy}{dx} + 4y = 0$

3. $\dfrac{d^2y}{dx^2} - \cot x \dfrac{dy}{dx} - y \sin^2 x = 0$

4. $\sin^2 x \dfrac{d^2y}{dx^2} + \sin x \cos x \dfrac{dy}{dx} + 4y = 0$

5. $\cos x \dfrac{d^2y}{dx^2} + \sin x \dfrac{dy}{dx} - 2y \cos^3 x = 2 \cos^5 x$

6. $\dfrac{d^2y}{dx^2} + \left(1 - \dfrac{1}{x}\right)\dfrac{dy}{dx} + 4x^2 e^{-2x} y = 4(1 + x)e^{-x}$

7. $x \dfrac{d^2y}{dx^2} + (4x^2 - 1)\dfrac{dy}{dx} + 4x^3 y = 2x^3$

8. $\dfrac{d^2y}{dx^2} + \tan x \dfrac{dy}{dx} + y \cos^2 x = 0$

9. $\dfrac{d^2y}{dx^2} - \dfrac{1}{x}\dfrac{dy}{dx} + 4x^2 y = x^4$

10. $\dfrac{d^2y}{dx^2} + (3 \sin x - \cot x)\dfrac{dy}{dx} + 2y \sin^2 x = e^{-\cos x} \sin^2 x$

11. $\dfrac{d^2y}{dx^2} - (1 + 4e^x)\dfrac{dy}{dx} + 3e^{2x} y = e^{2(x+e^x)}$

15.5 Method of Variation of Parameters

We shall now explain the method for finding the complete solution of a linear differential equation whose complementary function is known.

Consider the equation

$$\frac{d^2y}{dx^2} + P\frac{dy}{dx} + Qy = R \tag{1}$$

Let $y = Au + Bv$ be the complementary function of Eq. (1), where A and B are constants and u, v are functions of x.

Further, we assume that

$$y = Au + Bv \tag{2}$$

is the complete solution of Eq. (1), where A and B are no longer constants but functions of x to be so chosen that Eq. (1) will be satisfied.

Differentiating Eq. (2), we get

$$\frac{dy}{dx} = Au_1 + Bv_1 + A_1u + B_1v$$

We choose A and B such that

$$A_1u + B_1v = 0 \tag{3}$$

Then

$$\frac{dy}{dx} = Au_1 + Bv_1 \tag{4}$$

Differentiating Eq. (4), we get

$$\frac{d^2y}{dx^2} = Au_2 + Bv_2 + A_1u_1 + B_1v_1 \tag{5}$$

Substituting the values of y, $\frac{dy}{dx}$ and $\frac{d^2y}{dx^2}$ in Eq. (1), we get

$$Au_2 + Bv_2 + A_1u_1 + B_1v_1 + P(Au_1 + Bv_1) + Q\,(Au + Bv) = R$$

$$\Rightarrow \quad A(u_2 + Pu_1 + Qu) + B\,(v_2 + Pv_1 + Qv) + A_1u_1 + B_1v_1 = R$$

But

$$u_2 + Pu_1 + Qu = 0 \text{ and } v_2 + Pv_1 + Qv = 0$$

because u and v are solutions of the equation

$$\frac{d^2y}{dx^2} + P\frac{dy}{dx} + Qy = 0$$

Therefore, Eq. (1) will reduce to

$$A_1u_1 + B_1v_1 = R \tag{6}$$

From Eqs. (3) and (6), we get

$$\frac{A_1}{-v} = \frac{B_1}{u} = \frac{R}{uv_1 - u_1 v}$$

$$\Rightarrow \qquad A_1 = \frac{-vR}{uv_1 - u_1 v} \quad \text{and} \quad B_1 = \frac{uR}{uv_1 - u_1 v}$$

On integrating, we get A and B. Let

$$A = f(x) + c_1 \quad \text{and} \quad B = F(x) + c_2$$

Then, the complete solution is

$$y = uf(x) + vF(x) + c_1 u + c_2 v$$

Note. A and B, the functions of x, are called parameters.

Examples

1. Solve the equation by the method of variation of parameters

$$x^2 \frac{d^2y}{dx^2} + x\frac{dy}{dx} - y = x^2 e^x$$

Solution. First of all we find the solution of the equation

$$x^2 \frac{d^2y}{dx^2} + x\frac{dy}{dx} - y = 0$$

This is homogeneous equation. Putting $x = e^z$ and $D \equiv \dfrac{d}{dz}$, the equation reduces to

$$[D(D-1) + D - 1]\, y = 0$$

$$\Rightarrow \qquad (D^2 - 1)\, y = 0$$

Therefore
$$y = c_1 e^z + c_2 e^{-z} = c_1 x + \frac{c_2}{x}$$

Writing the given equation in the standard form, we have

$$\frac{d^2y}{dx^2} + \frac{1}{x}\frac{dy}{dx} - \frac{1}{x^2}\, y = e^x \tag{1}$$

Let $y = Ax + \dfrac{B}{x}$ be a solution of Eq. (1), where A and B are functions of x. Then

$$\frac{dy}{dx} = A - \frac{B}{x^2} + A_1 x + \frac{B_1}{x}$$

Choose A and B such that

$$A_1 x + \frac{B_1}{x} = 0 \tag{2}$$

Therefore $$\frac{dy}{dx} = A - \frac{B}{x^2}$$

and $$\frac{d^2y}{dx^2} = A_1 + \frac{2B}{x^3} - \frac{B_1}{x^2}$$

Substituting the values of y, $\frac{dy}{dx}$ and $\frac{d^2y}{dx^2}$ in Eq. (1), we get

$$\left(A_1 + \frac{2B}{x^3} - \frac{B_1}{x^2}\right) + \frac{1}{x}\left(A - \frac{B}{x^2}\right) - \frac{1}{x^2}\left(Ax + \frac{B}{x}\right) = e^x$$

$$\Rightarrow \qquad A_1 - \frac{B_1}{x^2} = e^x \tag{3}$$

Solving Eqs. (2) and (3), we get

$$A_1 = \frac{1}{2}e^x \quad \text{and} \quad B_1 = -\frac{1}{2}x^2e^x$$

On integrating, we get

$$A = \frac{1}{2}e^x + a \quad \text{and} \quad B = -\frac{1}{2}e^x(x^2 - 2x + 2) + b$$

Hence, the complete solution is

$$y = Ax + \frac{B}{x}$$

i.e. $$y = \left(\frac{1}{2}e^x + a\right)x + \frac{1}{x}\left[-\frac{1}{2}e^x(x^2 - 2x + 2) + b\right]$$

$$\Rightarrow \qquad y = ax + \frac{b}{x} + e^x - \frac{e^x}{x}$$

2. Solve the equation by the method of variation of parameters

$$x^2\frac{d^2y}{dx^2} - 2x(1 + x)\frac{dy}{dx} + 2(1 + x) = x^3$$

Solution. Writing the given equation in standard form, we have

$$\frac{d^2y}{dx^2} - \frac{2}{x}(1 + x)\frac{dy}{dx} + \frac{2(1 + x)}{x^2}y = x \tag{1}$$

Here $$P + Qx = -\frac{2}{x}(1 + x) + \frac{2}{x}(1 + x) = 0.$$

Therefore $y = x$ is a part of complementary function of Eq. (1). Putting $y = vx$ in the equation

$$\frac{d^2y}{dx^2} - \frac{2}{x}(1 + x)\frac{dy}{dx} + \frac{2}{x^2}(1 + x)y = 0$$

it reduces to

$$\left(x\frac{d^2v}{dx^2} + 2\frac{dv}{dx}\right) - \frac{2}{x}(1 + x)\left(v + x\frac{dv}{dx}\right) + \frac{2(1 + x)}{x^2}vx = 0$$

$\Rightarrow$
$$\frac{d^2v}{dx^2} - 2\frac{dv}{dx} = 0$$

Therefore
$$v = c_1 + c_2 e^{2x}$$

Hence complementary function is

$$y = vx = c_1x + c_2xe^{2x}$$

Now, let $y = Ax + Bxe^{2x}$ be a solution of Eq. (1), where A and B are functions of x. Then

$$\frac{dy}{dx} = A + B(e^{2x} + 2xe^{2x}) + A_1x + B_1xe^{2x}$$

Choose A and B such that

$$A_1x + B_1xe^{2x} = 0 \tag{2}$$

Therefore
$$\frac{dy}{dx} = A + B(1 + 2x)e^{2x}$$

and
$$\frac{d^2y}{dx^2} = A_1 + B_1(1 + 2x)e^{2x} + 4B(1 + x)e^{2x}$$

Substituting the values of y, $\frac{dy}{dx}$ and $\frac{d^2y}{dx^2}$ in Eq. (1), we get

$$[A_1 + B_1(1 + 2x)e^{2x} + 4B(1 + x)e^{2x}] - \frac{2(1 + x)}{x}[A + B(1 + 2x)e^{2x}]$$

$$+ \frac{2(1 + x)}{x^2}[Ax + Bxe^{2x}] = x$$

$\Rightarrow$
$$A_1 + B_1(1 + 2x)e^{2x} = x \tag{3}$$

Solving Eqs. (2) and (3), we get

$$A_1 = -\frac{1}{2} \quad \text{and} \quad B_1 = \frac{1}{2}e^{-2x}$$

On integrating, we get

$$A = -\frac{1}{2}x + a \quad \text{and} \quad B = -\frac{1}{4}e^{-2x} + b$$

Hence, the complete solution is

$$y = Ax + Bx\, e^{2x}$$

i.e

$$y = -\frac{1}{2}x^2 - \frac{1}{4}x + ax + bxe^{2x}$$

EXERCISES

Solve the following equations by the method of variation of parameters:

1. $\dfrac{d^2y}{dx^2} + y = \operatorname{cosec} x$

2. $\dfrac{d^2y}{dx^2} - y = \dfrac{2}{1+e^x}$

3. $(1-x)\dfrac{d^2y}{dx^2} + x\dfrac{dy}{dx} - y = (1-x)^2$

4. $(1-x^2)\dfrac{d^2y}{dx^2} - 4x\dfrac{dy}{dx} - (1+x^2)\,y = x$

5. $\dfrac{d^2y}{dx^2} + (1-\cot x)\dfrac{dy}{dx} - y\cot x = \sin^2 x$

6. $\dfrac{d^2y}{dx^2} + n^2y = \sec nx$

7. $\dfrac{d^2y}{dx^2} + 4y = 4\tan 2x$

MISCELLANEOUS EXERCISES

Solve the following equations:

1. $\sin^2 x\dfrac{d^2y}{dx^2} = 2y$, given that $y = \cot x$ is a solution of it.

2. $(x\sin x + \cos x)\dfrac{d^2y}{dx^2} - x\cos x\dfrac{dy}{dx} + y\cos x = 0$, given that $y = x$ is a solution of it.

3. $\dfrac{d^2y}{dx^2} + \left(1 + \dfrac{2}{x}\cot x - \dfrac{2}{x^2}\right)y = x\cos x$, given that $y = \dfrac{\sin x}{x}$ is a solution of it.

4. $x\dfrac{d^2y}{dx^2} - 2(x+1)\dfrac{dy}{dx} + (x+2)\,y = (x-2)\,e^x$

5. $\dfrac{d^2y}{dx^2} - (1+x)\dfrac{dy}{dx} + xy = x$

6. $\dfrac{d^2y}{dx^2} - \dfrac{2}{x}\dfrac{dy}{dx} + \left(n^2 + \dfrac{2}{x^2}\right)y = 0$

7. $\dfrac{d^2y}{dx^2} + \dfrac{1}{x^{1/3}}\dfrac{dy}{dx} + \left(\dfrac{1}{4x^{2/3}} - \dfrac{1}{6x^{4/3}} - \dfrac{6}{x^2}\right)y = 0$

8. $\dfrac{d^2y}{dx^2} - \dfrac{1}{x^{1/2}}\dfrac{dy}{dx} + \dfrac{1}{4x^2}(-8 + x^{\frac{1}{2}} + x)\,y = 0$

9. $x^2\dfrac{d^2y}{dx^2} + (x - 4x^2)\dfrac{dy}{dx} + (1 - 2x + 4x^2)\,y = 0$

10. $\dfrac{d^2y}{dx^2} - \cot x\,\dfrac{dy}{dx} - y\sin^2 x = \cos x - \cos^3 x$

11. $\dfrac{d}{dx}\left(\cos^2 x\,\dfrac{dy}{dx}\right) + y\cos^2 x = 0$

12. $(1 + x)^2\,\dfrac{d^2y}{dx^2} + (1 + x)\,\dfrac{dy}{dx} + y = 4\cos(\log(1 + x))$

13. $(1 - x^2)\dfrac{d^2y}{dx^2} + x\dfrac{dy}{dx} - y = x(1 - x^2)^{\frac{3}{2}}$

16

Simultaneous Differential Equations

16.1 Introduction

So far we have considered the differential equations having one independent variable and one dependent variable. Many a times, we come across with the linear differential equations having one independent variable and two or more dependent variables. Such equations are called *simultaneous differential equations*. The complete solution of such equations can only be obtained if the number of equations is equal to the number of dependent variables. We shall first consider such equations with constant coefficients.

16.2 Simultaneous Equations with Constant Coefficients

16.2.1 Symbolic Method

Let x, y be two dependent variables and t the independent variable. Consider the simultaneous equations

$$f_1(D)x + f_2(D)y = f(t) \tag{1}$$

and

$$F_1(D)x + F_2(D)y = F(t) \tag{2}$$

where D stands for the operator $\dfrac{d}{dt}$ and f_1, f_2, F_1, F_2 are rational integral functions with constant coefficients.

Eliminating one of the dependent variables, say y, from these equations by operating on both the sides of Eq. (1) with $F_2(D)$ and both the sides of Eq. (2) with $f_2(D)$, we get

$$F_2(D)f_1(D)x + F_2(D)f_2(D)y = F_2(D)f(t) \tag{3}$$

$$f_2(D)F_1(D)x + f_2(D)F_2(D)y = f_2(D)F(t) \tag{4}$$

Since the functions F_1 and F_2 have only constants in their coefficients, it follows that

$$F_2(D)f_2(D)y = f_2(D)F_2(D)y$$

Therefore, on subtracting Eq. (4) from (3), we get

$$[F_2(D)f_1(D) - f_2(D)F_1(D)]x = F_2(D)f(t) - f_2(D)\,F(t)$$

This equation being linear differential equation in x and t can be solved by the methods discussed earlier. Substituting the value of x either in Eq. (1) or in (2), the value of f can be obtained.

16.2.2 Method of Differentiation

Let the two given equations connect t with the four quantities x, y, $\frac{dx}{dt}$ and $\frac{dy}{dt}$. Differentiating them with respect to t, four equations containing x, y, $\frac{dx}{dt}, \frac{dy}{dt}, \frac{d^2x}{dt^2}, \frac{d^2y}{dt^2}$ are obtained and from these four equations y, $\frac{dy}{dt}$ and $\frac{d^2y}{dt^2}$ are eliminated. In this way an equation of the second order, in which x is the dependent and t the independent variable is obtained. Solving this equation we get the value of x in terms of t. From this solution calculate $\frac{dx}{dt}$ and substitute the values of x and $\frac{dx}{dt}$ in either of the given equations, to get the value of y in terms of t.

Examples

1. Solve the simultaneous equations

$$\frac{dx}{dt} + 4x + 3y = t$$

$$\frac{dy}{dt} + 2x + 5y = e^t$$

Solution. The given equations can be written as

$$(D + 4)\,x + 3y = t \tag{1}$$

$$2x + (D + 5)y = e^t \tag{2}$$

Eliminating y from these equations by operating on both the sides of Eq. (1) with $D + 5$ and both the sides of Eq. (2) with 3, we get

$$(D + 5)(D + 4)x + 3(D + 5)y = (D + 5)\,t \tag{3}$$

$$6x + 3(D + 5)y = 3e^t \tag{4}$$

Subtracting Eq. (4) from (3), we get

$$(D + 5)(D + 4)x - 6x = 1 + 5t - 3e^t$$

$$\Rightarrow \qquad (D^2 + 9D + 14)x = 1 + 5t - 3e^t \qquad (5)$$

This equation is linear differential equation with constant coefficients. The auxiliary equation is

$$m^2 + 9m + 14 = 0$$

$$\Rightarrow \qquad (m + 2)(m + 7) = 0 \quad \Rightarrow \quad m = -2, -7$$

Therefore $\qquad \text{C.F.} = c_1e^{-2t} + c_2e^{-7t}$

Further

$$\text{P.I.} = \frac{1}{(D^2 + 9D + 14)}(1 + 5t - 3e^t)$$

$$= \frac{1}{(D^2 + 9D + 14)} \cdot 1 + 5 \cdot \frac{1}{D^2 + 9D + 14} t - 3 \cdot \frac{1}{D^2 + 9D + 14} e^t$$

$$= \frac{1}{14}\left(1 + \frac{D^2 + 9D}{14}\right)^{-1}(1) + \frac{5}{14}\left(1 + \frac{D^2 + 9D}{14}\right)^{-1}(t) - 3 \cdot \frac{1}{1 + 9 + 14} e^t$$

$$= \frac{1}{14}\left(1 - \frac{D^2 + 9D}{14} + \ldots\right)(1) + \frac{5}{14}\left(1 - \frac{D^2 + 9D}{14} + \ldots\right)(t) - \frac{1}{8}e^t$$

$$= \frac{1}{14} + \frac{5}{14}\left(t - \frac{9}{14}\right) - \frac{1}{8}e^t = \frac{5}{14}t - \frac{31}{196} - \frac{1}{8}e^t$$

Hence, the general solution of Eq. (5) is

$$x = c_1e^{-2t} + c_2e^{-7t} + \frac{5}{14}t - \frac{31}{196} - \frac{1}{8}e^t \qquad (6)$$

Now $\qquad \dfrac{dx}{dt} = -2c_1e^{-2t} - 7c_2e^{-7t} + \dfrac{5}{14} - \dfrac{1}{8}e^t$

Substituting these values of x and $\dfrac{dx}{dt}$ in first of the given equations, we get

$$-2c_1e^{-2t} - 7c_2e^{-7t} + \frac{5}{14} - \frac{1}{8}e^t$$

$$+ 4\left(c_1e^{-2t} + c_2e^{-7t} + \frac{5}{14}t - \frac{31}{196} - \frac{1}{8}e^t\right) + 3y = t$$

$$\Rightarrow \qquad y = \frac{1}{3}\left[-2c_1e^{-2t} + 3c_2e^{-7t} + \frac{5}{8}e^t - \frac{3}{7}t + \frac{27}{98}\right] \qquad (7)$$

Equations (6) and (7) together give the required solution.

2. Solve the simultaneous equations

$$\frac{dx}{dt} + 5x + y = e^t \tag{1}$$

$$\frac{dy}{dt} - x + 3y = e^{2t} \tag{2}$$

Solution. Differentiating Eq. (1) w.r.t. t, we get

$$\frac{d^2x}{dt^2} + 5\frac{dx}{dt} + \frac{dy}{dt} = e^t \tag{3}$$

Eliminating $\frac{dy}{dt}$ from Eq. (3) and y with the help of Eqs. (1) and (2), we get

$$\frac{d^2x}{dt^2} + 8\frac{dx}{dt} + 16x = 4e^t - e^{2t} \tag{4}$$

This equation is linear differential equation with constant coefficients. The auxiliary equation is

$$m^2 + 8m + 16 = 0 \;\Rightarrow\; m = -4, -4$$

Therefore $\qquad \text{C.F.} = (c_1 + c_2t)\, e^{-4t}$

Further $\qquad \text{P.I.} = \frac{1}{(D+4)^2} 4e^t - \frac{1}{(D+4)^2} e^{2t}$

$$= \frac{4}{25} e^t - \frac{1}{36} e^{2t}$$

Hence, the general solution of Eq. (4) is

$$x = (c_1 + c_2t)\, e^{-4t} + \frac{4}{25} e^t - \frac{1}{36} e^{2t} \tag{5}$$

Now $\qquad \frac{dx}{dt} = -4(c_1 + c_2t)\, e^{-4t} + c_2e^{-4t} + \frac{4}{25} e^t - \frac{1}{18} e^{2t}$

Substituting these values of x and $\frac{dx}{dt}$ in Eq. (1), we get

$$y = -(c_1 + c_2 + c_2t)\, e^{-4t} + \frac{1}{25} e^t + \frac{7}{36} e^{2t} \tag{6}$$

Equations (5) and (6) together give the required solution.

3. Solve

$$\frac{d^2x}{dt^2} - 4\frac{dx}{dt} + 4x - y = 0$$

$$\frac{d^2y}{dt^2} + 4\frac{dy}{dt} + 4y - 25x = 16e^t$$

Solution. The given equations can be written as

$$(D^2 - 4D + 4)\,x - y = 0 \qquad (1)$$

$$-25x + (D^2 + 4D + 4)y = 16e^t \qquad (2)$$

Eliminating y from these equations by operating on both sides of Eq. (1) with $D^2 + 4D + 4$ and adding to Eq. (2), we get

$$[(D^2 + 4D + 4)(D^2 - 4D + 4) - 25]\,x = 16e^t$$

$$\Rightarrow \qquad (D^4 - 8D^2 - 9)\,x = 16e^t \qquad (3)$$

This equation is linear differential equation with constant coefficients. The auxiliary equation is

$$m^4 - 8m^2 - 9 = 0$$

$$\Rightarrow \qquad (m^2 - 9)(m^2 + 1) = 0$$

$$\Rightarrow \qquad m = 3, -3, \pm i$$

Therefore $\qquad \text{C.F.} = c_1e^{3t} + c_2e^{-3t} + c_3 \cos t + c_4 \sin t$

Further $\text{P.I.} = \dfrac{1}{(D^4 - 8D^2 - 9)}16e^t = 16 \cdot \dfrac{1}{(1 - 8 - 9)}e^t = -e^t$

Hence, the general solution of Eq. (3) is

$$x = c_1e^{3t} + c_2e^{-3t} + c_3 \cos t + c_4 \sin t - e^t \qquad (4)$$

Now $\qquad \dfrac{dx}{dt} = 3c_1e^{3t} - 3c_2e^{-3t} - c_3 \sin t + c_4 \cos t - e^t$

and $\qquad \dfrac{d^2x}{dt^2} = 9c_1e^{3t} + 9c_2e^{-3t} - c_3 \cos t - c_4 \sin t - e^t$

Substituting these values in Eq. (1), we get

$$y = c_1e^{3t} + 25e^{-3t} + (3c_3 - 4c_4) \cos t + (4c_3 + 3c_4) \sin t - e^t \qquad (5)$$

Equations (4) and (5) together give the required solution.

4. Solve

$$\frac{dx}{dt} + 2\frac{dy}{dt} - 2x + 2y = 3e^t$$

$$3\frac{dx}{dt} + \frac{dy}{dt} + 2x + y = 4e^{2t}$$

Solution. The given equations can be written as

$$(D-2)x+2(D+1)y=3e^t \tag{1}$$

$$(3D+2)x+(D+1)y=4e^{2t} \tag{2}$$

Multiplying Eq. (2) by 2 and substracting from Eq. (1), we get

$$(5D+6)x=8e^{2t}-3e^t$$

This equation is linear differential equation with constant coefficients. Therefore

$$\text{C.F.}=c_1e^{-\frac{6}{5}t}$$

Further

$$\text{P.I.}=\frac{1}{(5D+6)}(8e^{2t}-3e^t)=\frac{8e^{2t}}{10+6}-\frac{3e^t}{5+6}=\frac{1}{2}e^{2t}-\frac{3}{11}e^t$$

Hence
$$x=c_1e^{-\frac{6}{5}t}+\frac{1}{2}e^{2t}-\frac{3}{11}e^t \tag{3}$$

Now
$$\frac{dx}{dt}=-\frac{6}{5}c_1e^{-\frac{6}{5}t}+e^{2t}-\frac{3}{11}e^t$$

Substituting these values in Eq. (1), we get

$$-\frac{6}{5}c_1e^{-\frac{6}{5}t}+e^{2t}-\frac{3}{11}e^t-2c_1e^{-\frac{6}{5}t}-e^{2t}+\frac{6}{11}e^t+2(D+1)y=3e^t$$

$$\Rightarrow \qquad 2(D+1)y=\frac{16c_1}{5}e^{-\frac{6}{5}t}+\frac{30}{11}e^t$$

Therefore
$$\text{C.F.}=c_2e^{-t}$$

and
$$\text{P.I.}=\frac{1}{2(D+1)}\left(\frac{16c_1}{5}e^{-\frac{6}{5}t}+\frac{30}{11}e^t\right)=\frac{\frac{8c_1}{5}e^{-\frac{6}{5}t}}{-\frac{6}{5}+1}+\frac{\frac{15}{11}e^t}{1+1}$$

$$=-8c_1e^{-\frac{6}{5}t}+\frac{15}{22}e^t$$

Hence
$$y=c_2e^{-t}-8c_1e^{-\frac{6}{5}t}+\frac{15}{22}e^t \tag{4}$$

Equations (3) and (4) together give the required solution.

5. Solve

$$\frac{dx}{dt}+\frac{2}{t}(x-y)=1$$

$$\frac{dy}{dt} + \frac{1}{t}(x + 5y) = t$$

Solution. The given equations can be written as

$$t\frac{dx}{dt} + 2x - 2y = t \tag{1}$$

$$t\frac{dy}{dt} + x + 5y = t^2 \tag{2}$$

Differentiating Eq. (1) w.r.t. t, we get

$$t\frac{d^2x}{dt^2} + 3\frac{dx}{dt} - 2\frac{dy}{dt} = 1 \tag{3}$$

Eliminating y and $\frac{dy}{dt}$ from Eqs. (1), (2) and (3), we get

$$t\frac{d^2x}{dt^2} + 8\frac{dx}{dt} + 12\frac{x}{t} = 2t + 6$$

$$\Rightarrow \qquad t^2\frac{d^2x}{dt^2} + 8t\frac{dx}{dt} + 12x = 2t^2 + 6t$$

This is a homogeneous linear equation. Putting $t = e^z$ and $\frac{d}{dz} \equiv D$, we get

$$[D(D - 1) + 8D + 12]x = 2e^{2z} + 6e^z$$

$$\Rightarrow \qquad (D^2 + 7D + 12)\, x = 2e^{2z} + 6e^z$$

Therefore $\qquad$ $\text{C.F.} = c_1e^{-3z} + c_2e^{-4z} = c_1t^{-3} + c_2t^{-4}$

and $\text{P.I.} = \dfrac{1}{(D^2 + 7D + 12)}(2e^{2z} + 6e^z) = \dfrac{2e^{2z}}{30} + \dfrac{6e^z}{20} = \dfrac{t^2}{15} + \dfrac{3t}{10}$

Hence

$$x = c_1t^{-3} + c_2t^{-4} + \frac{t^2}{15} + \frac{3t}{10} \tag{4}$$

Now

$$\frac{dx}{dt} = -3c_1t^{-4} - 4c_2t^{-5} + \frac{2t}{15} + \frac{3}{10}$$

Substituting these values in Eq. (1), we get

$$y = -\frac{1}{2}c_1t^{-3} - c_2t^{-4} + \frac{2t^2}{15} - \frac{t}{20} \tag{5}$$

Equations (4) and (5) together give the required solution.

EXERCISES

Solve the following simultaneous differential equations:

1. $\frac{dx}{dt} = -\omega y, \frac{dy}{dt} = \omega x$

2. $\frac{dx}{dt} + \frac{dy}{dt} + 2x + y = 0,$
   $\frac{dy}{dt} + 5x + 3y = 0$

3. $\frac{dx}{dt} - 7x + y = 0,$
   $\frac{dy}{dt} - 2x - 5y = 0$

4. $\frac{dx}{dt} - 5x - y = 0,$
   $\frac{dy}{dt} + 4x - y = 0$

5. $\frac{dx}{dt} + 5x - 2y = e^t,$
   $\frac{dy}{dt} - x + 6y = e^{2t}$

6. $4\frac{dx}{dt} + 9\frac{dy}{dt} + 11x + 31y = e^t$
   $3\frac{dx}{dt} + 7\frac{dy}{dt} + 8x + 24y = e^{2t}$

7. $\frac{dx}{dt} + \frac{dy}{dt} + 3x = \sin t,$
   $\frac{dx}{dt} + y - x = \cos t$

8. $4\frac{dx}{dt} + 9\frac{dy}{dt} + 44x + 49y = t$
   $3\frac{dx}{dt} + 7\frac{dy}{dt} + 34x + 38y = e^t$

9. $\frac{d^2x}{dt^2} - 3x - 4y + 3 = 0$
   $\frac{d^2y}{dt^2} + x + y + 5 = 0$

10. $2\frac{d^2x}{dt^2} - \frac{dy}{dt} - 4x = 2t$
    $2\frac{dx}{dt} + 4\frac{dy}{dt} - 3y = 0$

11. $\frac{d^2y}{dt^2} + \frac{dy}{dt} - 2y = \sin t$
    $\frac{dx}{dt} + x - 3y = 0$

12. $t\frac{dx}{dt} + 2x = t,$
    $t\frac{dy}{dt} - tx - ty - 2x = -t$

13. $\frac{d^2x}{dt^2} - \frac{dy}{dt} - 8x = 8t$
    $\frac{dx}{dt} + 2\frac{dy}{dt} - 3y = 0$

16.3 Simultaneous Equations of the Form $\frac{dx}{P} = \frac{dy}{Q} = \frac{dz}{R}$

Consider the equation of the form

$$\frac{dx}{P} = \frac{dy}{Q} = \frac{dz}{R} \tag{1}$$

where P, Q, R are functions of x, y, z. To find the solution of such equations the following methods will be suitable.

1. Selecting any two members of Eq. (1), say

$$\frac{dx}{P} = \frac{dy}{Q}$$

and solving, a solution $f(x, y, z) = c_1$ is obtained.

Further, selecting any other two members of Eq. (1), say

$$\frac{dy}{Q} = \frac{dz}{R}$$

and solving, a solution $F(x, y, z) = c_2$ is obtained. The two solutions $f(x, y, z) = c_1$ and $F(x, y, z) = c_2$ together give the complete solution of Eq. (1).

2. The given equations can be written as

$$\frac{dx}{P} = \frac{dy}{Q} = \frac{dz}{R} = \frac{l\,dx + m\,dy + n\,dz}{lP + mQ + nR} = \frac{l'dx + m'dy + n'dz}{l'P + m'Q + n'R}$$

where l, m, n and l', m', n' may be functions of x, y, z or constants. The multipliers are chosen such that one of the above equations can be integrated. Mostly the numerators are differential of the denominators or else the denominators are zero and numerators are exact differential. Sometimes only one set of multipliers may serve our purpose.

Examples

1. Solve the equations $\dfrac{dx}{yz} = \dfrac{dy}{zx} = \dfrac{dz}{xy}$.

Solution. Taking $\dfrac{dx}{yz} = \dfrac{dy}{zx}$, we obtain $x\,dx = y\,dy$.

On integrating, we get

$$x^2 - y^2 = c_1 \tag{1}$$

Further, taking $\dfrac{dy}{zx} = \dfrac{dz}{xy}$, we obtain $y\,dy = z\,dz$.

On integrating, we get

$$y^2 - z^2 = c_2 \tag{2}$$

Equations (1) and (2) constitute the complete integral.

2. Solve the equations

$$\frac{dx}{z(x+y)} = \frac{dy}{z(x-y)} = \frac{dz}{x^2+y^2}$$

Solution. Multiplying by x, $-y$, $-z$, we obtain

$$\frac{dx}{z(x+y)} = \frac{dy}{z(x-y)} = \frac{dz}{x^2+y^2} = \frac{x\,dx - y\,dy - z\,dz}{xz(x+y) - yz(x-y) - z(x^2+y^2)}$$

$$= \frac{x\,dx - y\,dy - z\,dz}{0}$$

$$\Rightarrow \qquad x\,dx - y\,dy - z\,dz = 0$$

On integrating, we get

$$x^2 - y^2 - z^2 = c_1 \tag{1}$$

Further, multiplying by y, x, $-z$, we obtain

$$\frac{dx}{z(x+y)} = \frac{dy}{z(x-y)} = \frac{dz}{x^2+y^2} = \frac{y\,dx + x\,dy - z\,dz}{yz(x+y) + xz(x-y) - z(x^2+y^2)}$$

$$= \frac{y\,dx + x\,dy - z\,dz}{0}$$

$$\Rightarrow \qquad y\,dx + x\,dy - z\,dz = 0$$

On integrating, we get

$$2xy - z^2 = c_2 \tag{2}$$

Equations (1) and (2) constitute the complete integral.

3. Solve

$$\frac{dx}{x-y} = \frac{dy}{y-x-z} = \frac{dz}{z}$$

Solution. We have

$$\frac{dx}{x-y} = \frac{dy}{y-x-z} = \frac{dz}{z} = \frac{dx+dy+dz}{0} = \frac{dx-dy+dz}{2(x-y+z)}$$

Therefore $\qquad dx + dy + dz = 0$

On integrating, we get

$$x + y + z = c_1 \tag{1}$$

Further $\qquad \dfrac{dx - dy + dz}{x - y + z} = \dfrac{2\,dz}{z}$

On integrating, we get

$$\log|x - y + z| = \log z^2 + \log|c_2|$$

$$\Rightarrow \qquad |x - y + z| = |c_2|\,z^2 \tag{2}$$

Equations (1) and (2) constitute the complete integral.

4. Solve $\qquad \dfrac{dx}{(z^2 - 2yz - y^2)} = \dfrac{dy}{xy + zx} = \dfrac{dz}{xy - zx}$

Solution. We have

$$\frac{x\,dx}{x(z^2 - 2yz - y^2)} = \frac{dy}{x(y+z)} = \frac{dz}{x(y-z)}$$

$$\Rightarrow \qquad \frac{x\,dx}{z^2 - 2yz - y^2} = \frac{dy}{y+z} = \frac{dz}{y-z}$$

Multiplying by 1, y, z, we get

$$\frac{x\,dx}{z^2 - 2yz - y^2} = \frac{dy}{y+z} = \frac{dz}{y-z} = \frac{x\,dx + y\,dy + z\,dz}{(z^2 - 2yz - y^2) + y(y+z) + z(y-z)}$$

$$= \frac{x\,dx + y\,dy + z\,dz}{0}$$

$$\Rightarrow \qquad x\,dx + y\,dy + z\,dz = 0$$

On integrating, we get

$$x^2 + y^2 + z^2 = c_1 \qquad (1)$$

Further, taking

$$\frac{dy}{y+z} = \frac{dz}{y-z}$$

We obtain $\quad y\,dy - z\,dy = y\,dz + z\,dz$

$$\Rightarrow \qquad y\,dy - (y\,dz + z\,dy) - z\,dz = 0$$

On integrating, we get

$$y^2 - 2yz - z^2 = c_2 \qquad (2)$$

Equations (1) and (2) constitute the complete integral.

5. Solve $\qquad \dfrac{dx}{\dfrac{y-z}{yz}} = \dfrac{dy}{\dfrac{z-x}{zx}} = \dfrac{dz}{\dfrac{x-y}{xy}}$

Solution. The given equations can be written as

$$\frac{yz}{y-z}dx = \frac{zx}{z-x}dy = \frac{xy}{x-y}dz = \frac{yz\,dx + zx\,dy + xy\,dz}{y - z + z - x + x - y}$$

$$= \frac{yz\,dx + zx\,dy + xy\,dz}{0}$$

$$\Rightarrow \qquad yz\,dx + zx\,dy + xy\,dz = 0$$

On integrating, we get

$$xyz = c_1 \qquad (1)$$

Further, the given equations can be written as

$$\frac{xyz\,dx}{x(y-z)} = \frac{xyz\,dy}{y(z-x)} = \frac{xyz\,dz}{z(x-y)} = \frac{xyz\,(dx + dy + dz)}{x(y-z) + y(z-x) + z(x-y)}$$

$$= \frac{xyz\,(dx + dy + dz)}{0}$$

$$\Rightarrow \qquad dx + dy + dz = 0$$

On integrating, we get

$$x + y + z = c_2 \tag{2}$$

Equations (1) and (2) constitute the complete integral.

6. Solve $\dfrac{dx}{x^2 - yz} = \dfrac{dy}{y^2 - zx} = \dfrac{dz}{z^2 - xy}$

Solution. We have

$$\frac{dx - dy}{x^2 - yz - y^2 + zx} = \frac{dy - dz}{y^2 - zx - z^2 + xy} = \frac{dz - dx}{z^2 - xy - x^2 + yz}$$

$$\Rightarrow \quad \frac{dx - dy}{(x - y)(x + y + z)} = \frac{dy - dz}{(y - z)(x + y + z)} = \frac{dz - dx}{(z - x)(x + y + z)}$$

$$\Rightarrow \quad \frac{dx - dy}{x - y} = \frac{dy - dz}{y - z} = \frac{dz - dx}{z - x}$$

Now, taking

$$\frac{dx - dy}{x - y} = \frac{dy - dz}{y - z}$$

and integrating, we get

$$\log |x - y| = \log |y - z| + \log |c_1|$$

$$\Rightarrow \quad |x - y| = |c_1 (y - z)| \tag{1}$$

Further, taking

$$\frac{dz - dx}{z - x} = \frac{dy - dz}{y - z}$$

and integrating, we get

$$\log |z - x| = \log |y - z| + \log |c_2|$$

$$\Rightarrow \quad |z - x| = |c_2 (y - z)| \tag{2}$$

Equations (1) and (2) constitute the complete integral.

EXERCISES

Solve the following equations:

1. $\dfrac{dx}{z} = \dfrac{dy}{-z} = \dfrac{dz}{z^2 + (x + y)^2}$
2. $\dfrac{dx}{x} = \dfrac{dy}{y} = \dfrac{dz}{z}$
3. $\dfrac{dx}{y} = \dfrac{dy}{x} = \dfrac{dz}{z}$
4. $\dfrac{dx}{y + z} = \dfrac{dy}{z + x} = \dfrac{dz}{x + y}$
5. $\dfrac{dx}{mz - ny} = \dfrac{dy}{nx - lz} = \dfrac{dz}{ly - mx}$
6. $\dfrac{dx}{x(y - z)} = \dfrac{dy}{y(z - x)} = \dfrac{dz}{z(x - y)}$

7. $\dfrac{dx}{x(y^2 - z^2)} = \dfrac{dy}{y(z^2 - x^2)} = \dfrac{dz}{z(x^2 - y^2)}$

8. $\dfrac{dx}{x} = \dfrac{dy}{xz} = -\dfrac{dz}{y}$

9. $\dfrac{x\,dx}{y^2 z} = \dfrac{dy}{xz} = \dfrac{dz}{y^2}$

10. $\dfrac{dx}{x^2 - y^2 - z^2} = \dfrac{dy}{2\,xy} = \dfrac{dz}{2\,xz}$

11. $\dfrac{dx}{y^2 - z^2} = \dfrac{dy}{z^2 - x^2} = \dfrac{dz}{x^2 - y^2}$

12. $\dfrac{dx}{y} = \dfrac{dy}{x} = \dfrac{dz}{z(x^2 - y^2)}$

MISCELLANEOUS EXERCISES

Solve the following equations:

1. $\dfrac{dx}{dt} - 3x - 2y = 0;\ \dfrac{dy}{dt} + 5x + 3y = 0$

2. $\dfrac{dx}{dt} + 2x - 3y = t;\ \dfrac{dy}{dt} - 3x + 2y = e^{2t}$

3. $2\dfrac{dx}{dt} - \dfrac{dy}{dt} + 2x + 2y = 11t;\ 2\dfrac{dx}{dt} + 3\dfrac{dy}{dt} + 5x - 3y = 0$

4. $3\dfrac{dx}{dt} + 2\dfrac{dy}{dt} - 4x + 3y = 8e^{-3t};\ 4\dfrac{dx}{dt} + \dfrac{dy}{dt} + 3x + 4y = 8e^{-t}$

   given that $x = \dfrac{1}{5}$, $y = 0$ when $t = 0$.

5. $\dfrac{d^2x}{dt^2} + 4x + y = te^{3t};\ \dfrac{d^2y}{dt^2} + y - 2x = \cos^2 t$

6. $t\dfrac{dx}{dt} + y = 0,\ t\dfrac{dy}{dt} + x = 0$

7. $\dfrac{d^2x}{dt^2} + m^2y = 0,\ \dfrac{d^2y}{dt^2} - m^2x = 0$

8. $\dfrac{dx}{y^2} = \dfrac{dy}{x^2} = \dfrac{dz}{x^2y^2z^2}$

9. $\dfrac{dx}{x^2} = \dfrac{dy}{y^2} = \dfrac{dz}{n\,xy}$

10. $\dfrac{l\,dx}{mn(y - z)} = \dfrac{m\,dy}{nl(z - x)} = \dfrac{n\,dz}{lm(x - y)}$

11. $\dfrac{dx}{y^2 + yz + z^2} = \dfrac{dy}{z^2 + zx + x^2} = \dfrac{dz}{x^2 + xy + y^2}$

12. $\dfrac{dx}{x^2 + y^2 - yz} = \dfrac{dy}{-x^2 - y^2 - xz} = \dfrac{dz}{z(x - y)}$

13. $\dfrac{dx}{x(y^2 + z)} = \dfrac{dy}{-y(x^2 + z)} = \dfrac{dz}{z(x^2 - y^2)}$

14. $\dfrac{dx}{x^3 + 3xy^2} = \dfrac{dy}{y^3 + 3x^2 y} = \dfrac{dz}{2z(x^2 + y^2)}$

15. $\dfrac{dx}{2x^2 + y^2 + z^2 - 2yz - zx - xy} = \dfrac{dy}{x^2 + 2y^2 + z^2 - yz - 2zx - xy}$

$$= \dfrac{dz}{x^2 + y^2 + 2z^2 - yz - zx - 2xy}$$

16. $\dfrac{dx}{x} = \dfrac{dy}{y} = \dfrac{dz}{z - a\sqrt{x^2 + y^2 + z^2}}$

Appendix

Beta and Gamma Functions

1. **Beta Function.** Consider the integral

$$\int_0^1 x^{m-1}(1-x)^{n-1}\,dx$$

It is a proper integral for $m \geq 1$, $n \geq 1$. The points 0 and 1 are the only points of infinite discontinuity of the integrand. When $m < 1$, then 0 is the point of infinite discontinuity and when $n < 1$, then 1 is the point of infinite discontinuity. The integral converges at 0 when $m > 0$ and converges at 1 when $n > 0$. Hence the integral exists for positive values of m and n. The integral $\int_0^1 x^{m-1}(1-x)^{n-1}\,dx$, for m, $n > 0$, is called *beta function* denoted by $B(m, n)$. Thus, we write

$$B(m, n) = \int_0^1 x^{m-1}(1-x)^{n-1}\,dx,\; m > 0,\; n > 0$$

2. **Gamma Function.** Consider the integral

$$\int_0^\infty x^{m-1} e^{-x}\,dx$$

The integrand has infinite discontinuity at 0 if $m < 1$. The integral converges when $m > 0$. Hence the integral exists for positive values of m. The integral $\int_0^\infty x^{m-1} e^{-x}\,dx$, for $m > 0$, is called *gamma function* denoted by $\Gamma(m)$. Thus, we write

$$\Gamma(m) = \int_0^\infty x^{m-1} e^{-x}\,dx, \quad m > 0$$

I. *A Property*

On integrating $\int_0^\infty x^m e^{-x}\,dx$, $m > 0$, by parts, we have

$$\int_0^\infty x^m e^{-x}\,dx = [-x^m e^{-x}]_0^\infty + m\int_0^\infty x^{m-1} e^{-x}\,dx$$

$$\Rightarrow \qquad \int_0^\infty x^m e^{-x}\,dx = m\int_0^\infty x^{m-1} e^{-x}\,dx$$

$$\Rightarrow \qquad \Gamma(m+1) = m\,\Gamma(m)$$

This formula helps us in evaluation of gamma function. For example, if we put $m = 6$, we have

$$\Gamma(6) = 5\Gamma(5) = 5.4\Gamma(4) = 5.4.3.2.1\ \Gamma(1)$$

The value of $\Gamma(1)$ can be evaluated as

$$\Gamma(1) = \int_0^\infty x^{1-1} e^{-x}\,dx = \int_0^\infty e^{-x}\,dx = [-e^{-x}]_0^\infty = 1$$

Hence $$\Gamma(6) = 5.4.3.2.1 = \lfloor 5$$

Similarly, we can show that

$$\Gamma(m+1) = \lfloor m$$

when m is a positive integer.

For non-integral values of m, we have

$$\Gamma(6.5) = (5.5)\ \Gamma(5.5)$$

II. *Value of* $\Gamma\left(\frac{1}{2}\right)$

We have

$$\Gamma\left(\frac{1}{2}\right) = \int_0^\infty t^{-\frac{1}{2}} e^{-t}\,dt$$

Let $t = x^2$. Then, $dt = 2x\,dx$ and we have

$$\Gamma\left(\frac{1}{2}\right) = 2\int_0^\infty e^{-x^2}\,dx \qquad (1)$$

Replacing x by y in Eq. (1), we get

$$\Gamma\left(\frac{1}{2}\right) = 2\int_0^\infty e^{-y^2}\,dy \qquad (2)$$

On multiplying Eqs. (1) and (2), we get

$$\left(\Gamma\left(\frac{1}{2}\right)\right)^2 = 4\int_0^\infty\int_0^\infty e^{-(x^2+y^2)}\,dx\,dy$$

$$= 4\int_0^{\frac{\pi}{2}}\int_0^\infty e^{-r^2}\,r\,d\theta\,dr, \text{ on transforming to polars}$$

$$= 2\int_0^{\frac{\pi}{2}} [e^{-r^2}]_0^\infty \, d\theta = 2\int_0^{\frac{\pi}{2}} d\theta = \pi$$

Hence $$\Gamma\left(\frac{1}{2}\right) = \sqrt{\pi}$$

III. *Integral of* $\sin^{2m-1} x \cos^{2n-1} x$

We have

$$\Gamma(m) = \int_0^\infty t^{m-1} e^{-t} \, dt, \quad m > 0$$

Let $t = x^2$. Then, $dt = 2x\,dx$ and we have

$$\Gamma(m) = \int_0^\infty x^{2m-2} e^{-x^2} 2x\,dx = 2\int_0^\infty x^{2m-1} e^{-x^2} dx \tag{1}$$

Similarly $$\Gamma(n) = 2\int_0^\infty y^{2n-1} e^{-y^2} dy \tag{2}$$

On multiplying Eqs. (1) and (2), we get

$$\Gamma(m)\,\Gamma(n) = 4\int_0^\infty \int_0^\infty x^{2m-1} y^{2n-1} e^{-(x^2+y^2)} \, dx\,dy$$

$$= 4\int_0^{\frac{\pi}{2}} \int_0^\infty (r\cos\theta)^{2m-1} (r\sin\theta)^{2n-1} e^{-r^2} \, r\,d\theta\,dr,$$

on transforming to polars

$$= 4\int_0^{\frac{\pi}{2}} \cos^{2m-1}\theta \sin^{2n-1}\theta \, d\theta \times \int_0^\infty r^{2m+2n-1} e^{-r^2} dr$$

$$= 2\int_0^{\frac{\pi}{2}} \cos^{2m-1}\theta \sin^{2n-1}\theta \, d\theta \cdot \Gamma(m+n)$$

Hence $$\int_0^{\frac{\pi}{2}} \cos^{2m-1}\theta \sin^{2n-1}\theta \, d\theta = \frac{\Gamma(m)\Gamma(n)}{2\Gamma(m+n)}$$

Note. The above result is true for all positive values of m and n, integral as well as fractional. Thus, the above result can also be written as

$$\int_0^{\frac{\pi}{2}} \cos^m\theta \sin^n\theta \, d\theta = \frac{\Gamma\left(\frac{m+1}{2}\right)\Gamma\left(\frac{n+1}{2}\right)}{2\Gamma\left(\frac{m+n+2}{2}\right)}$$

where $m > -1$, $n > -1$.

IV. *Relation between Beta Function and Gamma Function*

We have

$$B(m, n) = \int_0^1 x^{m-1}(1-x)^{n-1}\, dx$$

Putting $x = \sin^2 \theta$ in the above integral, we get

$$B(m, n) = \int_0^{\frac{\pi}{2}} (\sin \theta)^{2m-2}(\cos \theta)^{2n-2}\, 2 \sin \theta \cos \theta\, d\theta$$

$$= 2\int_0^{\frac{\pi}{2}} \sin^{2m-1}\theta \cos^{2n-1}\theta\, d\theta = \frac{\Gamma(m)\,\Gamma(n)}{\Gamma(m+n)}$$

Hence $$B(m, n) = \frac{\Gamma(m)\,\Gamma(n)}{\Gamma(m+n)}$$

Answers to the Exercises

Pages 8 and 9

1. $\dfrac{(1+\sin x)^4}{4}$
2. $\log|\log\sin x|$
3. $\log|\sec(\sin^{-1}x)|$
4. $\dfrac{1}{\sqrt{2}}\log\left|\tan\dfrac{x}{2}\right|$
5. $\dfrac{1}{\sqrt{a^2+b^2}}\log\left|\tan\left(\dfrac{x-\theta}{2}+\dfrac{\pi}{4}\right)\right|$, where $\theta=\tan^{-1}\dfrac{b}{a}$
6. $\dfrac{1}{2}\log|2x+\sqrt{9+4x^2}|$
7. $\log|e^x+\sqrt{e^{2x}-1}|$
8. $-\dfrac{1}{2}$ $(\operatorname{cosec}x\cot x+\log|\operatorname{cosec}x+\cot x|)$
9. $\log|\sin(\log x)|$
10. $\log|\tan^{-1}x|$
11. $\dfrac{1}{m}e^{m\sin^{-1}x}$
12. $\dfrac{x^3}{3}\log x-\dfrac{x^3}{9}$
13. $\dfrac{x^3}{3}\left[(\log x)^2-\dfrac{2}{3}\log x+\dfrac{2}{9}\right]$
14. $\dfrac{1}{4}(2x^2-1)\sin^{-1}x+\dfrac{x}{4}\sqrt{1-x^2}$
15. $\dfrac{1}{2}(x^2\tan^{-1}x-x+\tan^{-1}x)$
16. $2\log 2-\dfrac{3}{4}$
17. $\pi-2$
18. $\dfrac{\pi}{8}$
19. $\dfrac{\pi}{2}-1$
20. 2
21. 1
22. $\dfrac{1}{4}$
23. $e^{\frac{\pi}{2}}$

Page 12

1. $\dfrac{1}{2}\log|x+1|+\dfrac{1}{2}\log|x+3|-\log|x+2|$
2. $\dfrac{1}{2}\log|x+1|-4\log|x+2|+\dfrac{9}{2}\log|x+3|$
3. $\log|x-2|-\log|x-1|$
4. $x+8\log|x-2|-4\log|x-1|$
5. $x+\dfrac{1}{2}\log|x-1|-8\log|x-2|+\dfrac{27}{2}\log|x-3|$
6. $\dfrac{1}{3}\log|x-1|+\log|x+1|-\dfrac{5}{6}\log|2x+1|$
7. $\dfrac{1}{6}\log|x|+\dfrac{1}{3}\log|x+3|+\dfrac{1}{2}\log|x-2|$

8. $\frac{1}{2}\log|x-1| + \frac{3}{2}\log|x-3| - 2\log|x-2|$

9. $\frac{1}{2}\log|x-1| + \frac{1}{18}\log|3x-1| - \frac{4}{9}\log|3x-2|$

10. $\frac{47}{4}\log|x-1| + \frac{5}{4}\log|x+3| - \frac{25}{2}\log|2x-1|$

Page 14

1. $\frac{5}{27}\log\left|\frac{x-2}{x+1}\right| + \frac{1}{3(x+1)^2} - \frac{4}{9(x+1)}$

2. $2\log\left|\frac{x-1}{x}\right| + \frac{2}{x} + \frac{1}{x^2} + \frac{1}{3x^3}$ 3. $\frac{1}{4}\log\left|\frac{x+1}{x-1}\right| - \frac{1}{2(x-1)} - \frac{1}{(x-1)^2}$

4. $2\log\left|\frac{x-2}{x-1}\right| + \frac{2}{x-1} + \frac{1}{2(x-1)^2}$

5. $\frac{1}{27}\log\left|\frac{x+2}{x-1}\right| - \frac{2}{9(x-1)} + \frac{1}{9(x+2)}$

6. $-\frac{3}{4}\log\left|\frac{x-1}{x+1}\right| - \frac{1}{x} - \frac{x}{2(x^2-1)}$

7. $\frac{2}{27}\log\left|\frac{x-1}{x+2}\right| - \frac{7}{9(x+2)} + \frac{5}{6(x+2)^2}$

8. $x - 36\log|x+2| + 28\log|x+1| + \frac{12}{x+1}$

9. $2\log|x| - \frac{1}{x} - \frac{1}{2x^2} - \frac{7}{4}\log|x-1| - \frac{1}{2(x-1)} - \frac{1}{4}\log|x+1|$

10. $\frac{1}{12}\log(27e^4)$

Pages 16 and 17

1. $\frac{1}{6}\log|x^2-x+1| - \frac{1}{3}\log|x+1| + \frac{1}{\sqrt{3}}\tan^{-1}\left(\frac{2x-1}{\sqrt{3}}\right)$

2. $\frac{1}{4}\log|x^2+1| - \frac{1}{2}\log|x+1| + \frac{1}{2}\tan^{-1}x$

3. $\log|x^2+2| - \frac{1}{2}\log|x^2+1|$

4. $-\frac{2}{25}\log|x-1| - \frac{1}{5(x-1)} + \frac{1}{25}\log|x^2+4| - \frac{3}{50}\tan^{-1}\frac{x}{2}$

5. $-\frac{1}{2}\tan^{-1}x + \frac{1}{4}\log\left|\frac{x-1}{x+1}\right|$

6. $x + \frac{18}{25}\log|x-1| + \frac{16}{25}\log|x^2+4| - \frac{1}{5(x-1)} - \frac{24}{25}\tan^{-1}\frac{x}{2}$

7. $\frac{1}{2}\log\left|\frac{x^2+x+1}{x^2+1}\right| + \frac{1}{\sqrt{3}}\tan^{-1}\left(\frac{2x+1}{\sqrt{3}}\right)$

8. $\frac{1}{2}\log|x-1| - \frac{1}{2(x-1)} - \frac{1}{4}\log|x^2+1|$

9. $-\frac{1}{4}\log\left|\frac{x^2-x+1}{x^2+x+1}\right| + \frac{1}{2\sqrt{3}}\tan^{-1}\left(\frac{x\sqrt{3}}{1-x^2}\right)$

10. $\frac{64}{17}\log\frac{5}{4} - \frac{27}{10}\log\frac{4}{3} - \frac{11}{340}\log 2 - \frac{7\pi}{680}$

Page 20

1. $\frac{1}{2}\log\left|\frac{x^2+1}{x^2+3}\right|$

2. $\frac{1}{8}\log\left|\frac{x^2(x^2+2)^3}{(x^2+1)^4}\right| - \frac{1}{4(x^2+2)}$

3. $\frac{1}{5}\log\left|\frac{x^5}{x^5+1}\right|$

4. $\frac{1}{2}\log\left|\frac{x^2}{x^2+1}\right| + \frac{1}{2(x^2+1)} + \frac{1}{4(x^2+1)^2}$

5. $\frac{1}{6}\log\frac{x^6}{x^6+1}$

6. $-\frac{1}{8\sqrt{2}}\tan^{-1}\frac{x}{\sqrt{2}} - \frac{x}{8(x^2+2)} - \frac{x}{2(x^2+2)^2}$

7. $\frac{1}{6}\log\frac{3}{2} + \frac{2}{3\sqrt{2}}\tan^{-1}\left(\frac{1}{4\sqrt{2}}\right)$

8. $\frac{3}{2}\tan^{-1}x - \frac{1}{2\sqrt{3}}\tan^{-1}\frac{x}{\sqrt{3}}$

9. $\frac{1}{2}\log\left|\frac{x^2-x+1}{x^2+x+1}\right|$

10. $\tan^{-1}\left(\frac{x-1}{x}\right)$

11. $\frac{x}{1-x^2}$

Page 22

1. $x + \log|1+e^x| - 2\log|1+2e^x|$

2. $\log|1+e^{-x}| - e^{-x}$

3. $\frac{1}{1-e^x} - \log|1-e^{-x}|$

4. $\log\left|2-\frac{1}{e}\right|$

5. $-\frac{3}{2}x + \frac{35}{36}\log|9e^{2x}-4|$

Pages 22 and 23

1. $\log|x| - \frac{1}{2}\log|1-x^2|$

2. $\Sigma\frac{a}{(a-b)(a-c)}\log|x-a|$

3. $\frac{5}{12}\log\left|\frac{x-2}{x+2}\right| - \frac{1}{3}\log\left|\frac{x-1}{x+1}\right|$

4. $-\frac{1}{12}\log|x-1| + \frac{13}{21}\log|x+2| + \frac{13}{28}\log|x-5|$

5. $2\log|x-1| - \log|x| - \frac{x}{(x-1)^2}$

6. $\frac{x}{1-x^2}$

7. $\frac{1}{48}\log|x-1| + \frac{8}{3}\log|x+2| - \frac{43}{16}\log|x+1|$

$-\frac{21}{8(x+1)} + \frac{7}{8(x+1)^2} - \frac{1}{6(x+1)^3}$

8. $\dfrac{1}{4\sqrt{2}} \log \left| \dfrac{x^2 + \sqrt{2}\,x + 1}{x^2 - \sqrt{2}\,x + 1} \right| + \dfrac{1}{2\sqrt{2}} \tan^{-1}\left(\dfrac{x^2 - 1}{x\sqrt{2}} \right)$

9. $\dfrac{3x - 4}{4(x-1)^2} + \dfrac{3}{8} \log |x - 1| - \dfrac{1}{24} \log |x + 1|$

$-\dfrac{1}{6} \log |x^2 - x + 1| + \dfrac{1}{\sqrt{3}} \tan^{-1}\left(\dfrac{2x - 1}{\sqrt{3}} \right)$

10. $-\dfrac{1}{2(x-1)} - \dfrac{1}{2} \log |x - 1| + \dfrac{1}{4} \log |x^2 + 1|$

11. $\dfrac{1}{6} \log \left| \dfrac{1 + x}{1 - x} \right| + \dfrac{1}{12} \log \left| \dfrac{1 + x + x^2}{1 - x + x^2} \right| + \dfrac{1}{2\sqrt{3}} \tan^{-1}\left(\dfrac{x\sqrt{3}}{1 - x^2} \right)$

12. $\dfrac{1}{2a^3} \tan^{-1} \dfrac{x}{a} + \dfrac{1}{2a^2} \left(\dfrac{x}{x^2 + a^2} \right)$

13. $\dfrac{1}{4\sqrt{2}} \log \left| \dfrac{x^2 - \sqrt{2}\,x + 1}{x^2 + \sqrt{2}\,x + 1} \right| + \dfrac{1}{2\sqrt{2}} \tan^{-1}\left(\dfrac{x^2 - 1}{\sqrt{2}\,x} \right)$

14. $\dfrac{1}{4} \log \left| \dfrac{x^2 - x + 1}{x^2 + x + 1} \right| + \dfrac{1}{2\sqrt{3}} \tan^{-1}\left(\dfrac{x^2 - 1}{\sqrt{3}\,x} \right)$

15. $\dfrac{1}{\sqrt{3}} \tan^{-1}\left(\dfrac{x^2 - 1}{\sqrt{3}\,x} \right) + \dfrac{2}{\sqrt{3}} \tan^{-1}\left(\dfrac{2x^2 + 1}{\sqrt{3}} \right)$

16. $\dfrac{1}{3\sqrt{2}} \tan^{-1} \dfrac{x}{\sqrt{2}} + \dfrac{1}{3\sqrt{2}} \tan^{-1} (\sqrt{2}\,x)$

17. $\dfrac{1}{(a^2 - b^2)} \left(\dfrac{1}{b} \tan^{-1} \dfrac{x}{b} - \dfrac{1}{a} \tan^{-1} \dfrac{x}{a} \right)$

18. $-\dfrac{3}{25} \log |x - 1| + \dfrac{1}{5(x-1)} + \dfrac{1}{8} \log |x - 2| - \dfrac{1}{400} \log |x^2 + 4| + \dfrac{7}{200} \tan^{-1} \dfrac{x}{2}$

19. $\dfrac{1}{\sqrt{3}} \tan^{-1}\left(\dfrac{x^2 - 1}{\sqrt{3}\,x} \right)$

20. $\dfrac{6}{5} \log |x + 1| + \dfrac{9}{10} \log |x^2 - 3x + 6| - \dfrac{1}{\sqrt{5}} \tan^{-1}\left(\dfrac{2x - 3}{\sqrt{15}} \right)$

21. $\dfrac{1}{2} \log |x^2 + 1| - \log |x + 1| + \dfrac{5}{2(x+1)} + \dfrac{3}{2} \tan^{-1} x$

22. $-\dfrac{10}{x - 1} + \dfrac{35}{6} \log |2x - 1| - \dfrac{2}{15} \log |5x - 1|$

23. $\log \frac{4}{3}$ 24. $\frac{1}{\sqrt{2}} \log(\sqrt{2}-1) + \frac{\pi}{2\sqrt{2}}$ 25. $\frac{1}{\sqrt{2}} \log(\sqrt{2}+1) + \frac{\pi}{2\sqrt{2}}$

Page 25

1. $\frac{2}{9}(x\sqrt{x-1}) + \frac{26}{9\sqrt{15}} \tan^{-1}\sqrt{\frac{3x-3}{5}}$

2. $\frac{2}{3}(x+2)^{\frac{3}{2}} + 2(x+2)^{\frac{1}{2}} + \frac{4}{\sqrt{3}} \log\left|\frac{\sqrt{x+2}-\sqrt{3}}{\sqrt{x+2}+\sqrt{3}}\right|$

3. $-\frac{\sqrt{x+1}}{x} + \frac{1}{2}\log\left|\frac{\sqrt{x+1}+1}{\sqrt{x+1}-1}\right|$ 4. $\frac{1}{2}\log\left|\frac{2\sqrt{x+1}-1}{2\sqrt{x+1}+1}\right|$

5. $2\log\left|\frac{\sqrt{x+1}-1}{\sqrt{x+1}+1}\right| - 6\sqrt{x+1}$

6. $\frac{2}{5}(x+2)^{\frac{5}{2}} - 2(x+2)^{\frac{3}{2}} + 6\sqrt{x+2} + \frac{1}{\sqrt{3}}\log\left|\frac{\sqrt{x+2}-\sqrt{3}}{\sqrt{x+2}+\sqrt{3}}\right|$

7. $\frac{3}{5}(x+12)(x-3)^{\frac{2}{3}}$

Pages 26 and 27

1. $\cos^{-1}\left(\frac{1-x}{x\sqrt{2}}\right)$ 2. $\sin^{-1}\left(\frac{x+1}{\sqrt{2}(x+2)}\right)$

3. $-\frac{1}{\sqrt{3}} \sinh^{-1}\left(\frac{\sqrt{3}(x+1)}{x-1}\right)$ 4. $\frac{1}{3}\sinh^{-1}\left(\frac{5x-1}{\sqrt{2}(2-x)}\right)$

5. $-\frac{1}{\sqrt{3}}\sinh^{-1}\left(\frac{5-x}{\sqrt{23}(x+1)}\right)$ 6. $\frac{\sqrt{3x^2+4x-5}}{x+2} + 4\sin^{-1}\left(\frac{4x+9}{19(x+2)}\right)$

Page 28

1. $-\frac{1}{\sqrt{2}}\sin^{-1}\sqrt{\frac{1-x^2}{1+x^2}}$ 2. $-\frac{1}{\sqrt{2}}\sinh^{-1}\sqrt{\frac{x^2+1}{x^2-1}}$

3. $\frac{1}{2\sqrt{2}}\log\left|\frac{x\sqrt{2}+\sqrt{x^2-1}}{x\sqrt{2}-\sqrt{x^2-1}}\right|$ 4. $\frac{1}{2\sqrt{33}}\log\left|\frac{x\sqrt{11}+\sqrt{3x^2-12}}{x\sqrt{11}-\sqrt{3x^2-12}}\right|$

5. $\frac{1}{2\sqrt{5}}\left(\log\left|\frac{\sqrt{x^2+9}-\sqrt{5}}{\sqrt{x^2+9}+\sqrt{5}}\right| + \tan^{-1}\left(\frac{x\sqrt{5}}{2\sqrt{x^2+9}}\right)\right)$

6. $\frac{2}{\sqrt{3}}\cos^{-1}\sqrt{\frac{3(x^2+x+2)}{7(x^2+x+1)}}$

Pages 28 and 29

1. $2\left[\frac{1}{7}(x+2)^{\frac{7}{2}} - (x+2)^{\frac{5}{2}} + 3(x+2)^{\frac{3}{2}} - 5(x+2)^{\frac{1}{2}}\right] + \frac{1}{\sqrt{3}}\log\left|\frac{\sqrt{x+2}-\sqrt{3}}{\sqrt{x+2}+\sqrt{3}}\right|$

2. $\tan^{-1}\sqrt{4x+5}$

3. $\frac{1}{4\sqrt{3}}\log\left|\frac{\sqrt{x+1}-\sqrt{3}}{\sqrt{x+1}+\sqrt{3}}\right| - \frac{1}{2}\tan^{-1}\sqrt{x+1}$

4. $\frac{2}{\sqrt{3}}\tan^{-1}\left(\frac{x}{\sqrt{3(x+1)}}\right)$ 5. $-\frac{1}{\sqrt{2}}\sinh^{-1}\left(\frac{x+1}{x-1}\right)$ 6. $\frac{1}{3}\log(3+\sqrt{8})$

7. $\frac{1}{\sqrt{2}}\cosh^{-1}(4x-3) + \frac{1}{\sqrt{15}}\cosh^{-1}\left(\frac{8-11x}{x+2}\right)$

8. $\frac{1}{2}\sqrt{2x^2+3x-4} - \frac{3}{4\sqrt{2}}\cosh^{-1}\left(\frac{4x+3}{\sqrt{41}}\right) + \frac{1}{\sqrt{2}}\sin^{-1}\left(\frac{5x+14}{\sqrt{41}(x+2)}\right)$

9. $\frac{1}{4\sqrt{2}}\log\left|\frac{\sqrt{4+2x}-\sqrt{2-x}}{\sqrt{4+2x}+\sqrt{2-x}}\right|$ 10. $\frac{1}{\sqrt{2}}\sinh^{-1}\left(\frac{1+x}{1-x}\right)$

11. $\sin^{-1}\left(\frac{3x+1}{\sqrt{5}(x+1)}\right)$ 12. $\sin^{-1}\left(\frac{x+3}{\sqrt{5}(x+1)}\right) + \cos^{-1}\left(\frac{2-x}{x\sqrt{5}}\right)$

13. $\frac{1}{2\sqrt{2}}\sinh^{-1}\left(\frac{1+x}{1-x}\right) + \sinh^{-1}\left(\frac{1-x}{1+x}\right)$

14. $\cosh^{-1}\left(\frac{x}{\sqrt{2}}\right) + \frac{2}{\sqrt{3}}\tan^{-1}\sqrt{\frac{x^2-2}{3}} + \frac{2}{\sqrt{3}}\sinh^{-1}\sqrt{\frac{x^2-2}{2x^2+2}}$

15. $\sqrt{x^2-3} - \frac{1}{\sqrt{5}}\tan^{-1}\sqrt{\frac{x^2-3}{5}} + \cosh^{-1}\left(\frac{x}{\sqrt{3}}\right) - \frac{1}{\sqrt{10}}\sinh^{-1}\sqrt{\frac{2x^2-6}{3x^2+6}}$

16. $\frac{1}{\sqrt{2}}\sin^{-1}\left(\frac{x\sqrt{2}}{1+x^2}\right)$ 17. $\frac{\sqrt{x^4+x^2+1}}{x}$

Page 34

1. $\frac{1}{\sqrt{7}}\log\left|\frac{\sqrt{7}+\tan\frac{x}{2}}{\sqrt{7}-\tan\frac{x}{2}}\right|$ 2. $\frac{2}{3}\tan^{-1}\left(\frac{1}{3}\tan\frac{x}{2}\right)$ 3. $\frac{4}{3}\tan^{-1}\left(\frac{4}{3}+\frac{5}{3}\tan\frac{x}{2}\right)$

4. $\frac{1}{\sqrt{3}}\log(2+\sqrt{3})$ 5. $\frac{1}{3}\log 2$ 6. $\frac{\pi}{4}$ 7. $\frac{1}{2}\log 3$

8. $\frac{\pi}{1-a^2}$ or $\frac{\pi}{a^2-1}$, according as $a<1$ or $a>1$.

Page 38

1. $\frac{x}{2} + \frac{1}{2} \log | \cos x + \sin x |$ 2. $\frac{\pi}{4}$

3. $\frac{1}{41} (3x - \log | 4 \cos x + 5 \sin x |)$

4. $\frac{6}{5} x + \frac{3}{5} \log | \sin x + 2 \cos x + 3 | - \frac{8}{5} \tan^{-1}\left(\frac{\tan \frac{x}{2} + 1}{2}\right)$

5. $\frac{3}{25} x + \frac{4}{25} \log | 3 \cos x + 4 \sin x |$ 6. $\frac{4}{5} x + \frac{3}{5} \log | 2 \sin x + \cos x |$

7. $\frac{3}{13} x - \frac{2}{13} \log | 2 \sin x + 3 \cos x |$ 8. $\frac{7}{2} x + \frac{1}{2} \log | \sin x + \cos x |$

Page 38

1. $\frac{\pi}{\sqrt{3}}$ 2. $\frac{1}{\sqrt{3}} \log \left|\frac{\tan \frac{x}{2} + 2 - \sqrt{3}}{\tan \frac{x}{2} + 2 + \sqrt{3}}\right|$ 3. $\frac{1}{2\sqrt{3}} \log \left|\frac{\tan \frac{x}{2} + 2 - \sqrt{3}}{\tan \frac{x}{2} + 2 + \sqrt{3}}\right|$

4. $\frac{7}{25} x - \frac{1}{25} \log | 3 \sin x + 4 \cos x + 1 | - \frac{7}{50\sqrt{6}} \cosh^{-1}\left(\frac{5 + \cos\left(x - \tan^{-1} \frac{3}{4}\right)}{1 + 5 \cos\left(x - \tan^{-1} \frac{3}{4}\right)}\right)$

5. $\frac{2}{\sqrt{3}} \tan^{-1}\left(\frac{1}{\sqrt{3}} \tan\left(\frac{x}{2} - \frac{\pi}{8}\right)\right)$ 6. $\frac{1}{5} \log \left|\tan\left(\frac{x}{2} + \frac{\pi}{4}\right)\right| - \frac{3}{10} \tan^{-1}\left(\frac{1}{2} \tan \frac{x}{2}\right)$

7. $\frac{2}{\sqrt{7}} \tan^{-1}\left(\frac{1}{\sqrt{7}}\right)$ 8. $\frac{1}{2 - \tan \frac{x}{2}}$ 9. $2x + \log | 2 \cos x + \sin x + 3 |$

10. $\frac{1}{5} \log \left|\frac{1 + 2 \tan \frac{x}{2}}{4 - 2 \tan \frac{x}{2}}\right|$ 11. $\frac{1}{13} \log \left|\frac{4 + 6 \tan \frac{x}{2}}{9 - 6 \tan \frac{x}{2}}\right|$ 12. $\frac{1}{6} \tan^{-1}\left(\frac{2}{3} \tan\left(\frac{3x - 4}{6}\right)\right)$

13. $\frac{12}{13} x - \frac{5}{13} \log | 3 \cos x + 2 \sin x |$ 14. $\frac{5\pi}{4}$

Page 42

1. $k(b - a)$ 2. $\frac{1}{2}(b^2 - a^2)$ 3. $\frac{1}{3}(b^3 - a^3)$ 4. $\frac{1}{a} - \frac{1}{b}$

5. $2(\sqrt{b} - \sqrt{a})$ 6. $\frac{2}{3}(b^{\frac{3}{2}} - a^{\frac{3}{2}})$ 7. $e^b - a^a$

8. $\sin b - \sin a$ 9. $\frac{1}{2}(b - a) - \frac{1}{4} (\sin 2b - \sin 2a)$

Page 50

1. $\frac{1}{2}$ 2. $\frac{3}{8}$ 3. $\frac{1}{3}\log 2$ 4. $\frac{1}{2}\tan 1$

5. $\frac{\pi}{2}$ 6. $\frac{\pi}{2}$ 7. $\frac{\pi}{3}$ 8. $\frac{1}{4}\log 2$

9. $\frac{\pi+2}{2}$ 10. $\frac{\pi}{4}$ 11. $\frac{4}{e}$ 12. $2e^{(\pi-4)/2}$

Page 56

1. $\frac{\pi}{4}$ 2. $\frac{\pi}{4}$ 3. $\frac{\pi}{4}$

4. $\frac{1}{\sqrt{2}}\log(\sqrt{2}+1)$ 5. $\frac{\pi}{8}\log 2$ 6. 0

7. 0 8. $-\frac{\pi^2}{2}\log 2$ 9. $\pi\log 2$

Page 57

1. $\cosh b - \cosh a$ 2. $\sinh b \cosh b - \sinh a \cosh a$

3. $\frac{\pi}{4}+\frac{1}{2}\log 2$ 4. $\frac{1}{\sqrt{2}}$ 5. $\frac{1}{10}(5-\sqrt{5})$

6. $\frac{1}{14}$ 7. $\frac{1}{3}$ 8. $\frac{2}{3}(2\sqrt{2}-1)$

9. $\frac{1}{3}$ 10. $\frac{1}{2}$ 11. 1

14. $\pi\log 2$ 15. $-\pi\log 2$ 17. π

Page 60

1. $\frac{1}{3}\cos^3 x - \cos x$

2. $-\frac{1}{4}\sin^3 x\cos x - \frac{3}{8}\sin x\cos x + \frac{3}{8}x$

3. $-\frac{1}{6}\sin^5\cos x - \frac{5}{24}\sin^3 x\cos x - \frac{5}{16}\sin x\cos x + \frac{5}{16}x$

4. $-\frac{1}{7}\cos x\left(\sin^6 x + \frac{6}{5}\sin^4 x + \frac{8}{5}\sin^2 x + \frac{16}{5}\right)$

5. $\frac{1}{3}\cos^2 x\sin x + \frac{2}{3}\sin x$ 6. $\frac{1}{4}\cos^3 x\sin x + \frac{3}{8}\cos x\sin x + \frac{3}{8}x$

7. $\frac{1}{7}\sin x\left(\cos^6 x + \frac{6}{5}\cos^4 x + \frac{8}{5}\cos^2 x + \frac{16}{5}\right)$

Page 62

1. $\frac{5\pi}{32}$ 2. $\frac{16}{35}$ 3. $\frac{35\pi}{256}$ 4. $\frac{128}{315}$ 5. $\frac{63\pi}{512}$

6. $\frac{5\pi}{64}$ 7. $\frac{35\pi}{128}$ 8. $\frac{5\pi}{96}$ 9. $\frac{64}{35}$

10. $\frac{5\pi}{32}$ 11. $\frac{3\pi a^4}{16}$ 12. $\frac{11\pi}{192}$ 13. $\frac{63\pi a^5}{8}$

Page 65

1. $\frac{1}{2}\tan^2 x + \log|\cos x|$
2. $\frac{1}{5}\tan^5 x - \frac{1}{3}\tan^3 x + \tan x - x$
3. $-\frac{1}{4}\cot^4 x + \frac{1}{2}\cot^2 x + \log|\sin x|$
4. $\frac{3\pi - 8}{12}$
5. $\frac{1}{2}\left(\log 2 - \frac{1}{2}\right)$
6. $\frac{3\pi - 8}{12}$

Page 67

1. $\frac{1}{3}\tan x\,(\sec^2 x + 2)$
2. $-\frac{1}{2}\operatorname{cosec} x \cot x + \frac{1}{2}\log\left|\tan\frac{x}{2}\right|$
3. $-\frac{1}{4}\operatorname{cosec}^3 x \cot x - \frac{3}{8}\operatorname{cosec} x \cot x + \frac{3}{8}\log\left|\tan\frac{x}{2}\right|$
4. $\frac{a^6}{48}\left(67\sqrt{2} + 15\log\left|\tan\frac{3\pi}{8}\right|\right)$

Page 74

1. $\frac{1}{48}\sin x\cos x\,(3 + 2\cos^2 x - 8\cos^4 x) + \frac{1}{16}x$
2. $\frac{8}{693}$
3. $\frac{2}{35}$
4. $\frac{5\pi}{4096}$
5. $\frac{\pi a^6}{32}$
6. $\frac{128 - 71\sqrt{2}}{1680}$
7. $\frac{3\pi}{128}$
8. $\frac{128}{1155}$
9. $\frac{7\pi}{8}$
10. $\frac{5\pi a^4}{8}$
11. $\frac{5\pi}{8}$
12. $\frac{8}{3}$
13. $\frac{\pi a^6}{32}$
14. $\frac{5\pi}{192}$
15. $\frac{8}{693}$

Page 81

6. $\frac{1}{2m+1}\left[1 + \frac{m+1}{2m-1} + \frac{(m+1)m}{(2m-1)(2m-3)} + \ldots\right]$
7. $\frac{\pi}{2}$

Page 84

1. $(3x^2 - 6)\sin x - (x^3 - 6x)\cos x$
2. $\frac{1}{n^4}(4n^2x^3 - 24x)\cos nx + \frac{1}{n^5}(n^4x^4 - 12n^2x^2 + 24)\sin nx$
3. $-\frac{6}{625}$
4. $5\left(\frac{\pi}{2}\right)^4 - 60\left(\frac{\pi}{2}\right)^2 + 120$
5. $-\frac{\pi^2}{12} + \frac{2}{27}$

Pages 88 and 89

1. $\frac{5\pi a^4}{128}$
2. $\frac{7\pi a^5}{8}$
3. $\frac{4a^5}{7}$
4. $\frac{5\pi a^4}{8}$
5. $\frac{x(a^2 + x^2)^{\frac{5}{2}}}{6} + \frac{5a^2 x}{24}(a^2 + x^2)^{\frac{3}{2}} + \frac{5a^2}{16}\left(x\sqrt{a^2 + x^2} + a^2\sinh^{-1}\frac{x}{a}\right)$

Pages 89 and 90

2. $\frac{1}{8}\sin x\left(-\cos^7 x + \frac{1}{6}\cos^5 x + \frac{5}{24}\cos^3 x + \frac{5}{16}\cos x + \frac{5}{128}x\right)$

3. $\frac{1}{6}\cos x\left(\sin^5 x - \frac{1}{4}\sin^3 x - \frac{3}{8}\sin x\right) + \frac{1}{16}x$ 4. $\frac{3\pi - 8}{32}$ 5. $\frac{5\pi}{32}$

10. (i) $\frac{\pi a^6}{32}$ (ii) $\frac{5\pi a^4}{128}$ (iii) $\frac{\pi}{32}$ (iv) $\frac{16}{105}$

(v) $\frac{3\pi}{128}$ (vi) $\frac{\Gamma(m+1)\Gamma(n+1)}{\Gamma(m+n+2)}$ (vii) $\frac{3\pi - 4}{192a^3}$

13. $\frac{q^n \lfloor n}{(p+1)(p+1+q)(p+1+2q)\dots(p+1+nq)}$

14. $\left(\frac{1}{13}x^4 - \frac{4}{143}x^2 + \frac{8}{1287}\right)(1+x^2)^{\frac{9}{2}}$

15. $-\frac{1}{4}x^3\sqrt{a^2 - x^2} - \frac{3}{8}a^2 x\sqrt{a^2 - x^2} + \frac{3}{8}a^4 \sin^{-1}\frac{x}{a}$

Page 95

1. 4π 3. πa^2 5. $\frac{4a^2}{3}$ 6. $\frac{\pi a^2}{8}$

9. $\frac{(4+\pi)a^2}{2}, \frac{(4-\pi)a^2}{2}$ 10. $(\pi - 2)\,a^2$ 11. πa^2

12. $3\sqrt{3}\,a^2$ 13. $\frac{3\pi a^2}{4}$ 14. πa^2 15. $3\pi a^2$

17. $\frac{2}{3}(3\pi - 2)a^2$ 18. $\frac{1}{3}(9\sqrt{3} + 4\pi)a^2$ 19. $4a^2$

20. $4\pi - 9\sin^{-1}\left(\frac{\sqrt{7}}{3\sqrt{3}}\right) - 8\sin^{-1}\left(\frac{\sqrt{7}}{2\sqrt{3}}\right)$

Page 97

4. $3\pi a^2$ 5. πab 7. $\frac{4\pi a^2}{3}$

Pages 100 and 101

1. $\frac{3\pi a^2}{2}$ 2. $\frac{\pi(2a^2 + b^2)}{2}$ 5. $\frac{\pi a^2}{2}$ 7. $\frac{\pi}{3}$

8. $\frac{\pi a^2}{4}$ 9. $\frac{5a^2}{2}$ 10. $(\pi - 1)\,a^2$

Pages 101 and 102

2. $\frac{32}{105}a^{-\frac{3}{2}}b^{\frac{7}{2}}$ 3. $\frac{a^2}{12}(3\pi - 8)$ 7. $(\pi + 2)\,a^2$ 8. $\frac{a^2}{2}(4 - \pi)$

10. $\dfrac{3\pi a^2}{8}$ 13. πa^2 15. $\dfrac{1}{2}(4-\pi)$ 22. $\dfrac{5\pi a^2}{4}$

23. $\left(\dfrac{15\sqrt{3}}{16}-\dfrac{\pi}{2}\right)a^2$ 24. $\dfrac{a^2}{3}(9\pi+16), \dfrac{9\pi+16}{9\pi-16}$

Pages 106 and 107

1. $a[\sqrt{2}+\log(1+\sqrt{2})]$ 2. $c\sinh\dfrac{x}{c}$ 5. $\dfrac{a}{27}(13\sqrt{13}-8)$

6. $\dfrac{3}{2}+\dfrac{1}{4}\log 2$ 7. $\dfrac{4a}{\sqrt{3}}$ 9. $4\sqrt{3}$ 10. $6a$

11. $2a[\sqrt{2}+\log(1+\sqrt{2})]$ 12. $a\sec\alpha\,(e^{\theta\cot\alpha}-1)$ 13. $\sqrt{2}\,(e^{\frac{\pi}{2}}-1)$

14. $\dfrac{1}{2}a\alpha^2$ 17. $a\sqrt{3}\left[\log|\sqrt{3}\cos\theta+\sqrt{3\cos^2\theta+1}|+\dfrac{\sqrt{1+3\cos^2\theta}}{\cos\theta}\right]_\alpha^\beta$

Pages 111 and 112

1. $s=a\,[\sec\psi\tan\psi+\log|\sec\psi+\tan\psi|]$
3. $s=a\log|\sec\psi+\tan\psi|$ 5. $s=a\,\psi$
7. $s=\sqrt{2}\,e^{\frac{\pi}{4}}(\cosh\psi-\sinh\psi-1)$
8. $s=4a\left(1-\cos\dfrac{\psi}{3}\right)$ 9. $s+a\psi=0$

Pages 112 and 113

1. $(r_2-r_1)\sec\alpha$ 3. 768π 8. $\dfrac{a^2+ab+b^2}{a+b}$ 11. $l[\sqrt{2}+\log(1+\sqrt{2})]$

12. $f\left(\dfrac{r_2}{a}\right)-f\left(\dfrac{r_1}{a}\right)$, where $f(\theta)=\dfrac{a}{2}[\theta\sqrt{1+\theta^2}+\log|\theta+\sqrt{1+\theta^2}|]$

16. $\dfrac{1}{3}[(4+t^2)^{\frac{3}{2}}-8]$

19. $a\left[(t-2)+\dfrac{\sqrt{3}}{2}\log\left|\dfrac{(t-\sqrt{3})(2+\sqrt{3})}{(t+\sqrt{3})(2-\sqrt{3})}\right|\right]$, where $t=\sqrt{\dfrac{4a-3x}{a-x}}$

Pages 121–123

1. $\dfrac{4}{3}\pi ab^2$ 2. $2\pi ah^2$ 3. $\dfrac{\pi c^2}{2}\left(x+\dfrac{c}{2}\sinh\dfrac{2x}{c}\right)$ 4. $2\pi a^3\left(\log 2-\dfrac{2}{3}\right)$

5. $\dfrac{1536\pi}{5}$ cu cm 6. $2\pi^2$

7. $\dfrac{8\pi a^3}{5}$ 8. $5\pi^2a^3$ 9. $\dfrac{32\pi a^3}{15}$

11. $\dfrac{\pi^2a^3}{4}$ 12. $\dfrac{2\pi b}{3a}\left[6a^2b-b^3-3ab\sqrt{a^2-b^2}-3a^2\sin^{-1}\dfrac{b}{a}\right]$

14. $\dfrac{2\pi}{15}$ 16. $\dfrac{32\pi}{15}$ 17. $\dfrac{8\pi a^3}{3}$

18. (i) $\dfrac{\pi a^3}{12}\left[\dfrac{3}{\sqrt{2}}\log(1+\sqrt{2})-1\right]$; (ii) $\dfrac{\pi^2 a^3}{4\sqrt{2}}$

19. $\dfrac{3\pi}{4}$ 20. $2\pi^2 a^3$ 21. $\dfrac{\pi a^3}{12}$

22. $\pi a^3(8\log 2-3)$ 23. $\dfrac{2\pi a^3}{3}$ 24. $\dfrac{1}{5}, \dfrac{5\pi}{28}$

Pages 128 and 129

1. $2\pi a^2$ 2. $\pi a^2[3\sqrt{2}-\log(1+\sqrt{2})]$

3. $\pi c\left[(b-a)+\dfrac{1}{2}\left(\sinh\dfrac{2b}{c}-\sinh\dfrac{2a}{c}\right)\right]$

4. (i) $32\pi\left[1+\dfrac{1}{4\sqrt{3}}\log\dfrac{2+\sqrt{3}}{2-\sqrt{3}}\right]$; (ii) $8\pi\left(1+\dfrac{4\pi}{3\sqrt{3}}\right)$

5. $\dfrac{64\pi a^2}{3}$ 6. $\dfrac{64\pi a^2}{3}$ 7. $4\pi a^2$ 8. $\dfrac{32\pi a^2}{5}$

9. $\dfrac{48\sqrt{2}\,\pi a^2}{5}$ 10. $4\pi a^2\left(1-\dfrac{1}{\sqrt{2}}\right)$ 11. 3π

13. $\dfrac{62\pi}{5}$ 14. $\dfrac{\pi a^2}{3}$

Pages 129 and 130

1. $\dfrac{5\pi}{28a^3}$ 2. $80\pi a^3$ 3. $\dfrac{2\pi a^3}{15\sqrt{5}}$

10. 4 : 5 11. 4π; $4\pi^2$ 13. $\dfrac{128}{1215}\pi a^2(125\sqrt{10}+1)$

17. $\pi^2, 2\pi^2$ 22. $\dfrac{23\pi a^3}{60}$

Page 134

1. $x\dfrac{dy}{dx}=y\log y$ 2. $a^2\left(\dfrac{d^2y}{dx^2}\right)^2=\left[1+\left(\dfrac{dy}{dx}\right)^2\right]^3$

3. $\dfrac{d^2y}{dx^2}-8\dfrac{dy}{dx}+15y=0$ 4. $x\dfrac{d^2y}{dx^2}+2\dfrac{dy}{dx}-xy+x^2-2=0$

5. $y\left[1-\left(\dfrac{dy}{dx}\right)^2\right]=2x\dfrac{dy}{dx}$ 6. $\dfrac{d^2y}{dx^2}-2\dfrac{dy}{dx}+2y=0$

7. $x\left[y\dfrac{d^2y}{dx^2}+\left(\dfrac{dy}{dx}\right)^2\right]-y\dfrac{dy}{dx}=0$

Page 137

1. $\log|y| = \frac{x^2}{2} + c$
2. $y^2 = ax^2 + c$
3. $xy = ce^{y-x}$
4. $\log|c(\text{cosec } y - \cot y)| = -\cos x$
5. $\log|y| = e^x + c$
6. $y - x = c(1 + xy)$
7. $\log\left|\frac{x}{y}\right| = \frac{y+x}{xy} + c$
8. $\log|x(1-y)^2| = \frac{1}{2}(x^2 - y^2) - 2y + c$
9. $x^2 + y^2 + 2(x - y) + 2\log|(x-1)(y+1)| = c$
10. $y = c(a + x)(1 - ay)$
11. $(y + 1)(e^x + 1) = ce^y$
12. $\sqrt{1+x^2} + \sqrt{1+y^2} = \sinh^{-1}\left(\frac{1}{x}\right) + c$
13. $(e^y + 1)\sin x = c$
14. $1 + e^y = c(1 + x^2)$
15. $\sqrt{1+x^2} + \sqrt{1+y^2} = c$
16. $e^{-x^2} + y^{-2} = c$
17. $x^2 + \tan^2 y = c$
18. $(x-1)e^x = \frac{1}{y} + \frac{1}{2y^2} + c$
19. $x\log x + \log|\log y| = x + c$
20. $\log(x^2 + 1) = y^2 - 2y + 4\log|y + 1| + c$
21. $r = ae^{\theta\cot\alpha}$
24. (i) $y = ke^{\frac{x}{c}}$ (ii) $r(k - \theta) = c$
26. $\frac{1}{r} = k + ce^{\theta}$
29. $r^2 = a^2 \sin 2\theta$
30. $cr(1 - \cos\theta) = 2$

Pages 141 and 142

1. $y = \frac{e^{mx}}{m+a} + ce^{-ax}$
2. $xy = \frac{1}{3}x^3 + \frac{3}{2}x^2 + 2x + c$
3. $x + y - 3 = ce^{-x}$
4. $(1 + x^2)y = \sin x + c$
5. $y = c(1 - x^2) + \sqrt{1 - x^2}$
6. $y = cx + x\log|\tan x|$
7. $x^2 y = (2 - x^2)\cos x + 2x\sin x + c$
8. $xy = c\cos x + \sin x$
9. $y = \sin x - 1 + ce^{-\sin x}$
10. $y = \sin x + c\cos x$
11. $x = ce^{-\tan^{-1}y} + \tan^{-1}y - 1$
12. $y = \tan x - 1 + ce^{-\tan x}$
13. $16x^2 y = 4x^4 \log x - x^4 + c$
14. $(1 + x^2)y = \frac{4}{3}x^3 + c$
15. $y\log x = (\log x)^2 + c$
16. $2y\sin x + \cos 2x + c = 0$
17. $x = y^3 + cy$
18. $xe^y = \tan y + c$
19. $2y = e^{\tan^{-1}x} + ce^{-\tan^{-1}x}$
20. $y(x - 1) = x^2(x^2 - x + c)$
21. $x = y - a^2 + ce^{-\frac{y}{a^2}}$
22. $xy^2 = 2y + c$
23. $(1 + x^3)y = \frac{x}{2} - \frac{1}{4}\sin 2x + c$
24. $y = \cos x - 2\cos^2 x;\ \frac{1}{8}$

Page 146

1. $(2 + cx)xy^2 = 1$
2. $\frac{1}{xy} + \log |x| = c$
3. $cx^5y^5 + \frac{5}{2}x^3y^5 = 1$
4. $x = y(1 + c\sqrt{x})$
5. $y(\log ex + cx) = 1$
6. $cy = (1 - y)\sqrt{1 - x^2}$
7. $\sec^2 x = cy - \frac{y}{3}\tan^3 x$
8. $y^2(c + 2x) = e^{x^2}$
9. $y^{-2} = 1 + x^2 + ce^{x^2}$
10. $(x + 1)^2 y^3 = \frac{1}{6}x^6 + \frac{2}{5}x^5 + \frac{1}{4}x^4 + c$
11. $e^{-x^2} = y(c - \cos x)$
12. $e^x = y(c - x^2)$
13. $\sin y = (1 + x)(e^x + c)$
14. $\sin x = (c + x)y$
15. $\frac{1}{x \log y} = c + \frac{1}{2x^2}$
16. $\tan y = ce^{-x^2} + \frac{1}{2}(x^2 - 1)$
17. $\sqrt{1 + x^2} = y(c + \sinh^{-1} x)$
18. $cy^2(1 - x^2) = 1 + y^2 + y^2(1 - x^2)\log |1 - x^2|$
19. $\frac{1}{y} = c\cos x + \sin x$
20. $\frac{1}{y^{n-1}} = 1 + ce^{(n-1)\frac{x^2}{2}}$
21. $\frac{1}{y^2} = -\left(\sin^2 x + \sin x + \frac{1}{2}\right) + ce^{2\sin x}$
22. $e^{-y} = x\int \frac{e^{-y}}{y}\,dy + cx$

Page 149

1. $\sqrt{x^2 + y^2}\, e^{\tan^{-1}\frac{y}{x}} = c$
2. $y^2 = x(x + c)$
3. $y^3 = ce^{\frac{x^3}{y^3}}$
4. $y = x + cy\sqrt{x}$
5. $cx^2 = y + \sqrt{x^2 + y^2}$
6. $x^2y = c(y + 2x)$
7. $\log |x + y| + \frac{2xy}{(x + y)^2} = c$
8. $x^2 + y^2 = cx$
9. $y = ce^{\frac{y}{x}}$
10. $\frac{x}{y} + \log |xy| = c$
11. $y^2 - x^2 = c(y^2 + x^2)^2$
12. $\cos\frac{y}{x} = \log |cx|$
13. $(y - x)^2 = cxy^2$

Page 153

1. $\tan^{-1}\left(\frac{2y + 1}{2x + 1}\right) = \log\left|c\sqrt{x^2 + y^2 + x + y + \frac{1}{2}}\right|$
2. $\log |\xi| + c = -\frac{1}{2}\log |2v^2 - 1| + \frac{1}{2\sqrt{2}}\log\left|\frac{v\sqrt{2} - 1}{v\sqrt{2} + 1}\right|$,

   where $\xi = x - 1$ and $v = \frac{y + 1}{x - 1}$

3. $\tan^{-1}\left(\frac{y+3}{x+2}\right) + \log|c\sqrt{(y+3)^2 + (x+2)^2}| = 0$

4. $x + y + \frac{4}{3} = ce^{3(x-2y)}$

5. $x + y - 2 = c(y - x)^3$

6. $\frac{3}{2}(x^2 + y^2) + 2xy - 5(x + y) = c$

7. $\log|x - y - 1| = x - 2y + c$

8. $4x + 8y + 5 = ce^{4(x-2y)}$

9. $x - 2y + \log|x - y + 2| = c$

10. $\frac{2}{7}(2x + 3y) - \frac{9}{49}\log|14x + 21y + 22| = x + c$

11. $\log|2x + y - 1| + x + 2y = c$

12. $x + 2y - 5 = c(2x - y)^2$

13. $3(x - 2y) + \log|3x + 3y + 2| = c$

14. $x^2 - 4xy + 4y^2 + 6x - 10y = c$

15. $(y - x + 1)^2(y + x - 1)^5 = c$

16. $(y - x - 2)^4 = c(x + 5y + 2)$

Pages 156

1. $x^3 - 3axy + y^3 = c$

2. $ax^2 + 2hxy + by^2 + 2gx + 2fy + c = 0$

3. $(e^y + 1)\sin x = c$

4. $x^4 - 6x^2y^2 - y^4 = c$

5. $x^4 + 6x^2y^2 + y^4 = c$

6. $a^2x - x^2y - xy^2 - \frac{1}{3}y^3 = c$

7. $x + ye^{\frac{x}{y}} = c$

8. $y(x + \log x) + x\cos y = c, (x > 0)$

9. $2(x + y) + \sin 2x + \sin 2y - 4\sin\alpha\sin x\sin y = c$

10. $x\sin xy = c$

11. $x^4 + 2x^2y^2 + y^4 = c$

12. $\sqrt{x^2 + y^2} + \log|xy| + \frac{x}{y} = c$

13. $x^3y + x^2 - y^2 = cxy$

14. $x^3 + y^3 - x^2 - xy + y^2 = c$

15. $xy(x^2 + y^2) = c$

16. $x^3y - 2x^2y^3 + 3x^2 - y = c$

17. $\frac{1}{2}\sin 2y + x\cos 2y - x^3y^2 = c$

18. $xy^2 - x^2y + 3x^2 - 2y = c$

Page 163

1. $x^2 + \frac{e^x}{y} = c$

2. $\frac{x}{y} + \log|xy| = c$

3. $ax^2y - cy + 2e^x = 0$

4. $\log\left|\frac{x^2}{y}\right| - \frac{1}{xy} = c$

5. $x^2 - y - 1 - x\cos y = cx$

6. $\frac{e^x}{y} + \frac{2}{3}x^3 - \frac{1}{2}y^2 = c$

7. $\frac{1}{xy} = \log\left|\frac{x}{y}\right| + c$

8. $\log|y| = \frac{x^3}{3y^3} + c$

9. $\frac{x}{y} + \log\left|\frac{y^3}{x^2}\right| = c$

10. $cy\cos xy = x$

11. $x^2y^2 + xy\log\left|\frac{cx}{y}\right| = 1$

12. $(x^2 + y^2)e^x = c$

13. $x^4y(3 + y^2) + x^6 = c$ 14. $x^2y^3(1 + 2xy) = c$

15. $3x^2y^4 + 6xy^2 + 2y^6 = c$

16. $e^{6y}\left[\frac{x^2y^2}{2} - \frac{x^3}{3} + \frac{y^2}{6} - \frac{y}{18} + \frac{1}{108}\right] = c$

17. $5x^{-\frac{36}{13}}y^{\frac{24}{13}} - 12x^{-\frac{10}{13}}y^{-\frac{15}{13}} = c$ 18. $x^2y^4(x + y^2) = c$

19. $4x^{\frac{1}{2}}y^{\frac{1}{2}} - \frac{2}{3}x^{-\frac{3}{2}}y^{\frac{3}{2}} = c$

Page 164

1. $\frac{x + y}{a} = \tan\left(\frac{y - c}{a}\right)$ 2. $x = \log\left|\tan\left(\frac{x + y}{2}\right) + 1\right| + c$

3. $a \log\left|\frac{x - y - a}{x - y + a}\right| = 2y + c$ 4. $y = \tan\left(\frac{x + y}{2}\right) + c$

5. $4x + y + 1 = 2\tan(2x + c)$ 6. $cx^2 + 2xe^{-y} = 1$

7. $e^y = ce^{-x} + e^x - 1$ 8. $x \log|y| = e^x(x - 1) + c$

9. $(b - a)\log|(x + y)^2 - ab| = 2(x - y) + c$

10. $\sin y = (e^x + c)(1 + x)$ 11. $\sqrt{x^2 + y^2} = a \sin\left(c + \tan^{-1}\frac{y}{x}\right)$

12. $y^2 + x\log|cx| = 0$ 13. $\sec y = x + 1 + ce^x$

14. $x + y - 4\log|2x + 3y + 7| = c$

Pages 165 and 166

1. $y = 1 - ce^{\frac{1}{x}}$ 2. $\log\left|\frac{x}{y}\right| - \frac{x + y}{xy} = c$

3. $y + \frac{2}{3}(x - 2a)\sqrt{a + x} = c$ 4. $xy = \frac{x^3}{3} + \frac{3}{2}x^2 + 2x + c$

5. $(1 + x^2)y = \tan^{-1}x - \frac{\pi}{4}$ 6. $y = \cos x + c\sec x$

7. $y = \tan x + c\sqrt{\tan x}$ 8. $y = cx\sqrt{x^2 - 1} + ax$

9. $c(x - y)^{\frac{2}{3}}(x^2 + xy + y^2)^{\frac{1}{6}} = e^{\frac{1}{\sqrt{3}}\tan^{-1}\left(\frac{x+2y}{x\sqrt{3}}\right)}$

10. $\log|x(3y^2 - 2xy + x^2)| + \sqrt{2}\tan^{-1}\left(\frac{3y - x}{x\sqrt{2}}\right) = c$

11. $\sec\frac{y}{x} = cxy$ 12. $(x - 2y^3)y^2 = c$ 13. $i = \frac{E}{R}(1 - e^{\frac{-Rt}{L}})$

14. $\frac{1}{xy} = c - \int\frac{\sin x}{x}\,dx$ 15. $x^3y^{-3} = 3\sin x + c$

16. $\frac{1}{y^2} = -1 + (c + x)\cot\left(\frac{x}{2} + \frac{\pi}{4}\right)$ 17. $(2x^3 - y)^2 = cyx^6$

18. $\frac{1}{2}y\cos 2x + \frac{3}{2}y + \frac{1}{3}y^3 = c$

19. $r^2 + 2r(\sin\theta - \cos\theta) = c$

20. $\log\left|\frac{x}{y}\right| - xy = c$

21. $x^3y^3 + x^2 = cy$

22. $x^3y^2 + 4x^2y^6 = c$

23. $x^2y^2(y^2 - x^2) = c$

24. $e^x + \frac{my^2}{2x^2} = c$

25. $\frac{y(x^2+1)}{x} = \frac{1}{2}x^2\log x - \frac{1}{4}x^2 + c$

26. $3y^2 = cx^2 - 2x^2 e^{-\frac{1}{x^3}}$

27. $x^2 + y^2 + 2a^2(y^2 - x^2) = c$

28. $xy = c\cos x + \sin x$

29. $y = (1 + cy + \log y)\cos x$

30. $\tan^{-1} e^x = \frac{1}{2}\sec^2 y + c$

31. $\sqrt{1-x^2} + \sqrt{1-y^2} = 1$

32. $y = xe^{1+cx}$

33. $2x = (x - y)\log|cx|$

34. $x^3\tan y + y^4 + \frac{y^3}{x^2} = c$

35. $\tan xy - \cos x - \cos y = c$

36. $x^4 - x^2y^2 - 4xy + 6x = c$

37. $xy(x^2 - 3) + 4y^2 = c$

38. $(xy - 2)^2 + (x + 3)^2 = 2y^2 + 15$

39. $y^2\sin 2x = c + 2x(3 + y + y^2)$

40. $e^{xy} = 2xy^3 + y^3 - 3$

41. $2y^5 - 2x^2y^3 + 3x = 0$

Page 169

1. $(y - 2x - c)(y - 3x - c) = 0$

2. $(y - 3x - c)(y - 6x - c) = 0$

3. $y^2 = cx^{(1\pm\sqrt{5})}$

4. $y(1 \pm \cos x) = c$

5. $(y - c)(xy + cy - 1)(y - ce^{\frac{1}{x}}) = 0$

6. $(2y + x^2 - c)(x + \log|y| - c) = 0$

7. $(2y - x^2 - c)(2y + 3x^2 - c) = 0$

8. $(2y - x^2 - c)(2x - y^2 - c) = 0$

9. $(y - cx^2)(yx^3 - c) = 0$

10. $(y - e^x - c)(y + e^{-x} - c) = 0$

11. $(y^2 - x^2 - c)(xy - c) = 0$

12. $(2y - x^2 - c)(y - ce^x)(y + x - 1 - ce^{-x}) = 0$

13. $(y - x + c)(xy + c) = 0$

14. $(xy - c)(yx^2 - c) = 0$

15. $(y - x + c)(x^2 + y^2 - c^2) = 0$

16. $\sin^{-1}\frac{y}{x} = \pm\log|cx|$

17. $(y - c)(y + x^2 - c)(xy + cy + 1) = 0$

18. $(y - e^x + c)\left(y + \frac{x^2}{2} + c\right) = 0$

19. $(2xy + x^2 - c)(x^2 + y^2 - c) = 0$

Page 171

1. $x = \frac{c}{p^2} + \frac{2}{3}p,\ y = \frac{2c}{p} + \frac{1}{3}p^2$

2. $xy = c + c^2x$

3. $\cos[(\sqrt{1 - c^2 + 2cx - x^2} - y)/(c - x)] = c - x$

4. $x = 2ap + \frac{3}{2}bp^2,\ y = ap^2 + bp^3$

5. $x = \sin p + c,\ y = p\sin p + \cos p$

6. $x = 2\tan^{-1}p - p^{-1} + c,\ y = \log(p^3 + p)$

7. $x = \frac{1}{2}(\log p + 1)^2 + c,\ y = p\log p$

8. $x = b \log p + 2cp + A,\ y = a + bp + cp^2$
9. $x = (\log p - p + c)(p - 1)^{-1},\ y = xp^2 + p$
10. $x = \log p + \sin p + p \cos p + c,\ y = p + p^2 \cos p$
11. $x = \log (p + 1) - \log (p - 1) + \log p + c,\ y = p - \log (p^2 - 1)$
12. $x = \tan p + c,\ y = p \tan y + \log \cos p$

Page 174

1. $x = ap + bp^2,\ y = \frac{1}{2} ap^2 + \frac{2}{3} bp^3 + c$
2. $x = \frac{1}{2}\left(p + \frac{1}{p}\right),\ y = \frac{1}{4}p^2 - \frac{1}{2}\log p + c$
3. $x = \log p + \sin p,\ y = p(1 + \sin p) + \cos p + c$
4. $x = p + \sin p,\ y = \frac{1}{2}p^2 + p\sin p + \cos p + c$
5. $xp^3 = a + bp,\ yp^2 = \frac{3}{2} a + 2bp + cp^2$
6. $x = a \cos^3 t,\ y = -a \sin^3 t,\ p = \tan t$
7. $p^3 - p(y + 3) + x = 0,\ y(1 - p^2)^{\frac{1}{2}} + (1 - p^2)^{\frac{3}{2}} = c$
8. $x = c + a \log \frac{p}{1 - p},\ x = y + a \log p$

Page 178

1. $x = \frac{c}{p^2} - \frac{1}{p},\ y = \log p + \frac{2c}{p} - 2$
2. $y = cx + \frac{a}{2c}$
3. $y = cx + ac(1 - c)$
4. $y = cx + \frac{a}{c}$
5. $y = cx + (1 + c^2)^{\frac{3}{2}}$
6. $y = cx + \frac{a}{c^2}$
7. $x = 2(1 - p) + ce^{-p},\ y = [2(1 - p) + ce^{-p}](1 + p) + p^2$
8. $y = cx + c^2$
9. $y = cx + \frac{c}{c - 1}$
10. $c = \log | cx - y |$
11. $y = cx + \sin^{-1}c$
12. $y = cx - \frac{c - 1}{c}$
13. $y^2 = cx^2 - \frac{2c}{1 + c}$
14. $\sin y = c \sin x + c^2$

Page 178

1. $x = \frac{e^{\frac{1}{p}}}{p^2},\ y = c + e^{\frac{1}{p}}\left(1 + \frac{1}{p}\right)$
2. $y^2 = 2cx + c^3$
3. $(y - x - c)(xy - c) = 0$
4. $y = cx + c^3$
5. $y^2 = cx^2 + c^2$
6. $x = p^2 - 2p + 2,\ y = c + \frac{2}{3} p^3 - p^2$
7. $x = \frac{c}{p^2} - \frac{\cos p}{p^2} - \frac{\sin p}{p},\ y = \frac{2c}{p} - \frac{2 \cos p}{p} - \sin p$

8. $x = (\log p - p + c)\,(p-1)^{-2},\ y = xp^2 + p$
9. $c^2y^2 + 2cx = 1$
10. $x = \dfrac{c}{p^3} - 2e^p\left(\dfrac{1}{p} - \dfrac{2}{p^2} + \dfrac{2}{p^3}\right),$

$y = \dfrac{3c}{2p^2} - 2e^p\left(1 - \dfrac{3}{p} + \dfrac{3}{p^2}\right)$

11. $x + c = 2\tan^{-1}p - \log\left|\dfrac{1+\sqrt{1-p^2}}{p}\right|,$

$y = \sin^{-1}p + \log(1+p^2)$

12. $(y-c)\,(y-x-c)\,(2y-x^2-c)\,(y-ce^{2x}) = 0$
13. $xy = c(3cx-1)$
14. $y^4 = c(x-c)$
15. $y = 4c(cxy+1)$
16. $xp = \frac{1}{3}p^3 + c,\ 9(y + xp\log p) = (2 + 3\log p)p^3$
17. $e^y = ce^x + c^3$
18. $x^2 + y^2 = cx$
19. $(x-a)^2 + (y+c)^2 = 1$
20. $y^2 = 2cx - c^2$
21. $y = 2c\sqrt{x} + f(c^2)$
22. $y^2 = cx + c^2$
23. $(y-cx)^2 = a^2(1+c^2)$
24. $(c-y)\,(1+p^2) = 1, p = \tan(x - p/(1+p^2))$
25. $(y-cx^2)\,(y^2+3x^2-c) = 0$
26. $y^2 = cx^2 - \dfrac{bc}{ac+1}$
27. $xy = cy + c^2$
28. $x = \tan p + c,\ y = p\tan p + \log\cos p$
29. $(y-cx+2c)\,(y-cx+1) = 0$
30. $[y - c(x+1)]\,[y + x\log|cx|] = 0$

Page 184

1. $x^2 - y^2 = a^2$
2. $x^2 + 2y^2 = a^2$
3. $2x^2 + 3y^2 = c^2$
4. $x^2 + y^2 + 2fy - c = 0$
8. $r = ce^{-\frac{\theta^2}{2}}$
9. $r^2 = ce^{\theta^2}$
10. $r = c(1-\cos\theta)$
11. $(\log r)^2 + \theta^2 = c^2$
12. $r^n = c^n\sin n\theta$
13. $r^n = c^n\cos n\theta$
14. $r^n\cos n\theta = c^n$
15. $r^2 = c\ \cos\theta$
16. $\sec 5\theta + \tan 5\theta = ce^{\frac{25}{r}}$
17. $r^2 = c(1+\cos^2\theta)$
18. $r = ce^{-\sin\theta}$
19. $r = 2c(\sin\theta - \cos\theta)$
20. $r^2\sin^3\theta = c(1+\cos\theta)$
21. $r = \dfrac{2b}{1-\cos\theta}$

Page 193

1. $y = c_1e^{-x} + c_2e^{-2x}$
2. $y = c_1 + c_2e^{-2x} + c_3e^{x/3}$
3. $y = c_1e^{-x} + c_2e^{2x} + c_3e^{3x}$
4. $y = c_1 + c_2e^{x} + c_3e^{-2x}$
5. $y = c_1e^{2x} + c_2e^{-x} + c_3e^{-3x}$
6. $y = c_1e^{x} + c_2e^{3x} + c_3e^{5x}$
7. $y = c_1e^{3x} + c_2e^{x} + c_3e^{-x}$
8. $x = 0$

9. $x = c_1 + c_2e^{-t} + c_3e^{3t}$
10. $y = c_1e^{-2x} + c_2e^{-\frac{x}{2}} + c_3e^{\frac{3x}{2}} + c_4e^x$
11. $y = e^{-4x}(c_1 \cos 3x + c_2 \sin 3x)$
12. $y = e^{2x}(c_1 \cos 3x + c_2 \sin 3x)$
13. $y = c_1e^x + c_2 \cos x + c_3 \sin x$
14. $y = c_1e^{-x} + e^{2x}(c_2 \cos 3x + c_3 \sin 3x)$
15. $y = c_1 + c_2x + e^{-x}(c_3 \cos 3x + c_4 \sin 3x)$
16. $y = (c_1 + c_2x) \cos 3x + (c_3 + c_4x) \sin 3x$
17. $y = (c_1 + c_2x) \cos 2x + (c_3 + c_4x) \sin 2x$
18. $y = c_1 \cos x + c_2 \sin x + c_3 \cos 2x + c_4 \sin 2x$
19. $y = e^x[(c_1 + c_2x) \cos 2x + (c_3 + c_4x) \sin 2x]$
20. $y = e^{-3x} \sin 2x$
21. $y = (c_1 + c_2x)e^{3x}$
22. $y = c_1 + (c_2 + c_3x)e^{2x}$
23. $y = (c_1 + c_2x)e^{4x}$
24. $y = (c_1 + c_2x + c_3x^2)e^{2x}$
25. $y = c_1e^{-x} + (c_2 + c_3x)e^{2x}$
26. $y = c_1 + c_2x + (c_3 + c_4x)e^{-x}$
27. $y = c_1 + c_2x + c_3e^{2x} + c_4e^{-\frac{x}{2}}$
28. $y = c_1e^{3x} + (c_2 + c_3x + c_4x^2)e^{-2x}$
29. $y = c_1 + c_2x + c_3x^2 + (c_4 + c_5x)e^x$
30. $y = (c_1 + c_2x)e^{-x} + c_3e^{2x} + c_4 \cos x + c_5 \sin x$

Page 197

1. $y = c_1e^{2x} + c_2e^{3x} + \frac{1}{2}e^{4x}$
2. $y = c_1e^x + c_2e^{2x} + \frac{1}{12}e^{5x}$
3. $y = (c_1 + c_2x)e^{-x} + \frac{2}{9}e^{2x}$
4. $y = c_1e^x + c_2e^{-x} + \frac{1}{2}xe^x$
5. $y = (c_1 + c_2x)e^x + \frac{1}{2}x^2e^x$
6. $y = c_1 \cos x + c_2 \sin x - \frac{1}{3} \cos 2x$
7. $y = c_1 \cos 2x + c_2 \sin 2x - \frac{1}{4} \cos 2x \log \left| \tan\left(x + \frac{\pi}{4}\right) \right|$
8. $y = c_1 \cos 3x + c_2 \sin 3x + \frac{1}{3} x \sin 3x + \frac{1}{9} \cos 3x \log | \cos 3x |$
9. $y = c_1e^{-2x} + c_2e^x - \left(x + \frac{1}{2}\right)$
10. $y = c_1 \cos x + c_2 \sin x + \sin x \log \left| \tan\left(\frac{x}{2} + \frac{\pi}{4}\right) \right| - 1$

Page 198

1. $y = (c_1 + c_2x)e^{2x} + e^x$
2. $y = c_1e^{-4x} + c_2e^x + \frac{1}{6}e^{2x}$
3. $y = c_1e^{-4x} + c_2e^x + \frac{1}{5}e^x$
4. $y = (c_1 + c_2x)e^{kx} + \frac{e^k}{(k-1)^2}$
5. $y = e^{-\frac{x}{2}}\left(c_1 \cos \frac{\sqrt{3}}{2}x + c_2 \sin \frac{\sqrt{3}}{2}x\right) + e^{-x}$
6. $y = c_1e^x + c_2e^{3x} + xe^{3x}$
7. $y = c_1e^{-\frac{7x}{3}} + c_2e^{2x} + xe^{2x}$
8. $y = (c_1 + c_2x)e^{3x} + (c_3 + c_4x)e^{-3x} + \frac{1}{2}x^2 e^{3x}$

9. $y = c_1e^{-x} + c_2 + c_3x + (c_4 + c_5x + c_6x^2)e^{-x} + \frac{1}{12}x^3e^x$

10. $y = \frac{7}{4}e^{3x} + \frac{3}{4}e^{-x} - \frac{1}{2}e^x$

Page 202

1. $y = c_1 \cos 2x + c_2 \sin 2x - \frac{1}{5}\cos 3x$

2. $y = c_1 \cos x + c_2 \sin x - \frac{1}{2}\cos x$

3. $y = c_1e^{3x} + c_2e^{-x} - \frac{2}{3}e^{2x} - \frac{3}{5}\sin x + \frac{3}{10}\cos x$

4. $y = (c_1 + c_2x)e^x + \frac{1}{2}e^{-x} - \frac{11}{25}\sin 2x - \frac{2}{25}\cos 2x$

5. $y = c_1 \cos 2x + c_2 \sin 2x + \frac{2}{5}e^{-x} + \frac{1}{3}\sin x + \frac{1}{2}x \sin 2x$

6. $y = c_1e^{2x} + c_2e^{3x} + \frac{5}{78}\cos 3x - \frac{1}{78}\sin 3x$

7. $y = e^{4x}(c_1 \cosh\sqrt{7}\,x + c_2 \sinh\sqrt{7}\,x) + \frac{25}{29}\cos 5x - \frac{10}{29}\sin 5x$

8. $y = c_1e^x + c_2e^{2x} + 3e^{3x} + 6\cos 2x - 2\sin 2x$

9. $y = e^x(c_1 \cos 2x + c_2 \sin 2x) + 3\cos 3x - \sin 3x$

10. $y = c_1 \cos 4x + c_2 \sin 4x - 3x \cos 4x$

11. $y = \frac{3}{2}e^{3x} + 2e^{-x} - \frac{1}{2}e^x + 2\sin x - \cos x$

12. $y = \frac{1}{29}\left[e^{-2x}\left(37\cos 3x - \frac{2}{3}\sin 3x\right) + 9\sin 2x - 8\cos 2x\right]$

Pages 204 and 205

1. $y = c_1e^{2x} + c_2e^{3x} + \frac{1}{6}\left(x + \frac{5}{6}\right)$ 2. $y = e^{-x}(c_1 \cos x + c_2 \sin x) + \frac{1}{2}(x - 1)$

3. $y = c_1 + c_2e^{-x} + \frac{1}{4}x^4 - \frac{1}{3}x^3 + x^2 - 2x$

4. $y = e^{-\frac{x}{2}}\left[c_1 \cos\frac{\sqrt{3}}{2}x + c_2 \sin\frac{\sqrt{3}}{2}x\right] + x^3 - x^2 - 4x + 6$

5. $y = c_1e^{2x} + c_2e^{-2x} - \frac{1}{4}x^2 - \frac{1}{8}$ 6. $y = (c_1 + c_2x)e^x + x + 1$

7. $y = c_1e^x + c_2e^{-2x} - \frac{1}{10}(\cos x + 3\sin x) - \frac{1}{4}(2x + 1)$

8. $y = c_1 + c_2e^{3x} + c_3e^{-2x} - \frac{1}{6}\left(\frac{1}{3}x^3 - \frac{1}{6}x^2 + \frac{25}{18}x\right)$

9. $y = c_1 + (c_2 + c_3x)e^{-x} + \frac{1}{18}e^{2x} + \frac{1}{3}x^3 - \frac{3}{2}x^2 + 4x$

10. $y = c_1e^x + c_2e^{2x} + x^2 + 3x + 4$

11. $y = e^{-2x}(c_1 \cos x + c_2 \sin x) + 10x - 8 + \frac{1}{2} e^{3x}$

12. $y = 1 + 2x - 3x^2 - e^{-3x}$

13. $x = 2(1 - e^{-2t} \cos t - 2e^{-2t} \sin t)$

Page 209

1. $y = c_1 e^x + c_2 e^{-2x} + \frac{1}{3} xe^x$

2. $y = (c_1 + c_2 x)e^x + \frac{1}{8} e^{3x} (2x^2 - 4x + 3)$

3. $y = e^x(c_1 \cos 2x + c_2 \sin 2x) - \frac{1}{10} e^{2x} (\cos x - 2 \sin x)$

4. $y = c_1 e^{-x} + c_2 e^x + 2e^x(x^2 - x)$

5. $y = c_1 e^x + (c_2 + c_3 x)e^{-2x} - \frac{1}{18} x^2 (x + 1)e^{-2x}$

6. $y = e^x (c_1 \cos \sqrt{3}\, x + c_2 \sin \sqrt{3}\, x) + \frac{1}{2} e^x \cos x$

7. $y = c_1 e^{-x} + c_2 e^{-2x} + c_3 e^{3x} - \frac{1}{12} e^{2x} \left(x + \frac{17}{12} \right)$

8. $y = c_1 e^{2x} + c_2 e^{-6x} + \frac{1}{32} e^{2x} (2x^2 - 3)$

9. $y = (c_1 + c_2 x)e^{2x} + \frac{1}{20} x^5 e^{2x}$

10. $y = e^x (c_1 \cos \sqrt{3}\, x + c_2 \sin \sqrt{3}\, x) + \frac{1}{2} e^x \sin x$

11. $y = c_1 e^x + c_2 e^{-x} - \frac{1}{39} e^x (2 \cos 3x + 3 \sin 3x)$

12. $y = c_1 e^x + c_2 e^{-x} + c_3 \cos x + c_4 \sin x - \frac{1}{5} e^x \cos x$

13. $y = (c_1 + c_2 x + c_3 x^2)e^{-x} + \frac{1}{60} x^5 e^{-x}$

14. $y = e^{-x} (c_1 \cos \sqrt{3}\, x + c_2 \sin \sqrt{3}\, x) + \frac{1}{73} e^x (3 \sin 2x - 8 \cos 2x)$

15. $y = c_1 e^x + c_2 e^{3x} - \frac{1}{8} e^x(\sin 2x + \cos 2x) - \frac{1}{30} (2 \sin 3x + \cos 3x)$

16. $y = (c_1 + c_2 x)e^{-x} + \frac{1}{2} (1 - x) \cos x + \frac{1}{2} \sin x$

17. $y = c_1 \cos 2x + c_2 \sin 2x + \frac{1}{3} x \sin x - \frac{2}{9} \cos x$

18. $y = c_1 e^x + c_2 e^{-x} + c_3 \cos x + c_4 \sin x + \frac{1}{4} \left(\frac{x^3}{3} - \frac{5}{2} x \right) \cos x - \frac{3}{8} x^2 \sin x$

19. $y = c_1 e^x + c_2 e^{2x} + \frac{1}{20} x (3 \sin 2x - \cos 2x) + \frac{3}{25} \cos 2x - \frac{7}{200} \sin 2x$

20. $y = (c_1 + c_2 x)e^x - e^x(2 \cos x + x \sin x)$

Pages 209 and 210

1. $y = c_1 e^{-x} + c_2 e^{4x}$

2. $y = c_1 e^{-2x} + c_2 e^{-\frac{x}{2}}$

3. $y = c_1 + c_2 e^{4x} + c_3 e^{-3x}$

4. $y = c_1 e^x + c_2 e^{-x} + c_3 e^{3x}$

5. $y = c_1 + c_2e^{2x} + c_3e^{-4x}$ 6. $x = c_1e^t + c_2e^{2t} + c_3e^{-3t}$

7. $y = c_1e^{-x} + c_2e^{\frac{x}{2}} + c_3e^{\frac{2x}{5}}$ 8. $y = c_1e^x + c_2e^{2x} + c_3e^{3x} + c_4e^{-4x}$

9. $y = c_1e^x + c_2e^{-2x} - \frac{1}{2} - x + 6\cos 2x - 2\sin 2x$

10. $y = c_1e^x + e^{-x}(c_2\cos 2x + c_3\sin 2x) +$
$\frac{1}{17}(2\cos 2x - 9\sin 2x) - 2x^2 - \frac{9x}{5} - \frac{82}{25}$

11. $y = c_1e^{2x} + c_2e^{-3x} + 2xe^{2x} - 3e^{3x} + x + 2$

12. $y = c_1e^{2x} + c_2e^{-x} + \frac{3}{2} - 3x - 2xe^{-x}$

13. $y = -6e^x + 2e^{3x} + 3x^2 + 8x + 10$ 14. $y = (2x + 6)\cos x - \sin x + 3(x^2 - 2)$

15. $y = c_1 + c_2x + c_3\cos x + c_4\sin x + \frac{1}{4}x^4 - 3x^2 + x\sin x + 2x\cos x$

16. $y = -\frac{1}{12}e^{-x} - \frac{2}{3}e^{2x} + \frac{1}{4}e^{3x} + \frac{1}{2}e^x$ 17. $y = e^{2x}\sin 5x$

18. $y = e^{3x}(2\sin 4x - 3\cos 4x)$ 19. $y = (c_1 + c_2x)e^{\frac{x}{2}} + (c_3 + c_4x)e^{-x}$

20. $y = c_1e^x + (c_2 + c_3x + c_4x^2)e^{2x}$ 21. $y = c_1e^x + c_2e^{2x} + (c_3 + c_4x)e^{-x}$

22. $y = 2 - e^{-x} - e^{-2x}$ 23. $y = (c_1 + c_2x + c_3x^2)e^x + c_4e^{-2x} + \frac{1}{18}x^3e^x$

24. $y = c_1\cos ax + c_2\sin ax + \frac{1}{2a}x\sin ax$

25. $y = (c_1 + c_2x + c_3x^2)e^{2x} + \frac{1}{24}x^4e^{2x} + (x^2 - 6x + 12)e^{3x}$

26. $y = (c_1 + c_2x)e^x + \frac{1}{2}(x\cos x + \cos x - \sin x)$

27. $y = e^{\frac{x}{\sqrt{2}}}\left(c_1\cos\frac{x}{\sqrt{2}} + c_2\sin\frac{x}{\sqrt{2}}\right) + e^{-\frac{x}{\sqrt{2}}}\left(c_3\cos\frac{x}{\sqrt{2}} + c_4\sin\frac{x}{\sqrt{2}}\right)$
$+ \frac{1}{2}[(x^2 - 2)\cos x - 4x\sin x]$

28. $y = (c_1 + c_2x)e^{2x} + (3\sin 2x - 2x^2\sin 2x - 4x\cos 2x)e^{2x}$

Pages 214 and 215

1. $y = c_1x^{-1} + c_2x^2 + \frac{1}{3}\left(x^2 - \frac{1}{x}\right)\log x$

2. $y = x(c_1 + c_2\log x) + 2\log x + 4$

3. $y = (c_1 + c_2\log x) + c_3x^2 + \frac{1}{4}x^3 - \frac{3}{2}x(\log x)^2$

4. $y = c_1 + c_2\log x + c_3(\log x)^2 + 3x^2$

5. $y = x^2[c_1\cos(\log x) + c_2\sin(\log x)] - \frac{1}{2}x^2\log x\cos(\log x)$

6. $y = (c_1 + c_2\log x)x + c_3x^{-1} + \frac{1}{4}x^{-1}\log x$

7. $y = c_1x^2 + c_2x^{-2} + \frac{1}{5}x^3$

8. $y = c_1x^2 + c_2x + \frac{1}{2}x^3 - x\log x + x^2\left[\frac{1}{2}(\log x)^2 - \log x\right]$

9. $y = c_1x^{-1} + x[c_2\cos(\log x) + c_3\sin(\log x)] + 5x + 2x^{-1}\log x$

10. $y = c_1x^2 + x^{\frac{5}{2}}(c_2x^{\frac{\sqrt{21}}{2}} + c_3x^{-\frac{\sqrt{21}}{2}}) - \frac{1}{5}x^3$

11. $y = x[c_1\cos(\log x) + c_2\sin(\log x)] + x\log x$

12. $y = x^{-3}[c_1\cos(\log x^2) + c_2\sin(\log x^2)] + \frac{1}{13}\log x - \frac{6}{169}$

13. $y = c_1x^{-1} + c_2x^{-2} + \frac{1}{6}x + \frac{1}{2}\log x - \frac{3}{4}$

Page 219

1. $y = c_1x^2 + c_2x^3 + \frac{1}{2}x$
2. $y = c_1x^2 + c_2x^{-2} + \frac{1}{4}x^2\log x$
3. $y = x^2[c_1 + c_2\log x + (\log x)^2]$
4. $y = c_1x^{-1} + c_2x^4 + \frac{1}{5}x^4\log x$
5. $y = c_1x^{-1} + c_2x^{-5} + \frac{1}{60}x^2$
6. $y = (c_1 + c_2\log x)x^{-2} + \frac{1}{36}x^4$
7. $y = c_1x^{-1} + c_2x^{-5} + \frac{2}{77}x^6$
8. $y = (c_1 + c_2\log x)x^{-1} + 1 - \frac{1}{2}x + \frac{1}{9}x^2$
9. $y = c_1x + c_2x^2 + \frac{1}{6x}$
10. $y = x^2(c_1 + c_2\log x) + \frac{x^m}{(m-2)^2}$

Page 223

1. $y = (2x-1)\left[c_1 + c_2(2x-1)^{\frac{\sqrt{3}}{2}} + c_3(2x-1)^{-\frac{\sqrt{3}}{2}}\right]$

2. $y = c_1(x+a)^2 + c_2(x+a)^3 + \frac{1}{6}(3x+2a)$

3. $y = (5+2x)^2[c_1(5+2x)^{\sqrt{2}} + c_2(5+2x)^{-\sqrt{2}}]$

4. $y = [c_1 + c_2\log(x+1)](x+1)^2 +$
$\frac{1}{4}[2(x+1)^2\{\log(x+1)\}^2 - 8x - 7]$

5. $y = [c_1 + c_2\log(1+2x)](1+2x)^2 + (1+2x)^2\{\log(1+2x)\}^2$

Pages 223 and 224

1. $y = c_1x + (c_2 + c_3\log x)x^{-2}$
2. $y = [c_1 + c_2\log x + c_3(\log x)^2]x^2$
3. $y = c_1\cos(\log x) + c_2\sin(\log x)$
4. $y = c_1x^{-1} + c_2x^{-2} + x^{-2}e^x$
5. $y = c_1x + c_2x^{-1} + \frac{x^m}{m^2-1}$
6. $y = c_1 + c_2x^{-1} + \frac{1}{2}(\log x)^2 - \log x$
7. $y = (c_1 + c_2\log x)x^{-2} + \frac{1}{9}x\left(\log x - \frac{2}{3}\right)$

8. $y = c_1 x^{-1} + \sqrt{x}\left[c_2 \cos\left(\frac{\sqrt{3}}{2}\log x\right) + c_3 \sin\left(\frac{\sqrt{3}}{2}\log x\right)\right] + \frac{1}{2}x + \log x$

9. $y = x[c_1 \cos(\sqrt{3}\log x) + c_2 \sin(\sqrt{3}\log x)]$
$+\frac{3}{5}\cos(\log x) - \frac{2}{5}\sin(\log x) + \frac{1}{2}x\sin(\log x)$

10. $y = x[c_1 + c_2 \log x + c_3(\log x)^2 + c_4(\log x)^3] + 4 + \log x$

11. $y = c_1 x^3 + c_2 x^{-2} + \frac{1}{50}(5\log x - 2)\,x^3 \log x$

12. $y = c_1(x+3)^2 + c_2(x+3)^3 + \frac{1}{2}(x+2)$

13. $y = c_1 + c_2 \log(x+1) + \{\log(x+1)\}^2 + x^2 + 8x$

14. $y = x^2(c_1 x^{\sqrt{3}} + c_2 x^{-\sqrt{3}})$
$+x^{-1}\left[\frac{(5\sin(\log x) + 6\cos(\log x))\log x}{61} + \frac{54\sin(\log x) + 282\cos(\log x)}{3721} + \frac{1}{6}\right]$

15. $y = c_1 x^{-1} + x^{-1}[c_2 \cos(\log x) + c_3 \sin(\log x)] + \frac{1}{20}x^2 + \frac{3}{10}x - 2$

16. $y = (c_1 + c_2 \log x)\cos(\log x) + (c_3 + c_4 \log x)\sin(\log x) + (\log x)^2 + 2\log x - 3$

Page 230

1. $y = -e^x - \frac{1}{4}c_1(2x+5) + c_2 e^{2x}$ 2. $y = e^x(c_1 \log|x| + c_2)$

3. $y = 1 + c_1 x \int e^{\frac{x^3}{3}} \cdot x^{-2}\,dx + c_2 x$ 4. $y = x^3 \log|x| + x^2 + c_1 x^3 + c_2 x$

5. $y = e^x(x^2 - x) + c_1 x e^x + c_2 x$ 6. $y = c_1 e^x + c_2 x - (1 + x^2)$

7. $y = c_1 e^x (x+1)^5 + c_2 e^x - \frac{1}{2}xe^x$ 8. $y = c_1 e^x + c_2 e^{3x}(4x^3 - 42x^2 + 150x - 183)$

9. $y = e^x \log|x| + c_1 e^x \int x^{-1} e^{-x}\,dx + c_2 e^x$

10. $y = c_2 x + \frac{1}{4}x\cos 2x - \frac{1}{2}x\sin 2x - c_1 \cos x$

Pages 233 and 234

1. $y = e^{x^2}(c_1 e^x + c_2 e^{-x} - 1)$

2. $y = e^{\frac{x^2}{2}}\left(c_1 \cos\sqrt{3}\,x + c_2 \sin\sqrt{3}\,x + \frac{1}{4}e^x\right)$

3. $y = e^{-x^2}(c_1 e^{x\sqrt{2}} + c_2 e^{-x\sqrt{2}})$ 4. $y = e^{\frac{1}{2}bx^2}(c_1 \cos\sqrt{b}\,x + c_2 \sin\sqrt{b}\,x)$

5. $y = \frac{1}{x}(c_1 \cos nx + c_2 \sin nx)$ 6. $y = e^{-\frac{x^2}{2}}(c_1 x + c_2) + x$

7. $y = x(c_1 \cos x + c_2 \sin x) + \frac{1}{2}xe^x$

8. $y = \sqrt{x}\,e^{-\frac{x^4}{8}}\left[c_1 \cos\left(\frac{\sqrt{3}}{2}\log x\right) + c_2 \sin\left(\frac{\sqrt{3}}{2}\log x\right)\right]$

9. $y = \frac{1}{x^2} e^{3x}\left(c_1 x^2 + c_2 x^{-1} + \frac{1}{3} x^2 \log x\right)$

10. $y = e^{-\frac{x^2}{2}}\left(c_1 \cos 2x + c_2 \sin 2x + \frac{1}{4} x\right)$

11. $y = e^{x^2} (c_1 \cos \sqrt{2}\, x + c_2 \sin \sqrt{2}\, x) + \frac{1}{2} e^{x^2}$

12. $y = e^{\frac{x}{2}} (c_1 x^3 + c_2)$ 13. $y = (c_1 \cos \sqrt{6}\, x + c_2 \sin \sqrt{6}\, x) \sec x$

Page 240

1. $y = c_1 \cos \frac{a}{x} - c_2 \sin \frac{a}{x}$ 2. $y = c_1 \cos (2 \tan^{-1} x) + c_2 \sin (2 \tan^{-1} x)$
3. $y = c_1 e^{-\cos x} + c_2 e^{\cos x}$
4. $y = c_1 \cos\left(2 \log \tan \frac{x}{2}\right) + c_2 \sin\left(2 \log \tan \frac{x}{2}\right)$
5. $y = c_1 e^{\sqrt{2} \sin x} + c_2 e^{-\sqrt{2} \sin x} + \sin^2 x$
6. $y = c_1 \cos (2e^{-x} (x + 1)) - c_2 \sin (2e^{-x}(x + 1)) + e^{-x}(x + 1)$
7. $y = (c_1 + c_2 x^2) e^{-x^2} + \frac{1}{2}$ 8. $y = c_1 \cos (\sin x) + c_2 \sin (\sin x)$
9. $y = c_1 \cos x^2 + c_2 \sin x^2 + \frac{1}{4} x^2$ 10. $y = c_1 e^{\cos x} + c_2 e^{2\cos x} + \frac{1}{6} e^{-\cos x}$
11. $y = c_1 e^{e^x} + c_2 e^{3e^x} - e^{2e^x}$

Page 245

1. $y = (a - x) \cos x + (b + \log | \sin x |) \sin x$
2. $y = [a - e^{-x} + \log | 1 + e^{-x}|]\, e^x + [b - \log | 1 + e^x |] e^{-x}$
3. $y = ax + be^x + (1 + x + x^2)$ 4. $y(1 - x^2) = a \cos x + b \sin x + x$
5. $y = a(\sin x - \cos x) + be^{-x} - \frac{1}{10}(\sin 2x - 2 \cos 2x)$.
6. $y = c_1 \cos nx + c_2 \sin nx + \frac{1}{n^2} \cos nx \log | \cos nx | + \frac{1}{n} x \sin nx$
7. $y = a \cos 2x + b \sin 2x - \cos 2x \log | \sec 2x + \tan 2x |$

Pages 245 and 246

1. $y = c_1 - c_1 x \cot x + c_2 \cot x$ 2. $y = c_1 \cos x + c_2 x$
3. $y = \frac{\sin x}{x}\left[\frac{1}{6} x^3 + \left(c_1 - \frac{1}{4}\right)(-x^2 \cot x + 2x \log | \sin x |\right.$

$\left. -2 \int \log \sin x dx\right) + c_2 \Big]$

4. $y = -\frac{1}{2} x^2 e^x + xe^x + \frac{1}{3} c_1 x^3 e^x + c_2 e^x$
5. $y = -1 + c_1 e^x \int e^{-x+\frac{x^2}{2}} dx + c_2 e^x$

6. $y = x(c_1 \cos nx + c_2 \sin nx)$

7. $y = (c_1 x^{-2} + c_2 x^3) e^{-\frac{4}{3}x^{2/3}}$

8. $y = (c_1 x^2 + c_2 x^{-1}) e^{\frac{x}{2}}$

9. $y = \sqrt{x}\,[c_1 \cos(\log x) + c_2 \sin(\log x)]$

10. $y = c_1 e^{-\cos x} + c_2 e^{\cos x} - \cos x$

11. $y = (c_1 \cos\sqrt{2}\,x + c_2 \sin\sqrt{2}\,x)\sec x$

12. $y = c_1 \cos(\log(1+x)) + c_2 \sin(\log(1+x)) + 2\log(1+x)\sin(\log(1+x))$

13. $y = c_1(\sqrt{1-x^2} + x\sin^{-1}x) + c_2 x - \frac{1}{9}x(1-x^2)^{\frac{3}{2}}$

Page 254

1. $x = c_1 \cos \omega t + c_2 \sin \omega t,\ y = c_1 \sin \omega t - c_2 \cos \omega t$

2. $x = c_1 \cos t + c_2 \sin t,\ y = -\frac{1}{2}(c_1 + 3c_2)\sin t + \frac{1}{2}(c_2 - 3c_1)\cos t$

3. $x = e^{6t}(c_1 \cos t + c_2 \sin t)$,
$y = e^{6t}[(c_1 - c_2)\cos t + (c_1 + c_2)\sin t]$

4. $x = (c_1 + c_2 t)e^{3t},\ y = (1 - 2t)(c_2 - 2c_1)e^{3t}$

5. $x = c_1 e^{-4t} + c_2 e^{-7t} + \frac{7}{40}e^t + \frac{1}{27}e^{2t}$
$y = c_1 e^{-4t} - c_2 e^{-7t} + \frac{1}{40}e^t + \frac{7}{54}e^{2t}$

6. $x = (c_1 + c_2 t)e^{-4t} + \frac{31}{25}e^t - \frac{49}{36}e^{2t}$
$y = (c_1 + c_2 t)e^{-4t} - \frac{11}{25}e^t + \frac{19}{36}e^{2t}$

7. $x = c_1 e^{-t} + c_2 e^{3t} + \frac{1}{5}(\cos t - 2\sin t)$
$y = 2c_1 e^{-t} - 2c_2 e^{3t} + \frac{1}{5}(8\cos t - \sin t)$

8. $x = c_1 e^{-t} + c_2 e^{-6t} + \frac{19}{3}t - \frac{56}{9} - \frac{29}{7}e^t$
$y = c_1 e^{-t} + 4c_2 e^{-6t} - \frac{17}{3}t + \frac{55}{9} + \frac{24}{7}e^t$

9. $x = (c_1 + c_2 t)e^t + (c_3 + c_4 t)e^{-t} - 23$
$y = (c_1' + c_2' t)e^t + (c_3' + c_4' t)e^{-t} + 18$

10. $x = (c_1 + c_2 t)e^t + c_3 e^{-\frac{3}{2}t} - \frac{1}{3}t$
$y = -2(c_1 + c_2 t - 3c_2)e^t - \frac{1}{3}c_3 e^{-\frac{3}{2}t} - \frac{1}{3}$

11. $x = \frac{3}{2}c_1 e^{2t} - 3c_2 e^{-t} - \frac{3}{10}e^t(\cos t - 2\sin t)$
$y = c_1 e^t + c_2 e^{-2t} - \frac{1}{10}(\cos t + 3\sin t)$

12. $x = \frac{1}{3}t + c_1 t^{-2},\ y = c_2 e^t - \frac{1}{3}t - c_1 t^{-2}$

13. $x = (c_1 + c_2 t)e^{3t} + c_3 e^{-2t} - t$
$y = (3c_2 - 2c_1 - 3c_2 t)e^{2t} - \frac{1}{3}c_3 e^{-3t} - \frac{1}{3}$

Pages 258 and 259

1. $x + y = c_1,\ x - \frac{1}{2}\log[z^2 + (x+y)^2] = c_2$
2. $x = c_1 z,\ y = c_2 z$
3. $x^2 - y^2 = c_1,\ |y + \sqrt{y^2 + c_1}| = |c_2 z|$
4. $|x - y| = |c_1(y - z)|,\ (x-y)^2\,|x + y + z| = |c_2|$
5. $x^2 + y^2 + z^2 = c_1,\ lx + my + nz = c_2$
6. $x + y + z = c_1,\ xyz = c_2$
7. $x^2 + y^2 + z^2 = c_1,\ xyz = c_2$
8. $y^2 + z^2 = c_1,\ \log|c_2 x| = \tan^{-1}\frac{y}{x}$
9. $x^2 - z^2 = c_1,\ x^3 - y^3 = c_2$
10. $|y| = |c_1 z|,\ x^2 + y^2 + z^2 = |c_2 z|$
11. $x + y + z = c_1,\ x^3 + y^3 + z^3 = c_2$
12. $x^2 - y^2 = c_1,\ \log|x + y| = (x^2 - y^2)\log|z| + c_2$

Pages 259 and 260

1. $x = c_1\cos t + c_2\sin t$

$$y = -\frac{1}{2}(c_1 - 3c_2)\sin t + \frac{1}{2}(c_2 - 3c_1)\cos t$$

2. $x = c_1 e^t + c_2 e^{-5t} + \frac{3}{7}e^{2t} - \frac{2}{5}t - \frac{13}{25}$

$$y = c_1 e^t + c_2 e^{-5t} + \frac{4}{7}e^{2t} - \frac{3}{5}t - \frac{12}{25}$$

3. $x = c_1 e^{-\frac{11}{8}t} + 3t - 2,\ y = 3 + 5t + \frac{6}{19}c_1 e^{-\frac{11}{8}t}$

4. $x = \frac{1}{10}e^{-t}(10\cos 2t - \sin 2t) - \frac{4}{5}e^{-3t}$

$$y = \frac{1}{10}e^{-t}(21\sin 2t - 8\cos 2t) + \frac{4}{5}e^{-3t}$$

5. $x = c_1\cos\sqrt{3}t + c_2\sin\sqrt{3}t + c_3\cos\sqrt{2}t + c_4\sin\sqrt{2}t$

$$-\frac{49}{1452}e^{3t} + \frac{5}{66}te^{3t} - \frac{1}{12} - \frac{1}{4}\cos 2t$$

$$y = -3c_1\cos\sqrt{3}t - c_2\sin\sqrt{3}t - 2c_3\cos\sqrt{2}t$$

$$-2c_4\sin\sqrt{2}t + \frac{1}{66}te^{3t} + \frac{1}{3} - \frac{23}{1452}e^{3t}$$

6. $x = c_1 t + c_2 t^{-1},\ y = -c_1 t + c_2 t^{-1}$

7. $x = e^{\frac{mt}{\sqrt{2}}}\left(c_1\cos\frac{mt}{\sqrt{2}} + c_2\sin\frac{mt}{\sqrt{2}}\right) + e^{-\frac{mt}{\sqrt{2}}}\left(c_3\cos\frac{mt}{\sqrt{2}} + c_4\sin\frac{mt}{\sqrt{2}}\right)$

$$y = e^{\frac{mt}{\sqrt{2}}}\left(c_1\cos\frac{mt}{\sqrt{2}} - c_2\sin\frac{mt}{\sqrt{2}}\right) + e^{-\frac{mt}{\sqrt{2}}}\left(c_4\cos\frac{mt}{\sqrt{2}} - c_3\sin\frac{mt}{\sqrt{2}}\right)$$

8. $x^3 - y^3 = c_1,\ x^8 + \frac{3}{z} = c_2$

9. $y - x = c_1 xy,\ z = \frac{nxy}{y - x}\log\left|\frac{y}{x}\right| + c_2$.

10. $l^2x + m^2y + n^2z = c_1,\ l^2x^2 + m^2y^2 + n^2z^2 = c_2$
11. $|y - x| = |c_1(z - x)|,\ |y - x| = |c_2(z - y)|$
12. $x + y + z = c_1,\ x^2 + y^2 = c_2z^2$
13. $xyz = c_1,\ x^2 + y^2 - 2z = c_2$
14. $\dfrac{1}{(x+y)^2} - \dfrac{1}{(x-y)^2} = c_1,\ xy = c_2z^2$
15. $|x - y| = |c_1(y - z)|,\ |z - x| = |c_2(y - z)|$
16. $|y| = |c_1x|,\ |x^{1-a}| = |c_2(z + \sqrt{x^2 + y^2 + z^2})|$

Index

Index